Smart Land Use Planning II

Smart Land Use Planning II

Guest Editors

Xufeng Cui
Walter T. de Vries
Fei Li
Basanta Paudel

MDPI

Basel • Beijing • Wuhan • Barcelona • Belgrade • Novi Sad • Cluj • Manchester

Guest Editors

Xufeng Cui
School of Business Administration
Zhongnan University of Economics and Law
Wuhan
China

Walter T. de Vries
Department of Aerospace and Geodesy
Technical University of Munich (TUM)
München
Germany

Fei Li
Research Center for Environment and Health
Zhongnan University of Economics and Law
Wuhan
China

Basanta Paudel
Institute of Geographic Sciences and Natural Resources Research
Chinese Academy of Sciences (CAS)
Beijing
China

Editorial Office

MDPI AG
Grosspeteranlage 5
4052 Basel, Switzerland

This is a reprint of the Special Issue, published open access by the journal *Land* (ISSN 2073-445X), freely accessible at: https://www.mdpi.com/journal/land/special_issues/7RM6R2QT2F.

For citation purposes, cite each article independently as indicated on the article page online and as indicated below:

Lastname, A.A.; Lastname, B.B. Article Title. *Journal Name* **Year**, *Volume Number*, Page Range.

ISBN 978-3-7258-6724-0 (Hbk)
ISBN 978-3-7258-6725-7 (PDF)
https://doi.org/10.3390/books978-3-7258-6725-7

Contents

Preface

With the rapid evolution of geographic information systems, remote sensing, data mining, and related technologies, concepts such as intelligent land management, refined planning, big data analytics, and smart monitoring have transitioned from theory to widespread practice. Under the influence of a new wave of scientific and technological revolution, a key issue has emerged: how to leverage new technologies to propel the smart transformation of land-use planning and achieve sustainable land use. It is foreseeable that future land-use planning will be network-based, software-platform-centered, data-driven, and security-guaranteed, integrating the Internet, geographic information systems (GISs), Internet of Things (IoT), artificial intelligence (AI), cloud computing, and other technologies to form a diversified and integrated intelligent system.

The purpose of this Special Issue is to gather insights from both academia and industry, inviting scholars and practitioners to share their valuable practical experiences and profound perspectives in the field of smart land-use planning. We aim to provide a platform for showcasing and discussing cutting-edge theories, innovative tools, and practical case studies that underpin this transformation. The research compiled herein reflects how the integration of multiple technologies empowers the entire planning cycle—from perception and analysis to decision-making and supervision—while also delving into the challenges encountered in practice and potential solutions.

We hereby commend this Reprint to all researchers, engineers, planners, and policy-makers interested in land science, urban and rural planning, geographic information science, resource management, and sustainable development. We sincerely thank all the contributors for dedicating their outstanding research to this Reprint and extend our gratitude to the reviewers for their invaluable time and professional expertise. It is our hope that this Reprint will stimulate further discussion and collaboration, jointly advancing land-use planning towards a more intelligent, precise, and sustainable future.

Xufeng Cui, Walter T. de Vries, Fei Li, and Basanta Paudel
Guest Editors

land

Editorial

Smart Land Use Planning: Hotspots and Prospects

Xufeng Cui [1], Huakun Huo [1], Fei Li [2], Walter Timo de Vries [3,*] and Basanta Paudel [4]

1 School of Business Administration, Zhongnan University of Economics and Law, Wuhan 430073, China; cxf@zuel.edu.cn (X.C.)
2 Research Center for Environment and Health, Zhongnan University of Economics and Law, Wuhan 430073, China
3 Department of Aerospace and Geodesy, School of Engineering and Design, Technical University of Munich (TUM), Arcisstrasse 21, 80333 München, Germany
4 Lumbini Research Center, Lumbini Buddhist University, Lumbini, Rupandehi 32990, Nepal
* Correspondence: wt.de-vries@tum.de

1. Introduction

Land use planning is a crucial tool for achieving the optimal allocation of land resources. However, traditional land use planning exhibits inadequacies in adaptability when confronting increasingly complex and diverse land use issues, such as resilient cities development [1], Urban Digital Twins (UDTs) [2], and smart city initiatives [3]. Therefore, it is imperative to transform land use planning from a traditional experience-based paradigm to a data-driven one and address current land use issues through Smart Land Use Planning. Smart land use planning is a diversified integrated intelligent system that is network-based, with software platforms as the core, data as the key element, and security as the fundamental guarantee. It integrates technologies including Geographic Information System (GIS), Internet of Things (IoT), and Artificial Intelligence (AI) [4]. It can provide core technical support for the sustainable utilization of land resources and the achievement of regional development goals.

2. Technology Hotspots: Core Links of Digital Empowerment

The key feature of smart land use planning lies in the integrated application of multi-source big data and diverse tools [4]. It emphasizes the integration of intelligent perception technology, intelligent decision-making technology, and intelligent operation technology throughout the entire life cycle of land use planning, forming a data-driven planning and management process. Ultimately, this enhances the rationality of land resource allocation in terms of spatial layout, functional configuration, and utilization efficiency, thereby facilitating the efficient allocation of land resources.

2.1. Intelligent Perception

Spatio-temporal data elements are essential components of smart land use planning, and accurate temporal information and geographical locations constitute the basic premise for sound land use planning. The process of integrating such multi-source heterogeneous data in real-time through multiple technologies is defined as intelligent perception. From the perspective of smart land use planning, intelligent perception technology encompasses two aspects: dynamic perception and systematic cognition. Specifically, it acquires fine-classification information of Land Use and Land Cover (LULC) via hyperspectral remote sensing satellites [5] and deploys Internet of Things (IoT) technology to monitor the dynamic changes in land use, thereby realizing the real-time collection, transmission, storage, and processing of multi-source data. Based on the acquired dynamic land use data, technologies such as big data and simulation can be employed for efficient data cleaning, summarizing the laws governing land use changes, establishing a systematic land use framework, and predicting future spatial patterns of land use to guide land use planning.

The process of intelligent perception technology significantly improves the efficiency and quantity of data acquisition for smart land use planning, thereby providing data support and a model foundation for subsequent smart land use decision-making and operations. The massive volume of accurate data obtained through intelligent perception technology can maximally support the application of land use planning in various scenarios. Ranging from smaller-scale village planning and urban planning to regional and even national-level planning, data matching the required resolution and coverage can be obtained, forming a complete technical chain. Thus, intelligent perception technology serves as the foundation of smart land use planning.

2.2. Intelligent Decision-Making

In the past, the decision-making subjects of land use planning were typically governments, public organizations, or individuals. However, smart land use planning breaks the limitations of a single and relatively isolated subject in the traditional planning process. Specifically, through intelligent decision-making technology, it innovates decision-making tools and platforms to achieve human–computer interaction and collaborative governance. Against the backdrop of intelligent decision-making, based on the results of data perception, smart land use planning introduces decision-making tools such as Artificial Intelligence (AI) [6] and Geographic Information System (GIS) as auxiliary technologies. These tools intelligently identify numerous land elements and conduct spatial correlation analysis, providing decision-makers with corresponding adjustment schemes for land use planning. Additionally, digital platform technology enables the integration of various departments into the public space of the network, realizing resource sharing and inter-departmental collaboration. This maximizes the compatibility between various types of land use planning and their "red lines," avoiding conflicts between different plans.

The systematic application of this series of intelligent decision-making technologies not only rapidly generates scientific decision-making and planning schemes for decision-makers to choose from but also allows for flexible adjustments based on practical needs. Furthermore, digital platforms can engage more stakeholders in the land use planning process and expand their participation channels. Therefore, intelligent decision-making technology constitutes the core part of smart land use planning. It is pivotal to the full and effective utilization of planning data elements and provides directions and goals for the smooth implementation of subsequent land use planning.

2.3. Intelligent Operation

After making sound decisions, smart land use planning enters the practical operation phase. The key to this phase is to ensure the effective implementation of decision content, which requires the support of intelligent operation technology. Specifically, through technologies such as Digital Twins (DT) [7], Information and Communications Technology (ICT) [8], and City Information Modeling (CIM) [9], physical entities of land use that incorporate factors like population, environment, and economy are constructed. This enables bidirectional interaction between the ecological, economic, and social goals of planning and digital models, forming a closed feedback loop: the decision schemes and intervention strategies formulated by smart land use planning through systematic analysis can be implemented in the physical operation of land use scenarios. The relevant implementation results can then be dynamically monitored by means of technologies such as DT, ICT, and

CIM, thereby supporting the closure of the loop. This continuous cycle enables iterative improvement and refinement, provides land governance strategies, and allows intelligent operation technology to quickly adapt to changing conditions.

By establishing an intelligent adaptive planning and operation system, intelligent operation technology effectively drives the full-process optimization of land use planning from implementation to dynamic monitoring. It not only ensures the efficiency of planning implementation but also achieves full coverage of the monitoring link and the sustainable development of the entire process. This multi-faceted capability connects elements such as land, economy, and ecology with strategic planning, creating an integrated framework. It enhances the adaptability and dynamic capabilities of the land use planning system, ensuring that planning plays a guiding role throughout the entire land use process and preventing the "absence" of planning.

In summary, in the context of smart land use planning, intelligent perception, intelligent decision-making, and intelligent operation constitute three core links. These three links fully cover the entire life cycle of smart land use planning. By embedding technical elements into the entire process of planning, approval, implementation, monitoring, and evaluation, they realize intelligent, efficient, and scientific development. This effectively overcomes the limitations of traditional land use planning in aspects such as data acquisition efficiency, decision-making scientificity, and implementation accuracy. More importantly, through the in-depth integration of digital technology and planning business, they significantly improve the efficiency of land resource allocation, ultimately providing technical support for smart land use planning to realize the digital empowerment process of "data-driven—scientific decision-making—efficient operation."

3. Practical Challenges: Bottlenecks in Technology Implementation

Despite the increasing advancement of technical tools, the full-scale implementation of smart land use planning in the field of land use planning, which is characterized by strong policy relevance and complex interest relationships, still faces numerous challenges. Currently, the most prominent difficulties in smart land use planning stem from three aspects: institutional environment, human resource environment, and technological environment. These three factors collectively determine the level of intelligence and effectiveness of land use planning.

3.1. Challenges from the Institutional Environment

The government is regarded as a key player in smart cities. Similarly, the stability and continuity of the institutional environment, including the government and policies, also determine the degree of realization of Smart Land Use Planning. In practice, the implementation of smart land use planning requires a relatively long-term process. The application of technology, coordination of interests, and adjustment of goals also demand that policies remain relatively stable over a certain period to ensure that the planning path aligns with the needs of social development. However, due to differences in conditions and diverse goals across countries or regions, coupled with the rapid development of the social economy and science and technology, the weight of concepts such as ecological protection, human settlement optimization, and social equity continues to rise. This leads to timely adjustments in specific national development planning requirements, which in turn affect the application level and effectiveness of smart land use planning. For example, India's smart city development relies on the top-down management model of the Special Purpose Vehicle (SPV) Model [10], which tends to lead to fragmented management in smart city development due to decentralized governance; the Chinese government implements the "integration of multiple plans," and the integration of multiple planning systems such as

Land Use Planning and urban-rural planning may lead to difficulties in historical data compatibility, affecting the role of technical coordination; Iran adopts exogenous Land Use Planning (LUP) [11], which emphasizes external perspectives and tends to be disconnected from the country's actual conditions, weakening the feasibility of technology application. While these policies align with national development needs, they also alter the application scope and value orientation of land use planning itself. This directly affects the goal-setting and scope of application of smart land use planning, thereby requiring it to possess greater resilience and adaptability.

3.2. *Challenges from the Human Resource Environment*

The relationship between land and people is an eternal topic in geography and many other disciplines. From the perspective of smart land use planning, challenges related to human resources also constitute an important issue that must be addressed, mainly reflected in two dimensions: the shortage of interdisciplinary talents for smart land use planning and the difficulty in coordinating multiple stakeholders. On the talent supply side, smart land use planning imposes interdisciplinary application requirements on practitioners' knowledge structures. Practitioners must absorb and draw on knowledge from disciplines such as landscape ecology, economics, and management in terms of concepts, master various digital technology planning tools in terms of skills, and achieve refinement in intelligent perception, precision in intelligent decision-making, and accurate calculation in intelligent operation in terms of goals. These requirements for interdisciplinary talents impose significant pressure on practitioners. Practitioners must invest more time and effort to master these technologies, which directly prolongs the process of practical implementation of advanced planning technologies and affects the intelligent transformation of land use planning. Meanwhile, the development of smart land use planning involves various stakeholders, such as governments, investors, and citizens. These stakeholders have distinct and strong interest demands, and their understanding of planning goals often varies, which easily leads to conflicts of interest demands. Current smart land use planning lacks efficient coordination mechanisms. Although smart planning technologies can provide data support for conflict resolution and expand channels for public participation in governance, they are unable to directly balance the demands of all parties. This goal of considering multiple demands in the process of technology application also poses challenges to smart land use planning.

3.3. *Challenges from the Technological Environment*

The most significant challenges faced by smart land use planning have always centered on the application of technology. To a certain extent, the intelligentization process is a process of technology application. Therefore, the feasibility of technology must be a key consideration. This feasibility not only encompasses the connotation of technology application but also requires that technology conforms to social value concepts, both of which constitute major technological environment constraints for smart land use planning. On one hand, the regional environments covered by land use planning exhibit significant heterogeneity. Land use planning technologies must be able to accurately adapt to the complex and diverse environments of different regions. For example, cloudy climates in basin areas affect remote sensing monitoring, which tends to reduce the accuracy of remote sensing images. This requires the exploration of more media for monitoring planning targets in terms of technology, but this will significantly increase the cost of applying land use planning technologies. On the other hand, intelligent technologies may trigger social concerns. For instance, the development of smart cities requires tracking vehicle information [12], and such data tracking may give rise to issues related to data privacy and security. The protection of privacy will restrict the application scope and scenarios of intelligent planning technologies. Alleviating the contradiction between data security and planning accuracy requires smart land use planning to explore diverse technical paths that balance security and practicality and meet the dual needs of technology application and social values.

In summary, the bottlenecks in the implementation of smart land use planning technologies are not caused by a single factor but result from the interweaving and mutual constraints of the three environments: institutional, human resource, and technological. Furthermore, these three aspects do not exist in isolation. Policy adjustments at the institutional level may exacerbate the difficulties in updating interdisciplinary talents, while issues such as privacy concerns at the technological level also require institutional guarantees and public collaboration for resolution. This implies that future research and practice must be based on the interconnections among these three environments and construct integrated solutions to effectively overcome the current predicament of technology implementation.

4. Future Prospects: Outlook on Smart Governance

To realize its core value and large-scale application, effective smart land use planning must systematically address the aforementioned institutional, human resource, and technological issues. The resolution of these issues requires relying on the collaborative response of the three rather than isolated policies. This will promote the normalization of smart land use planning, making it a core force supporting the coordinated development of urban and rural areas, efficient resource utilization, and the achievement of mutually beneficial outcomes in ecological protection.

To address the pain point of enhancing the resilient governance capacity of smart land use planning amid challenges from the institutional environment, a conceptual shift is necessary to avoid overemphasizing physical and technical aspects [13]. Policy changes stem from the dynamic allocation of resources, and the resilient governance of planning requires embedding value concepts such as social sustainability and social justice into the entire technical process, enabling smart land use planning to obtain reliable long-term guiding capabilities. For example, guided by the goal of sustainable development, Shenzhen, China integrates smart city development technologies [14] into urban planning; through the combination of technology and green development goals, the planning becomes more adaptable when facing ecological pressures; the African Great Green Wall (GGW) Initiative, as a form of land use planning [15], emphasizes the combination of long-term security and stability of Land tenure security with social sustainability, significantly enhancing the resilience of planning in response to dynamic changes in power. This approach of binding land resource utilization with multiple goals can maximize the balance of allocation requirements for different resources. In the future development of smart land use planning, emphasis should continue to be placed on resource allocation issues, with priority given to their implementation. This will enhance the ability of planning to address complex issues and provide rigid support for the sustainable use of resources and social justice.

Focusing on the issues of talent scarcity and meeting the needs of stakeholders amid challenges from the human resource environment, future smart land use planning must prioritize the cultivation of personnel skills and engage more interdisciplinary talents in the land use planning process. This is not only an effective way to leverage public wisdom to solve problems and make up for the shortcomings of existing methods but also enables timely responses to citizens' demands, emphasizing public participation and social inclusion. The form of Participatory Budgeting [16] can promote the integration of social capital, social inclusion, and mainstream pluralism in the decision-making process into the content of land use planning. These characteristics are conducive to the development of a democratic and diversified spatial pattern of land use planning, thereby facilitating the formulation of more inclusive and high-satisfaction land use planning schemes.

To address the current challenges faced by smart land use planning at the technological level, such as insufficient diversification and weak scenario adaptability, future efforts need to promote the diversified expansion of the technology system. Meanwhile, the iteration of land use planning tools has progressively accelerated [17], requiring continuous dynamic optimization in practical applications to adapt to the core scenarios of land use planning. In this process, the effective implementation of technology is only a basic requirement; the more core aspect lies in that technology application must carry social value orientations. This requires the construction of social participation mechanisms in the technical framework, integrating public demands into the technology embedding link throughout the entire planning process, and maximizing the engagement of multiple stakeholders in the formulation of land use planning schemes. Ultimately, technology will break through the attribute of single tool integration and become a core link connecting multiple values.

In the face of the three core challenges, namely institutional, human resource, and technological, future smart land use planning needs to construct systematic solutions through the aforementioned measures and promote the collaborative governance of the three. This in-depth coupling of "institution—human resource—technology" can not only effectively resolve the current pain points and difficulties in planning practice but also promote smart land use planning to form an integrated system with resilience, inclusiveness, and sustainability. Ultimately, it will provide planning guarantees for countries or regions to utilize land resources in a rational and efficient manner, achieving goals such as coordinating economic development and land resource allocation, and ensuring the sustainable use of land. Furthermore, the advancement of smart land use planning from theory to practice represents an inevitable trend in the integration of empirical scientific concepts of land resource management with advanced technologies. The human-land relationship follows complex laws; future smart land use planning must adopt more advanced technologies and concepts, coordinate the demands of different stakeholders in a more equitable manner, and plan the future of land resources in a more sustainable way.

Author Contributions: Conceptualization, X.C. and W.T.d.V.; Methodology, H.H.; Writing—original draft, H.H. and X.C.; Writing—review and editing, X.C., H.H., F.L., W.T.d.V. and B.P. All authors have read and agreed to the published version of the manuscript.

Funding: The National Social Science Fund of China (Grant No. 25BGL204).

Conflicts of Interest: The authors declare no conflicts of interest.

References

1. Wang, T.; Xu, T.; Wang, Z.; Wang, H.; Kang, J.; Qiu, L.; Xue, S.; Fang, Z.; Zhang, Y. Where Do Resilient Cities Grow? Exploring the Pathways and Mechanisms of Resilience Development. *Sustain. Cities Soc.* **2025**, *133*, 106856. [CrossRef]
2. De Jaeger, A.; Swerts, T. Right to the Digital Twin City? Citizen Participation and Limits-by-Design in Rotterdam's Urban Digital Twin. *Cities* **2026**, *168*, 106498. [CrossRef]
3. Liu, Y. Research on the construction and application of geological environment information platform for smart cities. *Shanghai Land Resour.* **2024**, *45*, 258–263.
4. Cui, X.; Li, F.; De Vries, W.T. Smart Land Use Planning: New Theories, New Tools and New Practice. *Land* **2023**, *12*, 1315. [CrossRef]
5. Li, J.; Jiang, W. Geospatial Information Technology Innovations: From Earth Monitoring to Urban Planning. *Engineering* **2025**, *47*, 1–2. [CrossRef]
6. Bibri, S.E.; Huang, J. Artificial Intelligence of Things for Sustainable Smart City Brain and Digital Twin Systems: Pioneering Environmental Synergies between Real-Time Management and Predictive Planning. *Environ. Sci. Ecotechnol.* **2025**, *26*, 100591. [CrossRef] [PubMed]

7. Wang, Q.-C.; Sun, M.; Liu, X.; Tao, F.; Yang, D.; Bardhan, R. Reflecting City Digital Twins (CDTs) for Sustainable Urban Development: Roles, Challenges and Directions. *Digit. Eng.* **2025**, *5*, 100035. [CrossRef]
8. Kong, J.; Hwang, J.; Kim, H. Building Smarter Cities Together: Government-to-Government Partnerships in the Development of Smart Cities. *Cities* **2025**, *156*, 105532. [CrossRef]
9. Peng, W. Construction and application of digital base for intelligent monitoring of urban rail transit. *Shanghai Land Resour.* **2023**, *44*, 126–133. [CrossRef]
10. Mishra, A.P.; Anand, S.; Batar, A.K. Optimizing Regional Development through Smart Cities: A Case Study of Lucknow City, India. *Reg. Sci. Policy Pract.* **2025**, *17*, 100226. [CrossRef]
11. Ramezani, S.; Nastaran, M.; Nooraie, H.; Otsuki, K. The Role of the Institutional Environment in Land Use Planning in Iran: A Conceptual Framework. *Land Use Policy* **2023**, *135*, 106942. [CrossRef]
12. Sun, Z.; Huang, Z.; Hao, P.; Ban, X.; Huang, T. Batch-Based Vehicle Tracking in Smart Cities: A Data Fusion and Information Integration Approach. *Inf. Fusion* **2024**, *102*, 102030. [CrossRef]
13. Alizadeh, H.; Sharifi, A. Societal Smart City: Definition and Principles for Post-Pandemic Urban Policy and Practice. *Cities* **2023**, *134*, 104207. [CrossRef]
14. Zhao, W. Smart City Technologies for Sustainable Urban Planning: Evidence and Equity Lessons from Shenzhen. *Sustain. Futures* **2025**, *10*, 101198. [CrossRef]
15. Orou Sannou, R.; Guenther, E. Exploring the Resource Nexus between Forest-Based Land Restoration and Food Security: The Case of the African Great Green Wall Initiative Countries. *Land Use Policy* **2025**, *151*, 107499. [CrossRef]
16. Kędra, A.; Maleszyk, P.; Visvizi, A. Engaging Citizens in Land Use Policy in the Smart City Context. *Land Use Policy* **2023**, *129*, 106649. [CrossRef]
17. Liu, Y.; Timo De Vries, W.; Zhang, G.; Cui, X. From Tradition to Smart: A Comprehensive Review of the Evolution and Prospects of Land Use Planning Tools. *Heliyon* **2024**, *10*, e40857. [CrossRef] [PubMed]

land

MDPI

Article

When the Map Does Not Tell the Whole Story: Integrating Community Voices into GIS Gentrification Analysis

Ivis García

Department of Landscape Architecture & Urban Planning, School of Architecture, Texas A&M University, College Station, TX 77843, USA; ivis.garcia@tamu.edu

Abstract: This exploratory case study examines the alignment between GIS-based displacement models and lived experiences of residents in Salt Lake City, addressing the benefits and limitations of spatial tools in capturing urban displacement complexities. By comparing the Urban Displacement Project's Estimated Displacement Risk (EDR) model with qualitative interviews from diverse neighborhoods, the research highlights discrepancies between predictive outputs and community narratives. The findings reveal that while GIS models effectively identify displacement hotspots, they often underestimate risks in areas with high homeownership or recent development. Conversely, resident interviews provide valuable insights into emerging displacement pressures that GIS may overlook. This study underscores the importance of integrating spatial analysis with community engagement to produce more equitable land-use planning strategies. The study contributes to urban governance and sustainable development by advocating for policies that prioritize the voices of vulnerable populations, fostering more resilient and inclusive cities.

Keywords: displacement; gentrification; Geographic Information Systems (GIS); qualitative data; quantitative data; Salt Lake City

1. Introduction

1.1. Problem Statement and Research Gap

Urban displacement driven by gentrification has become a critical challenge facing cities across the United States, yet significant gaps remain in how we understand and predict these processes. While Geographic Information Systems (GIS) have emerged as powerful tools for mapping displacement risk through quantitative modeling, these approaches often fail to capture the complex lived realities of affected communities. This limitation is particularly pronounced in mid-sized cities, where gentrification studies remain underrepresented compared to research focused on major metropolitan areas.

The Urban Displacement Project's (UDP) GIS-driven Estimated Displacement Risk (EDR) model exemplifies the current state of spatial modeling for displacement prediction, utilizing demographic and housing data to identify census tracts exhibiting characteristics correlated with low-income renter displacement [1]. However, such models, while valuable for visualizing risk patterns across geographic areas, cannot fully account for the personal, cultural, and social dimensions that shape residents' experiences of housing pressures, rising rents, and neighborhood change.

1.2. Research Contribution and Approach

This study addresses these limitations by employing research method approach that shows the UDP's GIS-based predictions and aligns them with narratives and lived experiences gathered through community interviews across various neighborhoods in Salt Lake City. This research makes a unique contribution to the displacement literature by validating quantitative displacement models against ground-level community knowledge in a mid-sized city context—an understudied urban typology in gentrification research.

The conceptual relationship between gentrification and displacement is central to the analysis. Gentrification—defined as the influx of higher-income residents and reinvestment in historically under-resourced neighborhoods—often triggers both direct displacement (through evictions) and indirect displacement (through rent increases or loss of cultural institutions). These interconnected processes alter neighborhood demographics and housing markets, making communities less accessible to low-income, long-standing residents. This study examines how these forces unfold in real-time through community narratives.

1.3. Research Objectives

This study seeks to (1) reveal discrepancies between model predictions and lived realities, (2) validate displacement patterns identified through quantitative analysis, and (3) highlight the strengths and limitations of GIS-driven tools in capturing the complexities of urban displacement. By capturing the voices of residents—many of whom have either experienced displacement or perceive themselves to be at risk—this comparison provides critical insight into the human dimensions of displacement. These include factors such as social networks, emotional ties to place, and barriers to relocation that GIS models cannot quantify. The analysis also helps identify areas where the model may underestimate or overestimate displacement risk due to local factors such as recent housing developments, specific policy interventions, or impacts from events like the COVID-19 pandemic.

1.4. Significance and Broader Context

The integration of spatial data in urban planning has grown significantly over the past decade, offering powerful tools for mapping socio-economic changes [2]. GIS applications in displacement research extend beyond basic risk mapping to encompass force displacement due to disasters [3], climate vulnerability mapping [4], and land-use suitability modeling [5]. Studies have demonstrated how layering datasets related to income, race, and land-use can reveal patterns of inequality and inform more equitable policy interventions [6]. However, researchers increasingly caution against over-reliance on quantitative models, emphasizing the importance of contextual, ground-level data to avoid misinterpretations [7]. This research contributes to a growing body of work that highlights how emerging tools—including remote sensing, big data analysis, and predictive modeling—can benefit from validation through on-the-ground experiences.

1.5. Paper Organization

This paper offers insights that can inform targeted policy interventions, resource allocation, and proactive community engagement strategies by revealing how lived realities contribute to GIS analysis. The findings underscore the need for adaptive urban governance frameworks that integrate both technological advancements and localized community knowledge to create more resilient, inclusive, and socially just cities. As cities continue to grapple with rapid urbanization, rising housing costs, and shifting demographic patterns, this research highlights the imperative of balancing data-driven approaches with participatory planning to ensure that urban development efforts genuinely reflect the needs and experiences of at-risk communities.

The paper proceeds as follows: Section 2 reviews the relevant literature on GIS-based displacement modeling and community-based research approaches; Section 3 outlines the methods methodology; Section 4 presents the comparative analysis of UDP model outputs and community narratives; Section 5 discusses the implications of the findings for urban planning practice; and Section 6 concludes with recommendations for integrating quantitative and qualitative approaches in displacement research.

2. Literature Review

Displacement due to gentrification and increased housing costs is a major challenge in cities across the world [8–10]. The issue calls for the use of better tools to help in the prediction of the neighborhoods at risk to be able to manage the effects. There is an increased application of GIS and predictive modeling for measuring the risk of displacement [2]. However, this article contributes to the existing literature that argues that understanding displacement cannot be left alone to GIS models but that inclusion of qualitative, community-based of narratives of gentrification is also necessary to help in the prediction and management of the effects. We need both quantitative and qualitative data to not only study the phenomenon but come up with equitable and smart land-use planning strategies.

2.1. GIS and Predictive Displacement Models

The GIS-based models have become more relevant as they help in processing spatial data and determining areas which are likely to be impacted by displacement [2,11,12]. The Urban Displacement Project (UDP), who were author collaborators, is a research initiative conducted by scholars from University of California, Berkeley and it can be considered as a pioneering work in this regard [13]. The UDP employs GIS to generate Estimated Displacement Risk (EDR) models, which use demographic data, housing market trends, and historical redlining patterns to predict displacement vulnerability [14]. These models visualize displacement hotspots, providing policymakers with critical tools to guide intervention strategies [15].

A notable 2010 framework is the Voorhees Center Gentrification Index, which combines 13 variables, such as racial composition, educational attainment, income levels, housing characteristics, and family structure, to assess gentrification and displacement risk [16]. Garcia applied this index with never before used data from 1970 to 2010 for Salt Lake City, recreating it to reflect the unique housing and demographic dynamics of the region, thereby providing localized insights that complement broader GIS-based approaches [2]. The findings revealed that while the effects of gentrification and community decline were pronounced in Chicago, in Salt Lake City, the impact of upgrading and decline was significantly less pronounced. Similarly, Chapple and Zuk's (2016) displacement typology, which combines socioeconomic indicators with housing market metrics to classify neighborhoods based on gentrification and displacement risk [17]. These models underscore the growing use of predictive analytics in addressing housing insecurity and urban inequality.

However, GIS models have limitations. As Easton et al. (2020) notes, predictive tools often fail to capture the nuanced, lived experiences of residents [18]. GIS models may underrepresent displacement risk in areas with high homeownership or where informal housing arrangements exist [1]. Additionally, static datasets can lag real-time developments, missing emergent displacement pressures linked to economic shocks or pandemics [19].

2.2. Qualitative Approaches to Displacement

Complementing GIS, qualitative methodologies such as community interviews, focus groups, and dialog provide critical insights into the human dimensions of displacement [20]. Research in New York by Newman and Wyly (2006) underscores the importance of resident narratives in identifying displacement precursors, including rising rents, evictions, and shifts in neighborhood demographics [21]. They concentrate in using interviews to uncovering experiences on the ground highlighting the social and emotional impacts of gentrification.

Studies conducted in rapidly gentrifying cities illustrate how qualitative research can validate or challenge GIS predictions. For example, Gould and Lewis (2017) found that while GIS models indicated low displacement risk in certain Brooklyn neighborhoods, interviews with residents highlighted significant housing instability and eviction threats, reflecting a mismatch between model outputs and lived realities [22]. Qualitative data often reveals early warning signs that GIS models may overlook, such as the loss of local businesses, cultural institutions, and social networks [23].

While participatory GIS and gentrification studies are well-established fields, this research addresses a specific gap: systematic validation of displacement prediction models through community narratives in mid-sized cities. Unlike previous studies that focus on major metropolitan areas or develop new participatory frameworks, this study evaluates an existing model's accuracy against ground-level experiences in Salt Lake City's unique demographic and housing context.

2.3. Integrating GIS with Community Narratives

Emerging scholarship advocates for the integration of GIS with qualitative data to produce more comprehensive displacement assessments. This hybrid approach not only enhances model accuracy but also empowers communities to influence urban planning processes [24]. The UDP's Salt Lake City analysis exemplifies this integrative approach. By comparing EDR model outputs with resident interviews, the project identifies discrepancies and contextualizes data within broader social and economic frameworks. Studies using GIS along with narratives contribute to more equitable land-use planning and displacement mitigation policies [21,25,26].

While many studies have employed participatory methods, this paper offers a unique contribution by systematically comparing GIS model outputs with qualitative narratives from residents in Salt Lake City neighborhoods. This approach advances the existing body of research by demonstrating how community insights can validate and refine spatial models, ultimately enhancing their policy relevance.

2.4. Displacement in Mid-Sized Cities: Insights and Future Directions

Despite advancements in GIS and participatory methodologies, gaps remain in addressing displacement dynamics across diverse urban contexts. Research often focuses on large metropolitan areas, with limited attention to mid-sized cities and rural communities experiencing similar pressures [27,28]. Nationally, cities with populations between 100,000 and 500,000 are often considered mid-sized [29]. Thus, Salt Lake City with a population of 209,593 in 2023 fits within this mid-sized category [30].

Unlike larger metropolitan areas, mid-sized cities face unique challenges that include fewer financial resources, limited planning capacities, and rapidly shifting demographics [31–34]. These constraints can amplify vulnerabilities to displacement and gentrification, as such cities often lack the robust housing policies or funding mechanisms seen in larger urban centers [35]. Research highlights that mid-sized cities frequently encounter rapid population growth or decline, creating pressure on existing infrastructure and housing markets [36,37]. For instance, Portland, OR, and Providence, RI, have shown how limited budgets and smaller planning departments can struggle to respond to gentrification and displacement trends, leaving gaps in housing equity [9,38]. Additionally, the limited diversity of industries in these cities can lead to economic volatility, further exacerbating housing insecurity for low-income populations [39].

Demographic changes in mid-sized cities, such as increased in-migration of affluent populations, further complicate urban planning efforts [33,40]. Studies have shown that these shifts often result in rising housing costs and displacement of long-standing communities, particularly those of color and low-income households [15,18,41]. In Salt Lake City, for example, these dynamics mirror the challenges observed in other mid-sized cities, where the lack of targeted policies has allowed speculative development and gentrification to disproportionately impact vulnerable groups [42].

This study builds upon previous research in cities such as Portland, OR, and Providence, RI, which have also explored the intersection of GIS-based displacement modeling and community narratives. In Portland, researchers found that while GIS identified high-risk neighborhoods, qualitative interviews revealed unique local challenges, such as zoning changes and public opposition to affordable housing, that GIS models did not account for [43]. Similarly, in Providence, studies highlighted the role of historical inequities and policy decisions in shaping displacement risks, which were underrepresented in GIS outputs [38]. These examples illustrate that mid-sized cities often face distinct displacement dynamics, shaped by their scale, governance frameworks, and socio-economic conditions. By comparing Salt Lake City's findings to these contexts, this study underscores the importance of tailoring displacement interventions to the specific characteristics of mid-sized urban areas.

Furthermore, this paper contributes to theoretical debates in urban studies by integrating structural urban political economy theories—such as Smith's rent gap theory and Logan and Molotch's urban growth machine—with participatory, community-centered frameworks like Asset-Based Community Development (ABCD) and Participatory GIS [44–47]. This theoretical integration allows to analyze displacement as a multi-scalar process, driven both by macroeconomic forces and local development politics, and mediated through the lived experiences of residents. The approach foregrounds the limitations of data-driven models when disconnected from on-the-ground realities.

This research fills a significant gap in the literature by offering a novel comparison of GI models with qualitative analysis in a mid-sized city, which is often underrepresented in gentrification and displacement studies. It demonstrates how community insights can validate, refute, and enrich predictive GIS models, enhancing their policy relevance and social equity impacts. This contributes to ongoing efforts to build more just and participatory urban planning practices.

3. Materials and Methods

3.1. Partnership Background Between Author and EDR Researchers

The author worked with the UDP who did a data analysis for SLC as part of the "Thriving in Place Project," which is a community-driven process to analyze and understand gentrification and displacement and then craft a plan of action [48]. UPD employs a combination of GIS modeling and data analysis to evaluate displacement risk across the city [1]. At the core of the study is the EDR model, a GIS-driven tool designed to estimate the likelihood of displacement for low-income renter households across all census tracts in Utah, with a particular focus on Salt Lake City [1].

3.2. EDR Model

The EDR model identifies census tracts with characteristics strongly correlated with low-income renter population loss between 2015 and 2019. The model analyzes migration patterns by comparing low-income renter populations who left neighborhoods versus those who moved in, using American Community Survey data as the primary data source. The model generates risk classifications through a framework where Low Data Quality indicates tracts with fewer than 500 total households or census margins of error greater than 15% of the estimate. Probable Displacement means the model estimates displacement is likely occurring, Elevated Displacement indicates the model estimates a small amount of displacement, and High Displacement shows the model estimates relatively high displacement levels. The model produces three distinct map layers including an Overall Displacement layer that shows the number of income groups experiencing any displacement risk, a 50–80% AMI layer showing displacement risk specifically for low-income households, and a 0–50% AMI layer that combines extremely low-income and very low-income households using the more extreme displacement scenario when predictions differ.

Key model variables include demographic characteristics, housing market conditions, and built environment features, though the technical report notes that specific variable weights and mathematical formulas are proprietary to the Urban Displacement Project methodology. Critical model limitations acknowledged by UDP include that the model focuses exclusively on renter displacement rather than homeowner displacement pressures, uses 2015–2019 data while missing pandemic and post-pandemic housing market changes, does not incorporate recent housing construction or infrastructure projects, may overestimate risk in areas with high student or military populations due to natural mobility patterns, and may underestimate risk in high-homeownership areas experiencing gentrification pressures.

To enhance the depth of the analysis, the EDR model incorporates additional map layers detailing demographic and economic conditions, such as the percentage of low-income renters, an Affordable Market Index, segregation patterns, and redlined zones from the 1930s. Historical redlining data, sourced from the University of Richmond's Mapping Inequality project, delineates neighborhoods that were redlined in the 1930s, underscoring the enduring effects of racial segregation and disparities in wealth and homeownership [49].

Overlay layers further contextualize the analysis by including city limits and city council district boundaries, which help define the geographic scope of governance and policymaking. Additionally, the analysis integrates demographic-specific layers to account for unique population characteristics, such as tracts with over 30% student populations or over 40% retired individuals. While students' low wages and high migration rates may lead to overestimations of displacement risk, low-income non-student residents in these areas may still face significant pressures. Similarly, retired populations, despite their low incomes, often have stable housing situations, which may also skew risk predictions. Military infrastructure, another key layer, captures areas with military bases or facilities, where high in- and out-migration rates may affect displacement assessments. Finally, the inclusion of roads and transit networks reflects the critical role of accessibility and mobility in shaping housing dynamics.

Together, these layers provide a comprehensive framework for evaluating displacement risk, balancing historical, demographic, and infrastructural influences, and ensuring a more nuanced understanding of the factors contributing to housing instability. More details of the EDR model applied in SLC could be found in the technical paper: Urban Displacement Project's Salt Lake City Displacement Data Analysis [1].

3.3. *Limitations of EDR Model*

Because the model assesses displacement risk for renters, it may overlook displacement pressures faced by homeowners, leading to potential underestimations in areas with higher homeownership rates. Furthermore, the model does not account for new housing developments or infrastructure projects initiated after 2019, which may influence current displacement dynamics. The use of pre-2019 data introduces substantial limitations given COVID-19's transformative impacts on housing markets, with remote work policies, eviction moratoriums, and historically low interest rates fundamentally altering displacement dynamics.

3.4. Interview Data Collection

This study involved comparing the GIS model's outputs with qualitative neighborhood assessments conducted by University of Utah students under the author's guidance. This qualitative data, drawn from 22 community interviews per neighborhood and narratives, offers on-the-ground insights that either validate or challenge the model's predictions. In this study, data for six neighborhoods is presented—Poplar Grove, Glendale, Ballpark, Central City, East Central, Fairpark—for a total of 132 interviews.

The selection of neighborhoods—Poplar Grove, Glendale, Ballpark, Central City, East Central, and Fairpark—was informed by their historical, socio-economic, and demographic significance. These areas represent a spectrum of displacement pressures, from high-risk zones identified by the EDR model to neighborhoods experiencing early signs of gentrification. By focusing on these specific neighborhoods, the study aimed to capture a diverse range of displacement dynamics, offering insights into how varying socio-economic factors interact with urban development and housing policies. Additionally, these areas provided opportunities to compare the model's outputs with rich, community-driven qualitative data, enabling the identification of discrepancies and validation of predictive patterns.

To ensure a diverse and representative sample, researchers employed multiple strategies to engage potential interviewees. Outreach was conducted in various public and private settings, including sidewalks, commercial establishments, and community events. A common practice involved researchers introducing themselves as graduate students from the University of Utah's City and Metropolitan Planning Master's Program. They explained that the interviews were part of a project for their Community Engagement in Planning course, aimed at understanding neighborhood changes.

3.5. Interview Procedures

Each interview lasted approximately 20 min and followed a semi-structured format, allowing participants to provide in-depth responses to key questions while enabling flexibility to explore emerging themes. Interviews were conducted in person, and participants were recruited through public outreach at community events, local businesses, and residential areas. Interviews were audio-recorded with participant consent and later transcribed verbatim for analysis by each student who conducted them. Participants could choose to use their names or not.

The interview process was structured to gather insights into participants' connections to the neighborhood and their views on its evolution. The interview questions were crafted to explore several key themes:

- Relationship to the Neighborhood:
  - How long have you lived or worked in this area?
  - Did you grow up here?
- Neighborhood Strengths:
  - What aspects of this neighborhood do you value most?
  - Which places hold personal significance for you?
- Changes and Losses:
  - Are there places that were important to you but no longer exist?
  - How have you observed the neighborhood change over the past 10–15 years?
- Community Connections:
  - Who are the key people, groups, or leaders that contribute to the neighborhood's character?
  - Are there business owners or community organizations that play an important role?

- Future Vision:
  - How would you like to see the neighborhood develop moving forward?
  - What changes or improvements would you prioritize?

The interview questions were designed to capture key dimensions of gentrification and displacement that quantitative data alone cannot reveal. Questions about participants' relationship to the neighborhood and their observations of changes provide insights into how gentrification impacts long-term residents and the social fabric of the area. From the perspective of Asset-based Community Development (ABCD), exploring neighborhood strengths and community connections highlights the cultural and social assets that define the neighborhood's identity, which are often threatened by displacement [46,50]. Lastly, asking about participants' future vision for the neighborhood helps identify their priorities and concerns, ensuring that community voices are considered in planning and policy decisions. These questions enable a comprehensive understanding of gentrification by uncovering the lived experiences behind broader socioeconomic trends.

3.6. Ethics of Qualitative Data Collection

To protect participant confidentiality, pseudonyms were used for most people unless they wanted to use their real name. Students were required to submit images of their engagement. The IRB allowed images and names of participants if they were interested in telling their stories. We found that business owners, non-profit, and advocates were interested in telling their stories using their names. If people choose anonymity the identifying information was removed from the transcripts. All participants provided informed consent under an Institutional Review Board (IRB)-approved protocol. A total of 12 students contributed to these interviews and they received training under the IRB, a committee responsible for ensuring the ethical treatment of human subjects in research. The IRB approval number is 00099240, titled "Gentrification and Neighborhood Change: Index and Affordability Strategies for Salt Lake City".

3.7. Limitations of Qualitative Data

While 132 interviews were conducted across six neighborhoods (22 per neighborhood), detailed demographic breakdowns of participants were not systematically collected, representing a limitation in assessing potential sampling bias. Despite the absence of explicit demographic information about all respondents, the reliability of the interview data can be supported by the methodological approach described in the documents, which focused on gathering in-depth perspectives from directly affected individuals within the neighborhoods. The research aimed to collect valuable information from people most affected by current development and displacement trends. The methodology emphasized engaging directly with residents to develop a deeper and more detailed story of the community and its assets, with collected information including personal experiences that were shared, which is crucial for understanding the qualitative impact of neighborhood changes from the perspective of those living them. By focusing on direct engagement and collecting unique stories, the interviews aimed to provide rich, contextualized data about neighborhood changes and community assets, which is a hallmark of qualitative research that prioritizes the depth and nuance of individual experiences over broad statistical representation. Therefore, while specific demographic breakdowns are not provided for all interviewees, the data's reliability stems from the direct, in-depth engagement with individuals significantly impacted by and knowledgeable about the neighborhood changes, ensuring the captured experiences are authentic and relevant to the study's objectives.

4. Background: Description of Salt Lake City and Neighborhoods

Data from the American Community Survey (ACS) for 2019 is presented in Table 1 to provide context for each neighborhood, offering a detailed snapshot of their demographic, economic, and housing characteristics. These metrics help illuminate the unique dynamics of neighborhoods like Poplar Grove, Glendale, Ballpark, Central City, East Central, and Fairpark, highlighting the socioeconomic conditions and housing trends that shape their identities. Figure 1 shows where Salt Lake City is located in the United States, and Figure 2 shows where the neighborhoods are situated within Salt Lake City.

Table 1. Demographic, Socioeconomic, and Housing Characteristics.

	Salt Lake City	Central City	Ballpark	East Central	Glendale	Fairpark	Poplar Grove
Population	210,314	10,152	5298	3175	20,899	6409	14,197
% Race (Use the "White alone" category)	73.9	73.1	66.3	80.6	42.5	68.0	36.5
% Ethnicity (Latino)	21.0	15.4	32.6	10.8	43.4	42.4	53.5
% Children under 18 years old	6.1	1.5	7.7	2.1	10.4	6.9	10.5
% Over 65 years old	11.4	10.6	7.3	7.8	7.0	12.3	7.3
% Female-headed families with children	9.3	5.2	15.9	1.4	18.8	17.5	15.9
% Family households	56.3	21.7	42.2	26.6	78.5	66.6	75.0
% Renter-occupied housing units	45.0	85.1	84.3	71.0	43.0	33.3	52.6
% Owner-occupied housing units	55.0	14.9	15.7	29.0	57.0	66.7	47.4
% College education (college degree of higher)	33.6	39.4	9.2	39.1	9.6	13.8	8.6
Median House Value	USD 342,316	USD 285,600	USD 244,000	USD 409,500	USD 190,033	USD 134,600	USD 174,067
Median Family Income	USD 91,332	USD 55,951	USD 51,534	USD 76,250	USD 62,447	USD 48,936	USD 43,576
% Persons Below Poverty	15.7	23.2	31.7	27.3	17.6	32.7	17.4
% Managerial Occupations	9.1	9.6	7.3	12.2	7.1	7.2	4.4

Figure 1. Where Salt Lake City, Utah, is located in the United States.

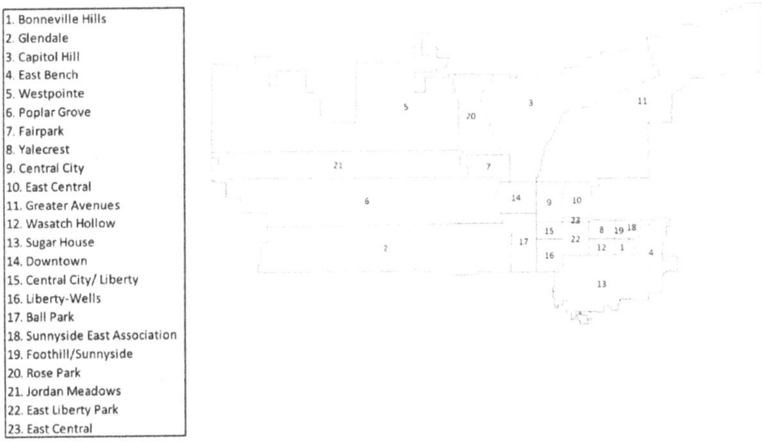

Figure 2. Where the neighborhoods are situated within Salt Lake City.

This contextual information is critical for understanding how gentrification manifests differently across neighborhoods. For example, indicators such as rising median house values, shifts in racial and ethnic composition, changes in educational attainment, and varying proportions of renter-occupied versus owner-occupied housing provide insights into the pressures and transformations that neighborhoods may experience. These data points are not only descriptive but also serve as key indicators of broader trends, helping to identify where gentrification and displacement pressures may be occurring and where targeted interventions might be needed to support vulnerable populations.

4.1. *Salt Lake City*

Salt Lake City, the capital of Utah, has a population of approximately 210,314. The city is predominantly composed of residents who identify as "White alone" (73.9%), with a significant Latino population (21%). It has a relatively younger demographic, with only 6.1% of its population under 18 years old and 11.4% over the age of 65. Family households comprise 56.3% of the total, with 9.3% being female-headed families with children. Housing in Salt Lake City is almost evenly split between renter-occupied units (45.0%) and owner-occupied units (55.0%). The city's median family income is USD 91,332, and 33.6% of residents have obtained a college degree or higher. Despite its relatively high median house value of USD 342,316, 15.7% of the population lives below the poverty line.

The city's demographic and economic makeup provides a unique lens for examining gentrification and displacement trends. Salt Lake City's mix of renters and homeowners indicates a population at risk for displacement, particularly as housing prices continue to rise. Neighborhoods with higher proportions of renters and vulnerable populations, such as families below the poverty line and Latino residents, often experience the earliest signs of gentrification. These signs include rising rents, new high-density developments, and an influx of higher-income residents. Such dynamics, combined with the city's relatively high percentage of college-educated individuals, suggest pressures for neighborhood upgrading and redevelopment. These changes risk displacing long-standing residents, disrupting social networks, and transforming the cultural character of vulnerable communities. Understanding these trends is essential for crafting policies that address housing inequities and promote inclusive urban development.

4.2. Central City

Central City exhibits strong indicators of gentrification, including the highest percentage of college graduates (39.4%) and one of the lowest percentages of family households (21.7%). Median house values are relatively high at USD 285,600, and the population is predominantly White (73.1%). With 85.1% renter-occupied housing and higher income levels (USD 55,951), the area reflects typical signs of urban gentrification, such as displacement of long-standing, low-income residents in favor of higher-income, educated newcomers.

4.3. Ballpark

Ballpark is characterized by high renter occupancy (84.3%) and a lower percentage of families (42.2%). The median house value is USD 244,000, and poverty levels are high (31.7%). The demographic composition includes a majority White population (66.3%) and 32.6% Latino residents. This neighborhood shows significant signs of gentrification, with low family household percentages and high renter occupancy, suggesting turnover and housing instability.

4.4. East Central

East Central stands out as the most affluent neighborhood in the group, with the highest median house value (USD 409,500) and median family income (USD 76,250). A significant portion of the population holds college degrees (39.1%), and only 10.8% of residents are Latino. High housing costs and a low percentage of family households (26.6%) suggest this area has undergone substantial gentrification, leading to reduced housing affordability for low-income populations.

4.5. Glendale

Glendale exhibits similar demographic characteristics to Poplar Grove, with 43.4% Latino residents and a slightly higher median house value (USD 190,033). Owner-occupied housing (57%) dominates, but renters still make up a significant portion (43%). The percentage of families below the poverty line is moderate (17.6%), and educational attainment remains low (9.6% with a college degree). While the signs of gentrification are subtle, its higher family household rates and homeownership could make it less vulnerable to displacement pressures.

4.6. Fairpark

Fairpark has a more balanced housing mix, with 33.3% renter-occupied units and 66.7% owner-occupied units. The median house value (USD 134,600) and income levels (USD 48,936) are among the lowest, and poverty rates are high (32.7%). While educational attainment (13.8%) and managerial occupation percentages (7.2%) remain low, the area shows fewer signs of gentrification due to its lower housing costs and stable family household percentages (66.6%).

4.7. Poplar Grove

Poplar Grove has a predominantly Latino population (53.5%), with relatively low educational attainment (8.6% college graduates) and a median house value of USD 174,067, among the lowest in the group. A significant portion of housing units are renter-occupied (52.6%), and poverty levels (17.4%) are moderate. Signs of gentrification are less apparent here, as low house values and income levels suggest limited economic changes compared to other neighborhoods. However, the area's proximity to more central locations may attract future development pressures.

According to the ACS data alone, Ballpark, Central City, and East Central show the strongest indicators of gentrification, with high renter occupancy, increasing house values, and higher levels of education and income. These patterns suggest displacement risks for low-income, minority, and family households. In contrast, neighborhoods like Fairpark, Glendale, and Poplar Grove retain more affordable housing and family-friendly characteristics, but they may face future pressures as development continues in adjacent areas.

5. Results: Alignment Between GIS Predictions and Lived Experiences

The results of this study reveal significant insights into the relationship between GIS-driven displacement models and the lived experiences of residents in Salt Lake City. By comparing the Urban Displacement Project's (UDP) Estimated Displacement Risk (EDR) model with qualitative interviews conducted across various neighborhoods, several key findings emerged, highlighting both the strengths and limitations of predictive GIS tools in capturing urban displacement dynamics. In Figure 3 the dark reddish color means that the zip code shows high displacement risk for two income groups (very low 0–50% AMI and low income 50–80%), the dark orange is elevated displacement for very low-income households (0–50% AMI) and light orange is probable displacement risk. This classification suggests that rising housing costs, demographic shifts, and increased development are contributing to the vulnerability of low-income households.

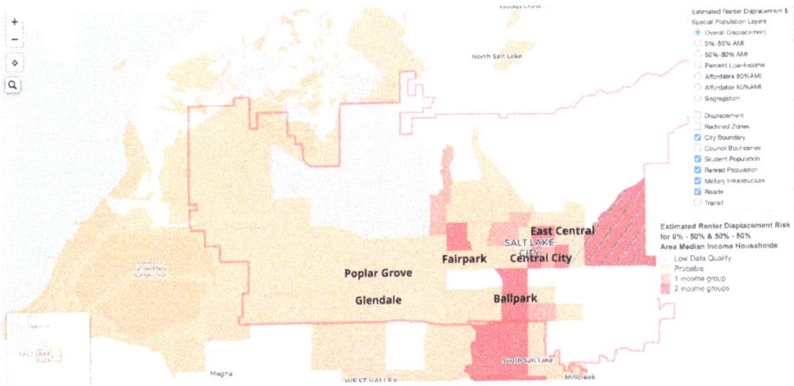

Figure 3. Urban Displacement Project's (UDP) Estimated Displacement Risk (EDR). View Map in Full Screen (https://www.urbandisplacement.org/maps/salt-lake-city-estimated-displacement-risk-model/, accessed on 20 November 2024).

5.1. Central City

The risk of displacement in Central City is evident through the voices of long-time residents and business owners. Bailey, who has lived in the neighborhood since her college years, notes that "the stability of the Central City neighborhood is threatened by the aggressive development of high-priced apartment buildings." She laments the demolition of historical buildings, replaced by luxury apartments, contributing to the erosion of the area's character and affordability. Similarly, Ken, a business owner for 25 years expressed concern that rising rents may force his closure, stating, "If my business is forced to close its current location, I will have nowhere to relocate to. Affordable space no longer exists in the valley."

Long-term residents like Gabriella, who lives in senior housing, worry about the vulnerability of their communities. She shares, "If folks from my building were displaced due to rising rents, I am unsure where they would go." Donnie, a homeowner, highlights the impact of short-term rentals, describing how "more than half my street is short-term rentals, making it feel like I have no neighbors." This shift toward short-term leasing reduces the availability of stable housing, contributing to displacement.

Even those involved in development acknowledge the issue. One developer suggests, "There should be a 2:1 replacement requirement for any affordable rental housing that is lost in development." The voices from Central City reflect a neighborhood in transition, with rising costs and changing demographics pushing out long-standing residents and businesses, underscoring the urgent need for policies to address displacement risks.

The map highlights a high risk of displacement in the area, which aligns with ongoing trends in the neighborhood. Historic homes have been demolished to make way for higher-density apartment buildings, reflecting increased development pressure. The area, previously redlined, is now seeing a surge in investment, contributing to rising property values and rents. With most residents being low-income renters, they are particularly vulnerable to these changes. Additionally, the neighborhood is attracting an influx of out-of-state residents, further intensifying competition for housing. The area's proximity to downtown and strong public transit access makes it desirable, accelerating the risk of displacement. However, the presence of affordable, deed-restricted units provides some safeguard, helping to stabilize households that meet the qualifications for these units.

5.2. Ballpark

The Ballpark neighborhood of Salt Lake City is undergoing significant redevelopment, sparking concerns about displacement and the erosion of community character as described by interviewees. Tracy, a resident since 2015, points out the stark juxtaposition of gentrification and persistent social issues, stating, "It's weird cause, like, I still have people smoking heroin on my front porch, but then I also have a bunch of yuppies that are like mad about it." This highlights the culture clash between long-time residents and wealthier newcomers, as rents in the area continue to rise, forcing many to downsize or leave.

Garrett, who rents a house with others, expresses frustration with the shifting landscape, noting that "I'm not a big fan of all the luxury apartments" replacing historic homes and businesses. Similarly, longtime resident Ron reflects on the physical transformation of the neighborhood, lamenting that "it's on its last legs" as apartment complexes replace single-family homes, diminishing the neighborhood's historic character and sense of community.

Matt, who has lived in Ballpark for 20 years, underscores the resilience of residents fighting to preserve the neighborhood's character, explaining, "We fight what we can fight, we're very careful to welcome those who are going to bring services to the community." However, he also recognizes the challenges posed by safety issues and the lack of local amenities, pointing out the need for more schools and better infrastructure. The voices from Ballpark paint a picture of a neighborhood at a crossroads, where redevelopment brings both opportunity and concern for those at risk of being displaced.

The map indicates elevated displacement risk for both income groups, which aligns with the area's history and current trends. The neighborhood was previously redlined, has a high concentration of low-income renters, and is experiencing rising rents alongside new apartment and townhome developments. Its proximity to transit further increases the likelihood of displacement.

In contrast, the map shows no displacement risk in other areas. This is consistent with the fact that these neighborhoods were historically rated as "best" or "still desirable" on redlining maps. They are characterized by large populations of wealthy White homeowners and high housing prices, making them exclusive and less susceptible to displacement pressures.

5.3. East Central

The interviews conducted in the East Central and East Liberty Park neighborhoods validate the mapping that indicates a high risk of displacement. Several residents expressed concerns about rising rents, new luxury developments, and the influx of wealthier individuals reshaping the area.

Kyle, a new resident, observed that "small single-family homes are being torn down and replaced by large, hideous luxury condos." He expressed fears that these new developments drive up housing costs without sufficiently increasing housing availability. Similarly, Ethan reflected on the changing character of the neighborhood, pointing out that "new apartment complexes are not creating more walkable neighborhoods" and are instead contributing to rising rents.

Long-term residents like Blair also highlighted the impact of new construction, sharing that "local businesses are getting torn down to make way for development," which he finds overwhelming. Gavin, planning to move out of state due to housing costs, emphasized that "rising rents are not just a possibility—they're happening now."

While some residents like Nikki have not yet experienced significant displacement, there is widespread acknowledgment that the increasing presence of luxury apartments and high property values threatens the neighborhood's stability. Dustin, a 12-year resident, noted that "larger apartments going up aren't the answer" and expressed concerns that people will have to relocate to less expensive areas further west if costs continue to rise.

Overall, the interviews consistently reflect anxiety about displacement, reinforcing the findings of the mapping analysis and painting a clear picture of a neighborhood in transition, facing the pressures of gentrification and increasing housing insecurity.

The map indicates high displacement risk in the northern part of the area, while the rest shows no significant risk. This aligns with local demographics, as the northern section near the university has a large student population, most of whom are renters with relatively low incomes. In contrast, the surrounding neighborhoods are more affluent, with a higher concentration of homeowners, contributing to greater stability and lower displacement risk.

5.4. Glendale

The interviews from Glendale reveal deep concerns about gentrification and displacement, reflecting a community at risk of significant change. Residents consistently highlight rising rents, increasing development, and demographic shifts as emerging threats to the neighborhood's stability.

Erin, a homeowner for two years, expressed apprehension about the influx of condominiums and rising housing prices, noting that "there are more rental properties and higher prices now than in previous years." She fears that if current trends continue, there will be no affordable options left within Salt Lake Valley for displaced residents. Similarly, Austin observed a decline in the racial diversity that once defined Glendale, attributing it to the arrival of young, wealthier newcomers. He remarked, "I was more of a minority when I moved here, but now there's been a shift."

Long-term residents like Kim, who has lived in Glendale for 18 years, voiced concerns about the neighborhood losing its character due to increased crime, homelessness, and development pressures. "You have to be rich now to live here," said James, underscoring the economic divide that gentrification exacerbates. Cheryl echoed these sentiments, warning that as more people move in, "they're going to start tearing a lot of places down soon."

Despite these challenges, many residents remain committed to preserving Glendale's community spirit. Cassy, a nine-year resident, values the neighborhood's diversity and hopes it will retain its cultural identity, explaining, "I've lived all over the valley and immediately felt the community here."

The collective narrative points to a strong sense of place and belonging, but also a growing unease about the future. Glendale residents recognize the need for affordable housing, improved infrastructure, and protections for long-term community members to mitigate displacement and ensure the neighborhood's inclusivity endures.

The map shows probable displacement risk for the lowest income group, while showing no significant risk for the 50–80% AMI group [1]. The area has a large Latinx population with flourishing businesses, along with other immigrants and many residents are lower-income households [1]. However, similar to other neighborhoods on the west side, the area has a lot of homeowners who are less likely to be displaced [1].

5.5. *Fairpark*

The Fairpark community in Salt Lake City reflects a neighborhood deeply concerned about the risk of gentrification and displacement. Interviews with residents reveal anxieties about the area's increasing development, rising property values, and the influx of new, wealthier residents. One long-time resident noted, "This place changes all the time... it's suddenly beginning to recover. This was just urban decay." While some see revitalization as positive, others fear it could lead to displacement of lower-income households.

Several interviewees pointed to the loss of local businesses and essential services, like the closure of a neighborhood Walgreens. One resident expressed frustration, saying, "There was no replacement for those businesses, they went away, they're gone." This lack of reinvestment in community-serving institutions raises concerns that new developments may not prioritize existing residents' needs.

Additionally, there is a notable divide between long-term residents and newcomers. A young resident reflected on the transformation, mentioning, "They see some economic growth and a fiscal dollar sign on it, now they're wanting to invest a lot more into where they overlooked for way too long." While development brings attention to the neighborhood, many fear that these changes could price out current residents, forcing them to relocate to more affordable areas outside of Salt Lake City.

The proximity of Fairpark to downtown and transit hubs makes it a prime target for redevelopment, and while residents acknowledge the benefits of investment, many worry about the neighborhood losing its cultural identity. Efforts to preserve affordability and community assets will be crucial in ensuring that revitalization does not lead to widespread displacement.

The map indicates probable displacement risk for both income groups, though the qualitative narrative suggests an even greater risk than the map reflects. Residents highlight the transformation of dilapidated homes into renovated properties, signaling increased investment and rising housing costs. The area's high poverty rate, large Latinx population, and the displacement of multigenerational families by younger renters further emphasize the pressures of gentrification. These firsthand accounts paint a picture of a neighborhood undergoing rapid change, where the threat of displacement looms larger than quantitative models may capture.

5.6. *Poplar Grove*

The Poplar Grove community in Salt Lake City reflects a growing concern about displacement and gentrification, as highlighted by residents' narratives and observations. While the map may show probable displacement risk, the interviews reveal deeper anxieties and lived experiences of change. Long-term residents like JT expressed frustration over the influx of new people, stating, "They are taking over the neighborhood and making it worse." He attributes rising costs, and neighborhood shifts to the arrival of wealthier newcomers and increasing development.

Other residents echoed similar sentiments. One resident, "P," noted how demand for housing is pushing prices up, sharing that his own home value increased by USD 80,000. He described receiving frequent cash offers for his property, which he sees as a sign of encroaching gentrification. P voiced concern over large-scale apartment developments, emphasizing that "I would like to see families move in, not apartment buildings with hundreds of people."

Steph and Dan, residents for nearly a decade, described an "interesting juxtaposition of new development and deterioration." They observed vacant businesses and deteriorating homes alongside new renovations and infrastructure improvements, reflecting the uneven nature of neighborhood change. While they appreciated enhancements to green spaces and trails, they were wary of the rising number of homeless individuals in parks, which they saw as a sign of economic strain and displacement.

Overall, the qualitative data reflects widespread community concern that gentrification is already underway, driven by increasing property values, new development, and shifting demographics. Residents express a strong desire to preserve the character and affordability of Poplar Grove, fearing that unchecked development could lead to the displacement of long-standing, lower-income households.

The map indicates probable displacement risk, with the northwest part of the area showing elevated risk. However, the qualitative narrative points to potentially greater displacement pressures. The neighborhood is highly diverse, with a population that is half Latinx, and new development is emerging along North Temple. Rising home prices, along with the area's green spaces and strong transportation access, make it increasingly appealing to higher-income residents, further heightening the risk of displacement.

6. Discussion

6.1. Summary Comparing EDR Model an Interview Data

The following neighborhood assessments represent exploratory comparisons based on 22 interviews per area. These findings are context-specific to Salt Lake City and should not be generalized beyond this case study without additional validation in similar urban contexts. Table 2 shows a detailed comparison of the EDR model's predicted displacement risk levels with the qualitative insights gathered from community narratives, highlighting key areas of alignment as well as discrepancies in neighborhoods across Salt Lake City. In many cases, the EDR model accurately identified areas experiencing displacement pressures, particularly in neighborhoods near downtown and the east side of the city. Residents in these areas reported increasing rent burdens, new high-density developments, and the loss of affordable housing stock—factors that closely aligned with the model's designation of elevated or high displacement risk. Neighborhoods such as Central City, Ballpark, and the East Central emerged as hotspots where both the model and community narratives highlighted ongoing displacement.

Table 2. Summary of Findings and Discrepancies between EDR Model and Community Narratives.

Neighborhood	EDR Model Risk Level	Community Narratives	Key Discrepancies
Central City	High	Increasing rents, high-density developments, and loss of affordable housing stock.	Aligns well with the model's predictions.
Ballpark	High	Rising housing costs and rapid gentrification observed.	Validates model output but adds nuanced cultural losses.
East Central	High	Residents noted significant commercial and residential displacement.	Model aligns; qualitative data adds psychological impacts.
Glendale	Moderate	Residents reported early gentrification signs, such as new developments and rising rents.	Model underestimates emerging pressures.
Fairpark	Moderate	Concerns about commercial displacement and loss of cultural networks.	Model does not reflect incremental changes.
Poplar Grove	Low	Early signs of gentrification and social network erosion identified.	Model underrepresents risks faced by homeowners.

6.2. EDR Analysis

While Table 2 presents a summary of alignment and discrepancies between the EDR model and community narratives, further analysis reveals key structural limitations in the model's predictive framework. For example, the model tends to underestimate displacement risk in neighborhoods like Glendale, Fairpark, and Poplar Grove due to its reliance on renter-focused metrics and lagging data sources. These areas, characterized by higher rates of homeownership and strong community ties, are experiencing what might be termed "cultural" or "anticipatory" displacement—forms of neighborhood change not readily captured by traditional indicators such as rent burden or income levels. In contrast, high-risk areas like Ballpark and Central City exhibit alignment between the model and resident accounts, but qualitative data reveals deeper impacts, including the erosion of social networks, loss of small businesses, and feelings of disconnection. This suggests that while GIS models are valuable for identifying displacement hotspots, they require complementary qualitative inputs to expose early warning signs, local nuances, and the broader emotional and social dimensions of displacement. Thus, the integration of spatial and narrative data does not merely validate the model but also exposes its blind spots, ultimately strengthening its policy relevance.

However, the study also found discrepancies between model predictions and ground-level experiences, particularly in the west side neighborhoods of Glendale, Fairpark, and Poplar Grove. While the EDR model suggested moderate or low displacement risk in these areas, interviews revealed significant concerns about rising housing costs, new developments, and the influx of higher-income residents. Community members described the early signs of gentrification, including commercial displacement and the gradual erosion of long-standing cultural and social networks. These findings suggest that the model may underrepresent displacement risk in areas with higher rates of homeownership, as it primarily focuses on renter displacement.

Several limitations of the GIS model became apparent through qualitative engagement. The model uses U.S. Census American Community Survey (ACS), Housing Urban Development (HUD) Fair Market Rent and the Bureau of Labor Statistics (BLS) Consumer Price Index data from Salt Lake City 2015–2019. Some of the limitation of this data is that it might now reflect recent upzoning changes [51], demolition of naturally occurring affordable housing, and other displacement pressures that have been taking place in the city [50]. Also, changes related to COVID-19 like the lowering of interest rates which increase home prices during the pandemic [52–55]. In addition, the overall inflation that the economy experience is also reflected in increases in rent [52,54,55].

6.3. Qualitative Data Analysis

Nonetheless, this research should be understood as a methodological exploration rather than a comprehensive validation study. The sample size of 132 interviews, while substantial for qualitative research, cannot support statistical generalization to other urban contexts. The findings are specific to Salt Lake City's unique demographic composition, housing market conditions, and policy environment during the study period.

In west side neighborhoods, interviewees noted that while displacement was not yet as widespread as in downtown areas, it was occurring incrementally, often affecting vulnerable populations first. These patterns, driven by speculative development and increasing land values, were not fully reflected in the model's risk assessments. The discrepancy underscores the need for real-time data integration and continuous model updates to reflect the rapidly evolving housing landscape.

Qualitative data plays a critical role in validating GIS-based models by providing a nuanced understanding of the lived experiences that quantitative metrics may fail to capture. Specifically, in the context of content validity, qualitative insights ensure that the variables used in the model reflect the realities on the ground. For example, interviews and community narratives can reveal socio-cultural factors, emotional ties, and informal networks that are not easily quantified but significantly impact displacement dynamics. By incorporating qualitative data, researchers can identify gaps or biases in the model, leading to more robust and comprehensive evaluations.

6.4. What We Can Learn

From a theoretical perspective, this study highlights the critical intersection of structural and human-centered approaches to understanding displacement. While GIS models provide valuable predictive insights, they must be contextualized through the lived experiences of residents to avoid perpetuating inequities in urban development. Practically, the findings underline the need for adaptive policymaking that combines data-driven strategies with community engagement to create equitable solutions. For example, policymakers should consider mechanisms to safeguard affordable housing, incentivize inclusive development, and support displaced populations with resources such as relocation assistance and social network rebuilding initiatives.

The focus on Salt Lake City as a mid-sized urban area provides a unique lens to understand displacement dynamics that differ from those in large metropolitan areas, such as New York or Los Angeles. As noted in the literature review, mid-sized cities often face distinct challenges, including fewer financial resources, limited planning capacities, and rapidly changing demographics [36,37]. The findings align with studies from other mid-sized cities, such as Portland and Providence, where discrepancies between GIS model predictions and community narratives have revealed early signs of displacement pressures [38,43]. However, the qualitative data further emphasized the emotional and psychological toll of displacement—factors that GIS models cannot easily quantify.

Residents expressed deep ties to their neighborhoods, with many citing generational connections and the importance of local social networks in providing stability. Displacement not only disrupted housing but also affected access to schools, jobs, and essential services, contributing to broader economic and social instability. This human dimension of displacement adds critical context to the model's findings, reinforcing the need for policies that prioritize community preservation alongside housing development.

Fostering participatory planning processes is essential for aligning development projects with the needs and aspirations of residents. Community engagement can provide valuable insights and build trust, ensuring that urban development is equitable and inclusive. Finally, there is a pressing need to develop real-time monitoring tools to better reflect current market and socio-economic conditions, particularly in the post-COVID-19 era. Investing in real-time data collection and regular model updates will allow policymakers to respond more effectively to rapidly changing housing dynamics and mitigate displacement pressures in a timely manner.

By integrating predictive models with community narratives, this study advances both the theory and practice of urban planning, emphasizing the importance of inclusive, data-informed policies to address displacement. Future research should explore how these approaches can be applied to other mid-sized cities, further enhancing their relevance and scalability.

Based on this case study's exploratory findings, future research might consider developing systematic approaches to incorporate community feedback into displacement models. This could include regular community surveys, real-time development tracking, and cultural displacement indicators, though such integration would require extensive validation across multiple urban contexts before implementation

6.5. Future Research

Longitudinal studies tracking displacement over time are needed to assess the long-term effectiveness of policy interventions [56]. Future research should prioritize cross-disciplinary collaborations, integrating insights from urban planning, sociology, and public health, among others [57]. Finally, expanding the accessibility of GIS tools so that residents could use them and fostering community partnerships will be essential in developing responsive, inclusive urban policies that safeguard vulnerable populations against displacement [58].

7. Conclusions

This exploratory study demonstrates the potential value of comparing GIS displacement models with community narratives while acknowledging significant methodological limitations. The case study approach in Salt Lake City suggests that community voices can provide valuable context for understanding model predictions, particularly in identifying areas where quantitative tools may under- or over-estimate displacement risks.

However, the findings cannot be generalized beyond Salt Lake City without additional research. The study's contribution lies in demonstrating a methodological approach for model validation rather than establishing universal principles for GIS-community integration.

Future research should expand this approach through multi-city comparative studies with larger sample sizes, longitudinal tracking of model accuracy over time, and THE systematic development of metrics for quantifying community-reported displacement pressures. Only through such expanded research can the field develop reliable frameworks for integrating technological tools with participatory planning approaches.

Funding: This research received no external funding.

Data Availability Statement: Reports and data could be found in https://www.slc.gov/can/thriving-in-place/, accessed on 16 July 2025.

Acknowledgments: This article was made possible through the GIS data and insights provided by the Urban Displacement Project's Salt Lake City Displacement Data Analysis, led by Tim Thomas and Julia Greenberg at the University of California, Berkeley. The dedicated efforts of students are also acknowledged: Poplar Grove—Skyler Barton and Jennifer Leslie; Glendale—Tyler Torres and Jenna Benson; Ballpark—Anthony Biamont and Jacob Klopfenstein; Central City—Kristin Reidelberger and Meghan Burrows; East Central/East Liberty Park—Matt Ryan and Will Goodreid; Fairpark—Justyna Kaniewska and Faria Afrin Zinia.

Conflicts of Interest: The author declares no conflicts of interest.

Abbreviations

The following abbreviations are used in this manuscript:

EDR	Estimated Displacement Risk
GIS	Geographic Information Systems
UDP	Urban Displacement Project
SLC	Salt Lake City
AMI	Area Median Income
HUD	Housing Urban Development

References

1. Thomas, T.; Greenberg, J. Urban Displacement Project's Salt Lake City Displacement Data Analysis 2022. Available online: https://www.urbandisplacement.org/maps/salt-lake-city-estimated-displacement-risk-model/ (accessed on 16 July 2025).
2. García, I.; Baker, J. *The Socioeconomic Change of Salt Lake City Community Council Districts 1970–2010*; Metropolitan Research Center: Salt Lake City, UT, USA, 2017.

3. Tomaszewski, B. *Geographic Information Systems (GIS) for Disaster Management*, 2nd ed.; Routledge: New York, NY, USA, 2020; ISBN 978-1-351-03486-9.
4. de Sherbinin, A.; Bukvic, A.; Rohat, G.; Gall, M.; McCusker, B.; Preston, B.; Apotsos, A.; Fish, C.; Kienberger, S.; Muhonda, P.; et al. Climate Vulnerability Mapping: A Systematic Review and Future Prospects. *WIREs Clim. Change* **2019**, *10*, e600. [CrossRef]
5. Malczewski, J. GIS-Based Land-Use Suitability Analysis: A Critical Overview. *Prog. Plan.* **2004**, *62*, 3–65. [CrossRef]
6. Yeh, A.G.O. From Urban Modelling, GIS, the Digital, Intelligent, and the Smart City to the Digital Twin City with AI. *Environ. Plan. B Urban Anal. City Sci.* **2024**, *51*, 1085–1088. [CrossRef]
7. Martha, B. The Impact of Gentrification on Urban Communities. *J. Sociol.* **2023**, *1*, 40–49.
8. Levy, D.K.; Comey, J.; Padilla, S. In the Face of Gentrification: Case Studies of Local Efforts to Mitigate Displacement. *J. Afford. Hous. Community Dev. Law* **2007**, *16*, 238–315.
9. Brown-Saracino, J. *The Gentrification Debates: A Reader*; Routledge: New York, NY, USA, 2010; ISBN 978-0-415-80165-2.
10. García, I. No Se Vende (Not for Sale). An Anti-Gentrification Grassroots Campaign of Puerto Ricans in Chicago. *Am. Crit.* **2019**, *3*, 35–61.
11. Assaad, R.H.; Jezzini, Y. Green Gentrification Vulnerability Index (GGVI): A Novel Approach for Identifying at-Risk Communities and Promoting Environmental Justice at the Census-Tract Level. *Cities* **2024**, *148*, 104858. [CrossRef]
12. Johnson, G.D.; Checker, M.; Larson, S.; Kodali, H. A Small Area Index of Gentrification, Applied to New York City. *Int. J. Geogr. Inf. Sci.* **2022**, *36*, 137–157. [CrossRef]
13. Mujahid, M.S.; Sohn, E.K.; Izenberg, J.; Gao, X.; Tulier, M.E.; Lee, M.M.; Yen, I.H. Gentrification and Displacement in the San Francisco Bay Area: A Comparison of Measurement Approaches. *Int. J. Environ. Res. Public Health* **2019**, *16*, 2246. [CrossRef] [PubMed]
14. Zuk, M.; Bierbaum, A.H.; Chapple, K.; Gorska, K.; Loukaitou-Sideris, A. Gentrification, Displacement, and the Role of Public Investment. *J. Plan. Lit.* **2018**, *33*, 31–44. [CrossRef]
15. Chapple, K.; Song, T. Urban Displacement and Predictive Modeling: Insights from the Urban Displacement. *Urban Stud. J.* **2021**, *58*, 511–531.
16. García, I.; Nolan, L.; Smith, J.; Sonomez, Z.; Zelalem, Y. *The Socioeconomic Change of Chicago's Community Areas (1970–2010)*; Nathalie P. Voorhees Center for Neighborhood and Community Improvement at the College of Urban Planning and Policy (CUPPPA) and University of Illinois at Chicago (UIC): Chicago, IL, USA, 2014.
17. Chapple, K.; Zuk, M. Forewarned: The Use of Neighborhood Early Warning Systems for Gentrification and Displacement. *Cityscape* **2016**, *18*, 109–130.
18. Easton, S.; Lees, L.; Hubbard, P.; Tate, N. Measuring and Mapping Displacement: The Problem of Quantification in the Battle against Gentrification. *Urban Stud.* **2020**, *57*, 286–306. [CrossRef]
19. Schafran, A. Pandemic and Displacement: Rethinking Housing Vulnerability. *City Community* **2020**, *19*, 671–679.
20. Bernstein, A.G.; Isaac, C.A. Gentrification: The Role of Dialogue in Community Engagement and Social Cohesion. *J. Urban Aff.* **2023**, *45*, 753–770. [CrossRef]
21. Newman, K.; Wyly, E.K. The Right to Stay Put, Revisited: Gentrification and Resistance to Displacement in New York City. *Urban Stud.* **2006**, *43*, 23–57. [CrossRef]
22. Gould, K.; Lewis, T. *Green Gentrification: Urban Sustainability and the Struggle for Environmental Justice*; Routledge: Oxfordshire, UK, 2017.
23. Fullilove, M.T. *Root Shock: How Tearing Up City Neighborhoods Hurts America, and What We Can Do About It*; One World/Ballantine: New York, NY, USA, 2005; ISBN 978-0-345-45423-2.
24. Corburn, J. *Street Science: Community Knowledge and Environmental Health Justice*; MIT Press: Boston, MA, USA, 2005.
25. Davis, B.; Foster, K.A.; Pitner, R.O.; Wooten, N.R.; Ohmer, M.L. Innovating Methodologies for Examining Gentrification-Induced Social and Cultural Displacement: An Illustration of Integrating Photovoice into Story Map. *Urban Aff. Rev.* **2024**, *60*, 367–386. [CrossRef]
26. Freeman, L. Displacement or Succession? Residential Mobility in Gentrifying Neighborhoods. *Urban Aff. Rev.* **2005**, *40*, 463–491. [CrossRef]
27. Immergluck, D. Large Redevelopment Initiatives, Housing Values and Gentrification: The Case of the Atlanta Beltline. *Urban Stud.* **2009**, *46*, 1723–1745. [CrossRef]
28. García, I.; Miller, S.; Holmes, T. Rural Communities Challenges and ResilientSEE: Case Studies from Disasters in Florida, Puerto Rico, and North Carolina. *Soc. Sci. Humanit. Open* **2023**, *7*, 100412. [CrossRef]
29. Raetz, H. Housing Characteristics of Small and Mid-Sized Cities. Available online: https://furmancenter.org/thestoop/entry/housing-characteristics-of-small-and-mid-sized-cities (accessed on 27 December 2024).
30. U.S. Census Bureau. U.S. Census Bureau QuickFacts: Salt Lake City, Utah; United States. Available online: https://www.census.gov/quickfacts/fact/table/saltlakecitycityutah,US/PST045219 (accessed on 24 November 2020).
31. Tassonyi, A.T. The Context and Challenges for Canada's Mid-Sized Cities. *SPP Res. Pap.* **2017**, *10*, 1–24. [CrossRef]
32. Friedman, A. *Planning Small and Mid-Sized Towns: Designing and Retrofitting for Sustainability*; Routledge: New York, NY, USA, 2014; ISBN 978-0-203-10781-2.

33. Bereitschaft, B. Neighbourhood Change among Creative–Cultural Districts in Mid-Sized US Metropolitan Areas, 2000–2010. *Reg. Stud. Reg. Sci.* **2014**, *1*, 158–183. [CrossRef]
34. Chettry, V. Dynamics of Urban Growth in Mid-Sized Cities Using Census Data. EBSCOhost. Available online: https://openurl.ebsco.com/contentitem/doi:10.22034%2FIJHCUM.2024.03.11?sid=ebsco:plink:crawler&id=ebsco:doi:10.22034%2FIJHCUM.2024.03.11 (accessed on 1 February 2025).
35. Cafferky, P. Planning for Anti-Displacement Development: An Affordable Housing Study in Central Falls. Master's Thesis, Massachusetts Institute of Technology, Cambridge, MA, USA, 2021.
36. Nicol, P.; Biggar, J. Optimizing Urban Density: Developer Positions on Densification in Two Mid-Sized Cities. *Plan. Pract. Res.* **2025**, *40*, 306–325. [CrossRef]
37. Mallach, A. *Smaller Cities in a Shrinking World: Learning to Thrive Without Growth*; Island Press: Washington, DC, USA, 2023; ISBN 978-1-64283-227-3.
38. RKG Associates Anti-Displacement and Comprehensive Housing Strategy. 2020. Available online: https://www.providenceri.gov/planning/comprehensive-housing-strategy/, (accessed on 16 July 2025).
39. Shelton, T. Mapping Dispossession: Eviction, Foreclosure and the Multiple Geographies of Housing Instability in Lexington, Kentucky. *Geoforum* **2018**, *97*, 281–291. [CrossRef]
40. Brewer, K.; Grant, J.L. Seeking Density and Mix in the Suburbs: Challenges for Mid-Sized Cities. *Plan. Theory Pract.* **2015**, *16*, 151–168. [CrossRef]
41. Delgado, E.; Swanson, K. Gentefication in the Barrio: Displacement and Urban Change in Southern California. *J. Urban Aff.* **2021**, *43*, 925–940. [CrossRef]
42. García, I. Evictions and Housing Instability among Latina and Immigrant Mothers in Salt Lake City. *J. Hous. Built Environ.* **2024**, *39*, 769–786. [CrossRef]
43. Bates, L.K. Gentrification and Displacement Study. 2013. Available online: https://www.portland.gov/sites/default/files/2020-01/2-gentrification-and-displacement-study-05.18.13.pdf (accessed on 16 July 2025).
44. Smith, N. Gentrification and the Rent Gap. *Ann. Assoc. Am. Geogr.* **1987**, *77*, 462–465. [CrossRef]
45. Logan, J.R.; Molotch, H. *Urban Fortunes: The Political Economy of Place*; University of California Press: Berkeley, CA, USA, 1988; ISBN 978-0-520-06341-9.
46. García, I. Asset Based Community Development (ABCD): Core Principles. In *Research Handbook on Community Development*; Phillips, R., Trevan, E., Eds.; Edward Elgar Publishing Company: Cheltenham, UK, 2020; pp. 67–75.
47. Brown, G.; Schebella, M.F.; Weber, D. Using Participatory GIS to Measure Physical Activity and Urban Park Benefits. *Landsc. Urban Plan.* **2014**, *121*, 34–44. [CrossRef]
48. Thriving in Place. Available online: https://www.slc.gov/can/thriving-in-place/ (accessed on 27 December 2024).
49. Nelson, R.; Winling, L. Mapping Inequality: Redlining in New Deal America. 2022. Available online: https://dsl.richmond.edu/panorama/redlining/ (accessed on 16 July 2025).
50. García, I.; Biamont, A.; Klopfenstein, J. A Case Study of Story Mapping, Neighborhood Change, and Community Assets of Ballpark, Salt Lake City. *Land* **2024**, *13*, 1573. [CrossRef]
51. Park, K.; Ewing, R.; Sabouri, S.; Choi, D.; Hamidi, S.; Tian, G. Guidelines for a Polycentric Region to Reduce Vehicle Use and Increase Walking and Transit Use. *J. Am. Plann. Assoc.* **2020**, *86*, 236–249. [CrossRef]
52. Wang, B. How Does COVID-19 Affect House Prices? A Cross-City Analysis. *J. Risk Financ. Manag.* **2021**, *14*, 47. [CrossRef]
53. Yiu, C.Y. Why House Prices Increase in the COVID-19 Recession: A Five-Country Empirical Study on the Real Interest Rate Hypothesis. *Urban Sci.* **2021**, *5*, 77. [CrossRef]
54. Boesel, M.; Chen, S.; Nothaft, F.E. Housing Preferences during the Pandemic: Effect on Home Price, Rent, and Inflation Measurement. *Bus. Econ. Clevel. Ohio* **2021**, *56*, 200–211. [CrossRef] [PubMed]
55. Aladangady, A.; Aneerg, E.; Garcia, D. House Price Growth and Inflation During COVID-19. 2022. Available online: https://www.federalreserve.gov/econres/notes/feds-notes/house-price-growth-and-inflation-during-covid-19-20221117.html?trk=organization_guest_main-feed-card_feed-article-content (accessed on 16 July 2025).
56. Lees, L. Gentrification and Social Mixing: Towards an Inclusive Urban Renaissance? *Urban Stud.* **2008**, *45*, 2449–2470. [CrossRef]
57. Thurber, A.; Krings, A.; Martinez, L.S.; Ohmer, M. Resisting Gentrification: The Theoretical and Practice Contributions of Social Work. *J. Soc. Work* **2021**, *21*, 26–45. [CrossRef]
58. Pánek, J.; Glass, M.R.; Marek, L. Evaluating a Gentrifying Neighborhood's Changing Sense of Place Using Participatory Mapping. *Cities* **2020**, *102*, 102723. [CrossRef]

land

MDPI

Article

Examining Municipal Procurement and Cooperation Networks in Smart Land Use Planning: The Yangtze River Delta Case

Gangjian Lin [1,2] and Yuanshuo Xu [1,2,*]

1 School of Public Affairs, Zhejiang University, Hangzhou 310002, China; gangjian_lin@zju.edu.cn
2 China Institute of Urbanization, Zhejiang University, Hangzhou 310002, China
* Correspondence: xuyuanshuo@zju.edu.cn

Abstract: Smart Land Use Planning (SLUP) has gained increasing attention in urban development, yet few studies examine its implementation from an urban governance perspective. This study investigates municipal SLUP project characteristics, their spatial distribution, and intercity cooperation networks by analyzing 3689 SLUP government procurement contracts in China's Yangtze River Delta urban agglomeration. Using the Latent Dirichlet Allocation model, this study identified four main SLUP project types: real estate management, land resource protection, land use planning, and geographic information services. Spatial analysis revealed significant imbalances across cities, with SLUP projects concentrated in central cities while other cities heavily depend on intercity cooperation for technical support and services. Network analysis showed a core–periphery structure, with industrial structure and institution similarities significantly facilitating cooperation, while geographic distance and cultural similarity had limited impact. Future research should expand data sources to enable cross-regional comparative analysis. This study offers empirical evidence for policymaking in the implementation of SLUP and regional coordinated development.

Keywords: government procurement; local government network; smart city; intercity cooperation

1. Introduction

With increasing urbanization and digital advancement, Smart Land Use Planning (SLUP) has become an important tool for optimizing land resource allocation and enhancing urban governance effectiveness. SLUP uses artificial intelligence algorithms such as such as machine learning (ML) and deep learning (DL) to process data, integrating GIS and big data analysis to address urban land scarcity, irrational land use, and environmental sustainability challenges [1–7]. It has significant value for improving land use efficiency and achieving smart city and sustainable development goals [8–10].

Existing studies on SLUP mainly focus on three aspects: technological applications, policy frameworks, and effectiveness assessment. Technological application studies focus on how to utilize remote sensing, big data, and artificial intelligence technologies to support land use decisions [11–13]. Policy framework studies explore institutional design and policy instruments that promote SLUP implementation [14–16]. Effectiveness assessment studies primarily examine SLUP's impact on land use efficiency, environmental quality, and social equity factors [17–20]. However, few studies discuss the government's role in SLUP implementation from an urban governance perspective, especially government behaviors and intercity cooperation at the regional scale. In other words, how do governments within a region implement SLUP?

Government is the key driver of smart city and SLUP implementation [21], and its investment directly affects the progress and effectiveness of smart city and SLUP [1,22]. Particularly under China's government-led urban development model, government procurement has become an important policy tool for promoting SLUP [8]. Government procurement refers to the act of government obtaining necessary goods, works, and services from other organizations using fiscal funds through statutory procedures [23,24]. It is an important means for local governments to provide public services. Through government procurement, governments can overcome their own capability limitations and improve service efficiency and performance by leveraging the strengths of other governments and enterprises [25,26]. Specifically in the SLUP context, since governments themselves lack sufficient digital technology capabilities, cooperation with enterprises through government procurement is an important pathway for SLUP implementation. Governments may adopt digital technologies in several types of public service delivery to achieve SLUP. For example, in land property rights management, digital registration systems improve the reliability and transparency of property information [27,28]; in land resource protection, intelligent monitoring technologies enable governments to more effectively monitor land use conditions [29]; in land use planning, data analysis and simulation tools provide scientific basis for decision-makers [30–32]. However, adopting these technologies may also generate challenges, including insufficient communication and integration between different systems and administrative levels, creating data silos, inefficiencies, and coordination difficulties in governance [33,34]. Investments in these areas reflect local governments' policy preferences and innovation willingness in SLUP, which is significant for understanding SLUP implementation, yet few studies have discussed this.

Moreover, due to differences in development stage within a region, government procurement in SLUP among different cities may exhibit significant variations, leading to spatially uneven distribution of SLUP development levels. This spatial imbalance may affect the overall regional SLUP development level, thereby hindering regional sustainable development goals [3,4,16]. In this context, cooperation between cities within a region plays an important role in promoting regional SLUP development [35,36]. Since government procurement involves cooperative relationships between city governments and other organizations, cities lacking digital technology capabilities can achieve SLUP by purchasing services from more capable cities within the region. Procurement contract relationships among cities constitute the intercity cooperation network in SLUP. Therefore, understanding the intercity cooperation mechanisms in SLUP government procurement is important for helping lagging cities implement SLUP projects, thereby enhancing regional SLUP development levels.

City networks provide an important perspective for understanding cooperative relationships between cities within a region. City networks are complex systems constituted by various flow relationships between cities (such as capital, information, technology) [37–39]. Castells describes them as social forms constituted by spaces of flows [40], while Taylor and Derudder emphasize that city networks are concrete manifestations of functional connections between cities in the context of globalization [41]. City network research examines various relationship types (Table 1): production networks explore economic value chains and capital flows between enterprises [42]; infrastructure networks investigate how transportation systems like aviation and railways affect urban development [43,44]; and innovation networks analyze academic collaboration and technology diffusion pathways [45,46]. In the field of public administration, government-involved city networks also receive attention. Shrestha and Feiock proposed the concept of local government networks [47]. These networks are collaborative structures where local governments interact with diverse actors—including government agencies, corporations, nonprofits, and citizens—to solve cross-jurisdictional problems, share policy expertise, and align regional development strategies [48,49]. In local government networks, intercity cooperation may be influenced by multiple factors, including geographic distance, size similarity, institutional similarity, and others [50,51]. However, existing research has not systematically revealed the formation mechanisms of intercity SLUP cooperation networks, especially in rapidly developing urban agglomeration regions.

Table 1. Summary of city network concepts and research themes.

Theme	References	Brief Summary
Production Networks	[34]	Networks that explore economic value chains and capital flows between enterprises.
Infrastructure Networks	[35,36]	Networks examining how transportation systems (aviation, railways) influence urban development.
Innovation Networks	[37,38]	Networks analyzing academic collaboration and technology diffusion pathways.
Local Government Networks	[39–41]	Collaborative structures involving local governments and diverse actors addressing cross-jurisdictional issues, sharing policy knowledge, and aligning regional development strategies.

The Yangtze River Delta urban agglomeration (YRD), one of China's most economically vibrant regions, offers an ideal case study for examining the characteristics of SLUP projects and cooperation networks across cities. Based on government procurement contracts, this study employs text analysis and social network analysis methods to explore the spatial distribution in SLUP government procurement projects and city cooperation networks across the YRD. Specifically, the study focuses on three questions: (1) What are the typological characteristics of SLUP government procurement projects in the YRD? (2) How are government SLUP projects spatially distributed across different cities in the region? (3) What are the structural characteristics of the intercity SLUP cooperation network, and which factors influence the formation of cooperative relationships?

For the first question, the study collects 3689 SLUP-related government procurement contracts and applies Latent Dirichlet Allocation (LDA) analysis, identifying four main project types. For the second question, the study analyzes the procurement scale and spatial distribution characteristics of SLUP-related government procurement projects across the 41 YRD cities. For the third question, the study develops an intercity SLUP cooperation network for the YRD region and employs Multiple Regression Quadratic Assignment Procedure (MRQAP) to examine the influence of city homogeneity and proximity on SLUP cooperative relationships. This paper makes three main contributions. First, it reflects government actions and preferences in the SLUP implementation process based on actual government procurement data. Second, it reveals the spatial distribution characteristics of SLUP government investment in the YRD region, enriching the regional research perspective on SLUP. Finally, it introduces social network analysis methods into SLUP research, revealing the structural characteristics and formation mechanisms of SLUP city cooperation networks, providing a new analytical framework for understanding regional cooperative actions and promoting SLUP development.

2. Materials and Methods

2.1. Study Area

The YRD region encompasses 41 cities across Shanghai Municipality, Jiangsu Province, Zhejiang Province, and Anhui Province, with a population over 220 million and covering an area of 358,000 square kilometers [52]. The YRD holds a crucial strategic position in China's regional development landscape and serves as a focal point for the country's digital economy growth. In 2023, the YRD's digital economy added value surpassing CNY 12 trillion, accounting for over 40% of the total GDP across the three provinces and one municipality. Meanwhile, major differences in digital development persist among cities in the YRD. For instance, in the 2022 Digital Financial Inclusion Index [53], seven cities in the YRD ranked among China's top ten in 2022, while Huaibei City was positioned only 140th nationally, highlighting the digital development gap between core and peripheral cities. In recent years, the integrated development of the YRD has been proposed as a national strategy. Cities within this region have established cooperation across multiple domains, with smart city initiatives, including SLUP, emerging as a key area of cooperation. However, systematic research is still lacking, especially on how much cities invest in SLUP and what factors promote or hinder intercity cooperation of SLUP. Therefore, focusing on the YRD as a research area helps understand local government behavior during regional integration processes and provides theoretical and practical insights for promoting more widespread and effective implementation of SLUP.

2.2. Data Source

This study utilizes government procurement data obtained from online government procurement contracts. Current social science research increasingly leverages government procurement data to analyze government service delivery [23,25]. In China, the Ministry of Finance mandates the disclosure of government procurement contract information on the China Government Procurement Website [54] and provincial procurement websites since 2017. We collected contracts related to government digitalization projects by using keywords including "system", "platform", "digital", "information", and "smart" on the China Government Procurement Network and the provincial procurement websites of Zhejiang, Jiangsu, Shanghai, and Anhui provinces. Data retrieval occurred in June 2023, yielding a total of 203,996 procurement contracts. We limited our search scope to land-related government agencies, including Bureau of Land and Resources, Bureau of Planning, and Bureau of Planning and Natural Resources, covering procurement data from 2018 to 2022. Python web crawlers (Python version 3.10.9) were used to extract the relevant data. For the collected contracts, we applied regular expressions to identify key contract information from unstructured web data, including project names, dates, administrative regions, purchasers, suppliers, and contract amounts. Subsequently, we matched supplier information from these contracts with enterprise registration data obtained from Tianyancha [55], one of China's largest enterprise databases, to identify supplier addresses. After manual verification of the extracted data, and excluding SLUP projects procured by central and provincial governments and simple goods procurement contracts (e.g., purchasing computers and servers, which typically do not involve complex collaboration processes), we obtained 3689 government procurement projects related to SLUP. Figure 1 illustrates the data acquisition and filtering procedure.

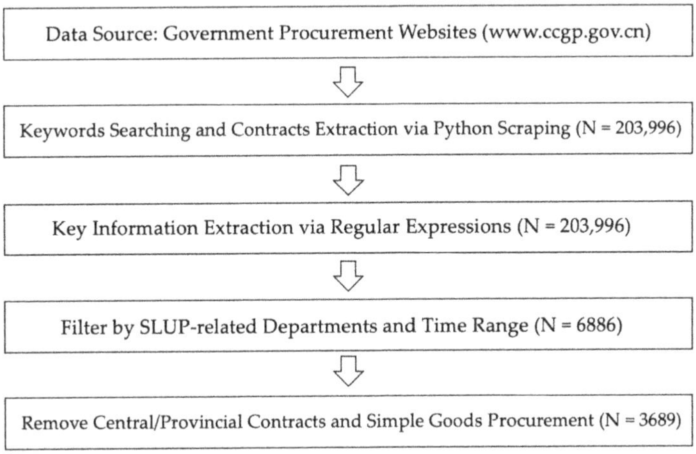

Figure 1. Data Acquisition and Filtering Procedure.

2.3. *Methods*

2.3.1. Latent Dirichlet Allocation Topic Model Analysis

We use LDA to identify the main types of SLUP projects from government procurement project titles. Latent Dirichlet Allocation (LDA) is a generative probabilistic topic model designed to discover latent thematic structures in text collections [56]. LDA assumes each document consists of multiple topics mixed in different proportions, with each topic representing a probability distribution over words in the vocabulary. The generative process of LDA is based on two multinomial distributions: topic–word distribution and document–topic distribution, both governed by Dirichlet priors. Implementing LDA analysis typically involves three steps: text preprocessing, model training, and topic inference. Preprocessing includes tokenization, stopword removal, and vocabulary construction. Model training generally employs methods such as variational Bayes or Gibbs sampling to estimate model parameters. Topic inference calculates the topic distribution for each document and word distribution for each topic. The key parameter in LDA is the number of topics K, typically determined by optimizing perplexity or coherence metrics. The Gensim library in Python is employed to perform LDA topic analysis.

2.3.2. Social Network Analysis

Social Network Analysis is a methodological approach for studying the relational structures between social entities by describing and analyzing connection patterns between nodes using graph theory and network theory [57]. In social networks, nodes represent actors while edges represent relationships or interactions between actors. In this study, social network analysis is used to construct and analyze the structural characteristics of intercity collaboration networks in SLUP across the YRD region. We used intercity SLUP procurement amounts as the weights for edges in the cooperation network. For example, if the government of City A procured SLUP projects worth CNY 1 million from a supplier registered in City B, the weight of the edge from City A to City B would be 1 (unit: million CNY). This study employs the following core metrics.

Average Weighted Degree Reflects the average value of weighted connections for nodes. The formula is as follows:

$$\bar{s} = \frac{1}{n}\sum_{i=1}^{n}\sum_{j=1}^{n} A_{ij} \tag{1}$$

In Equation (1), A_{ij} represents the weight of the connection between nodes i and j. The average weighted degree reflects the average strength of collaborative relationships between cities.

Network Density measures the closeness of network connections. The formula is as follows:

$$D = \frac{L}{n(n-1)/2} \tag{2}$$

In Equation (2), L is the actual number of edges and n is the number of nodes. Network density reflects the prevalence of intercity collaboration.

Average Clustering Coefficient is the mean of local clustering coefficients for all nodes in the network, used to evaluate the degree of close connections within each node's neighborhood. The formula is as follows:

$$CC = \frac{1}{n}\sum_{i=1}^{n}\frac{2d_i}{k_i(k_i-1)} \tag{3}$$

In Equation (3), k_i is the number of neighbors for node i, and d_i is the number of actual connections between these neighbors.

Centralization measures the imbalance in the distribution of centrality across nodes in the entire network, indicating whether the network exhibits concentrated characteristics dependent on a few core nodes. If the degree of a small number of nodes is significantly higher than other nodes, the centralization is high, suggesting that the network may be overly dependent on these key nodes for information control or resource allocation. The formula is as follows:

$$C_D = \frac{\sum_{i=1}^{n}\left(C_D^{\max} - C_D^{i}\right)}{(n-1)(n-2)} \tag{4}$$

In Equation (4), C_D^{max} is the maximum degree centrality value in the network, and C_D^i is the degree centrality of node i.

2.3.3. Multiple Regression Quadratic Assignment Procedure

Multiple Regression Quadratic Assignment Procedure (MRQAP) is a statistical method for analyzing relational data that effectively addresses autocorrelation issues in network data [58]. In this study, MRQAP is used to examine how factors such as city homogeneity and proximity influence the formation of smart land use planning collaboration networks. Compared to traditional linear regression, MRQAP offers significant advantages in network data analysis: First, linear regression assumes observations are independent, while relationships in network data are typically non-independent—MRQAP resolves this through quadratic assignment permutation tests. Second, MRQAP processes relational data directly in matrix form, preserving the integrity of network structures. Third, MRQAP provides more robust statistical inference for spatial autocorrelation and multicollinearity issues common in urban networks. In this study, intercity cooperation is influenced by multiple dimensions including geographic proximity and economic structural similarity, with complex interactions between these factors. MRQAP can accurately assess the net effect of each factor, revealing the intrinsic mechanisms of urban collaboration network formation. The basic regression model for MRQAP is

$$Y_{ij} = \alpha + \beta_1 X_{1ij} + \beta_2 X_{2ij} + \cdots + \beta_k X_{kij} + \varepsilon_{ij} \tag{5}$$

In Equation (5), Y_{ij} represents the relationship value between nodes i and j in the dependent variable matrix, X_{kij} represents the relationship value between nodes i and j in the kth independent variable matrix, and β_k is the regression coefficient.

3. SLUP Procurement Project Characteristics and Their Spatial Distribution

3.1. SLUP Procurement Project Characteristics

From 2018 to 2022, governments of 41 cities in the YRD region invested in 3689 SLUP projects, with a total expenditure of CNY 7.22 billion (approximately USD 1 billion). This substantial investment reflects the high priority these cities place on SLUP. Figure 2 shows the financial trends across this period, indicating a consistent upward trajectory in procurement spending. This growth demonstrates that SLUP in the YRD region is in a rapid development phase, maintaining momentum even during the pandemic. The sustained growth aligns with national digital development strategies, particularly the 2016 "National Informatization Development Strategic Outline", which provided policy support for SLUP initiatives. Additionally, China's reform from traditional land use planning to a comprehensive spatial planning system has created more application scenarios for smart technologies, driving both technological innovation and deeper implementation.

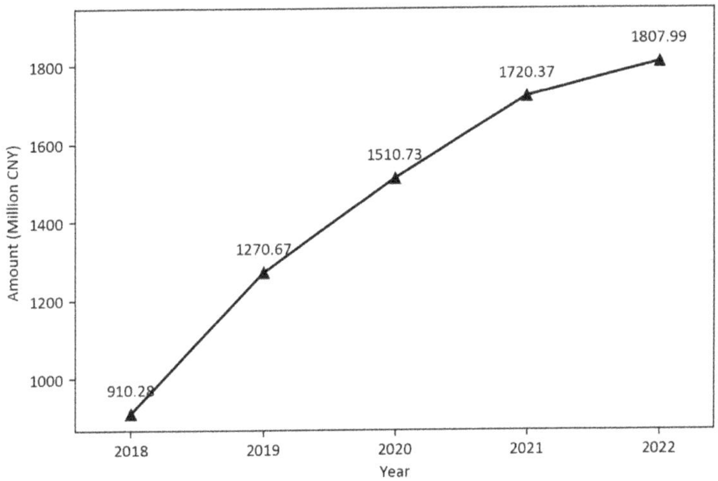

Figure 2. 2018–2022 annual SLUP project procurement amount changes (million CNY).

To understand government project characteristics in SLUP, we conducted LDA topic analysis on procurement project titles to identify distinct project categories. We first evaluated LDA models with 2–20 topics and plotted their topic-coherence scores (Figure 3). Topic coherence measures the semantic consistency of the top words within each topic—the higher the score, the easier the topic is for humans to interpret and the clearer its separation from other topics. The 11-topic solution achieved the highest coherence (0.5749), providing the best balance between detail and interpretability. Based on topic coherence evaluation, we determined 11 as the optimal number of topics, which were then manually categorized into four main groups: Real Estate Management, Land Use Planning, Land Resource Protection, and Geographic information Services. Table 2 lists each group's concise concept definition and its constituent sub-topics. Based on classification coherence evaluation, we determined 11 as the optimal number of topics, which were then manually categorized into four main groups: Real Estate Management, Land Use Planning, Land Resource Protection, and Geographic information Services.

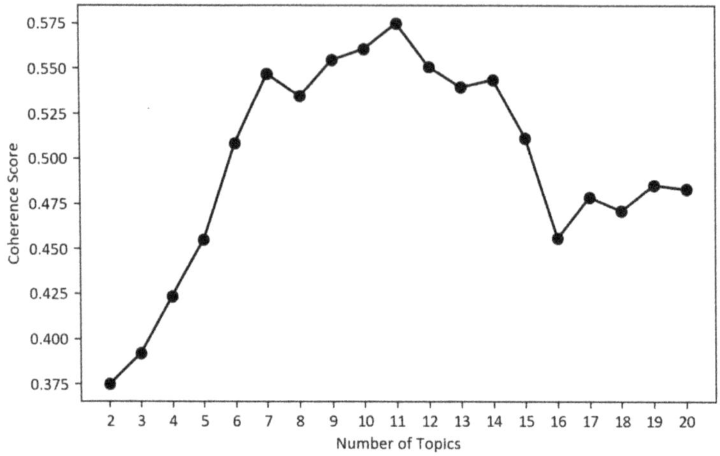

Figure 3. Topic-coherence Scores.

Table 2. Groups definition and sub-topics.

Group	Definition	Sub-Topics & Key Terms
Real Estate Management	Uses digital tools to record property rights, handle registrations, track transactions, and keep archives in order.	Topic 1: real estate/registration/maps/coordination Topic 11: real estate registration/archives/system upgrades
Land Resource Protection	Watches over land and the environment with sensors, 3-D models, and surveys to spot fires, erosion, or other risks and keep land records up to date.	Topic 2: monitoring/forest/fire prevention Topic 7: land resources/3D/supervision Topic 10: land/survey/database
Land Use Planning	Pulls different spatial plans into one shared map so regions can make long-term, well-coordinated decisions about land and infrastructure.	Topic 4: integration/unified planning/infrastructure Topic 5: one map/spatial planning/information systems Topic 6: planning/natural resources/14th Five-Year Plan
Geographic Information Services	Runs the mapping platforms and online geodata services the other groups rely on, and keeps that data secure and easy to share.	Topic 3: data/mapping/geographic information Topic 8: service projects/land/internet Topic 9: integration/data security/server

The first category is Real Estate Management, encompassing property registration, rights confirmation, and transaction management. It includes Topic 1 and Topic 11. Topic 1 focuses on optimizing registration processes and inter-governmental coordination, with keywords such as "real estate/registration/maps/coordination". Topic 11 addresses digital archiving and system upgrades, featuring keywords like "real estate registration/archives/system upgrades". This category demonstrates SLUP applications in real estate management, emphasizing property rights confirmation, information sharing, and process optimization.

The second category, Land Resource Protection, concentrates on dynamic land resource monitoring, ecological protection, and disaster prevention. It comprises Topics 2, 7, and 10. Topic 2 highlights forest fire prevention and state-owned land security monitoring, with keywords "monitoring/forest/fire prevention". Topic 7 emphasizes 3D spatial supervision and rural land remediation, with keywords "land resources/3D/supervision". Topic 10 focuses on land surveys and database development, with keywords "land/survey/database". This category reflects SLUP applications in land resource protection, particularly in ecological security and land surveys.

The third category is Land Use Planning, including spatial planning development, data integration, and multi-planning collaborative management. It contains Topics 4, 5, and 6. Topic 4 emphasizes cross-regional planning coordination, with keywords "integration/unified planning/infrastructure". Topic 5 reflects "one map" implementation and system integration, with keywords "one map/spatial planning/information systems". Topic 6 focuses on strategic planning, with keywords "planning/natural resources/14th Five-Year Plan". This category represents core SLUP content, especially unified planning and spatial planning system development.

The fourth category is Geographic information Services, covering digital government services, geographic information mapping, geographic information services, and data protection. It includes Topics 3, 8, and 9. Topic 3 focuses on basic geographic data management and smart scenario development, with keywords "data/mapping/geographic information". Topic 8 involves "Internet+ government services", with keywords "service projects/land/internet". Topic 9 emphasizes system integration and data security protection, with keywords "integration/data security/server". Unlike the first three groups, which focus on how the government manages land or sets policy, Geographic Information Services looks outward. It gives businesses and citizens the maps, location data, and secure online portals they need. Because it serves as a public utility—rather than a planning or regulatory tool—mixing it with the other groups would hide this special role. We therefore keep it as its own group. This category demonstrates the integration trend between geographic information technology and digital government services in the context of smart cities and SLUP development.

Figure 4 illustrates the procurement distribution across SLUP project categories and their temporal trends. In terms of procurement scale, Geographic information Services accounts for 30.8% of total investment, ranking first, followed by Land Use Planning at 29.9%. Land Resource Protection and Real Estate Management represent 25.6% and 13.6%, respectively. This distribution pattern reflects the priorities and characteristics of SLUP development in the YRD region. The highest proportion in Geographic Information Services indicates that geographic information infrastructure and data services are focal areas in smart land use planning. The similar proportion of Land Use Planning demonstrates that planning development and implementation remain core tasks. From a temporal perspective, Geographic Information Services shows rapid growth, increasing from CNY 217.32 million in 2018 to CNY 859.7 million in 2022. Land Use Planning exhibits a rise-then-decline pattern, peaking at CNY 568.36 million in 2019 before decreasing to CNY 368 million in 2022. This fluctuation likely results from China's planning cycle, as 2019–2020 marked the period for developing the 14th Five-Year Plan and a new round of spatial planning. This result reflects the cyclical nature of land use planning work, with concentrated investments during the planning phase followed by relatively reduced investments during implementation. Additionally, Land Resource Protection investments grew from CNY 321.36 million in 2018 to CNY 415.95 million in 2022. This increase closely aligns with national ecological civilization construction and sustainable development strategies. After 2020, as spatial ecological restoration projects gained momentum, localities increased smart investments in land resource protection, particularly in ecological monitoring, resource surveys, and disaster prevention. Real Estate Management investments remained generally stable, possibly because real estate management systems were already well-established before 2018, with current focus mainly on maintenance and upgrades.

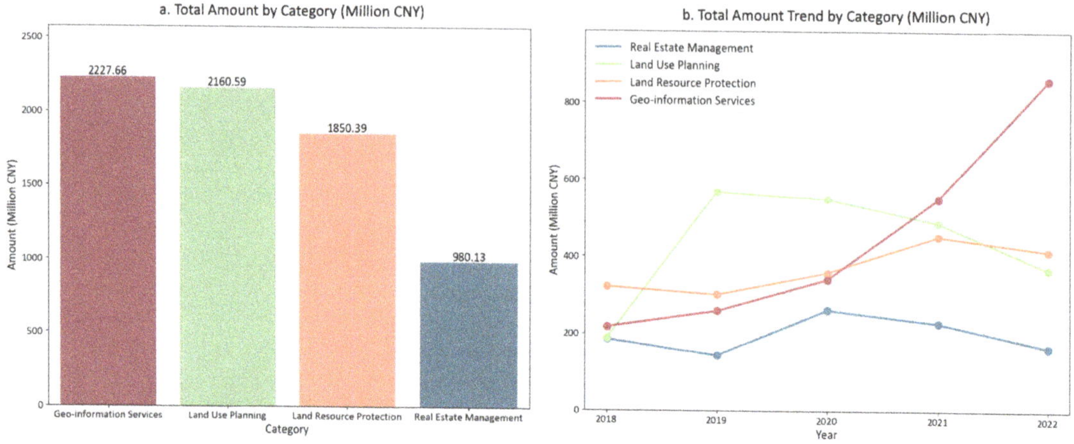

Figure 4. Procurement amounts and annual changes of SLUP projects by category (million CNY).

3.2. Spatial Distribution Patterns

Figure 5 illustrates the spatial distribution of procurement amounts and cross-regional procurement ratios in SLUP projects. The procurement amount distribution exhibits a clear "core-periphery" structure, with Shanghai and surrounding cities forming the center of SLUP project procurement in the YRD region. Shanghai ranks highest at CNY 915.1 million, followed by Suzhou (CNY 584.7 million) and Hangzhou (CNY 563.9 million). Beyond this central area, provincial capitals such as Hefei and Nanjing also show relatively high investment levels, creating a spatial pattern with central cities as cores and surrounding cities in a gradient distribution. Peripheral cities, mainly in the western and northern parts of the YRD region, exhibit significantly lower investments; for example, Chizhou in Anhui Province contracted only CNY 18.5 million, while Huaibei in Jiangsu Province contracted CNY 22.3 million, demonstrating an unbalanced spatial distribution. This pattern correlates strongly with urban economic development levels, administrative status, and innovation capacity, reflecting SLUP's resource endowment dependency.

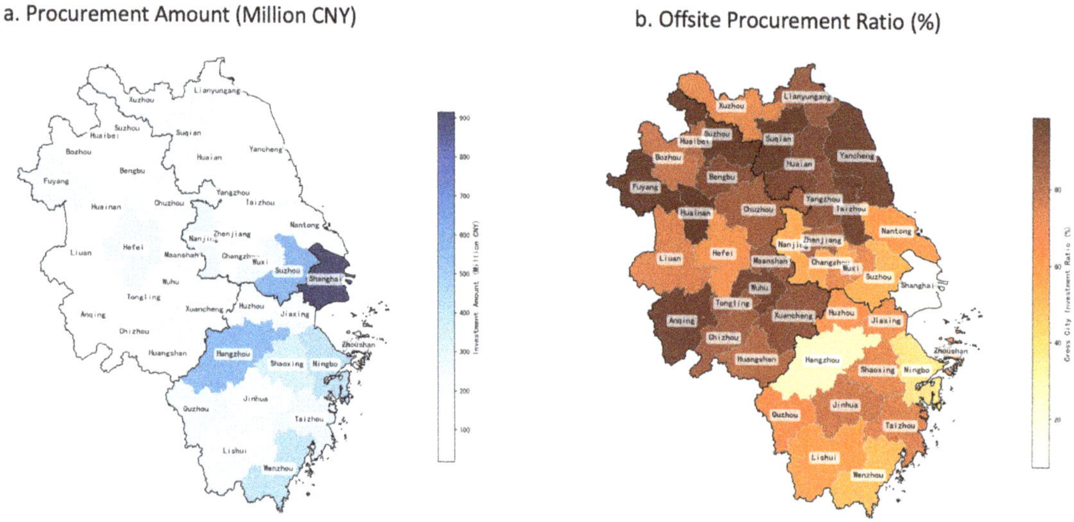

Figure 5. Spatial distribution of SLUP procurement amounts and cross-regional ratios.

The spatial distribution of cross-regional procurement ratios shows an opposite pattern to procurement amounts. From 2018–2022, cross-regional SLUP procurement in the YRD reached CNY 4.2 billion, accounting for 57.7% of total procurement amount. Generally, cities with higher economic development levels and administrative status demonstrate lower cross-regional procurement ratios. For instance, Shanghai, as the only municipality and most developed city in the YRD region, has a cross-regional procurement ratio of only 7.3%. Cities like Hangzhou and Ningbo also show relatively low cross-regional procurement ratios, indicating strong digital technology capabilities that can meet government SLUP project requirements. Peripheral cities, however, almost entirely depend on cross-regional procurement for SLUP projects. For example, Huainan in Anhui Province sourced 98.7% of its SLUP projects from outside vendors, while most cities in the northern and western parts of the region had cross-regional procurement ratios more than 80%. This emphasizes the importance of cooperation for peripheral cities in implementing SLUP. Overall, the spatial distribution of SLUP projects in the YRD is influenced by urban economic strength and administrative status, as well as city functional positioning and resource capability. These spatial disparities reflect the imbalanced development stages and digital technology capabilities within the region, while simultaneously creating conditions and necessities for intercity cooperation.

4. Intercity Cooperation Networks and Influencing Factors in SLUP

4.1. Characteristics of Cooperation Networks

Figure 6 displays the weighted cooperation networks in SLUP government procurement across the YRD region, alongside networks for different project types, rendered using Gephi Software (version 0.10) with the OpenOrd layout algorithm. The OpenOrd algorithm was configured with the following parameters: Edge Cut = 0.8, Scaling = 1.0, Iterations = 1000, and Gravity = 0.1, optimizing the visualization for clustering and spatial distribution of nodes based on network structure. Nodes represent cities, with nodes of the same color representing cities from the same province, and node size indicating degree. Edges represent procurement amounts between cities, with wider edges indicating larger amounts. Table 3 displays relevant descriptive metrics for the network.

Results show that the overall network comprises 41 nodes and 180 edges, with a network density of 0.22, indicating that cities in the YRD region have established only about 22% of potential cooperation relationships, displaying a relatively sparse network structure. The average weighted degree is 1360.77, reflecting the overall intensity of intercity cooperation. The average clustering coefficient is 0.031; this low value indicates few triangular cooperation relationships in the network, with city cooperation rarely exhibiting the closed structure of "friends of friends are also friends". The centralization index is 0.65; this relatively high value indicates that the overall cooperation network has a distinct "core-periphery" structure, with a few core cities dominating the network and controlling key resources and information channels. The network visualization in Figure 4 also reveals that cities from each province primarily form three sub-networks around their provincial capitals—Hefei, Nanjing, and Hangzhou—while Shanghai occupies the central position connecting these three sub-networks.

Examining networks by category reveals significant differences in network metrics. Regarding network connectivity, the Land Use Planning network has the highest number of edges (113) and network density (0.14), followed by the Land Resource Protection network with 97 edges and a density of 0.12. The Geographic Information Services network has 87 edges and a density of 0.11, slightly higher than the Real Estate Management network with 85 edges and a network density of 0.10. This indicates that in the land use planning field, cities have formed more extensive cooperation relationships, while cooperation in Land Resource Protection is also relatively active. City cooperation in Geographic Information Services and Real Estate Management fields is comparatively limited.

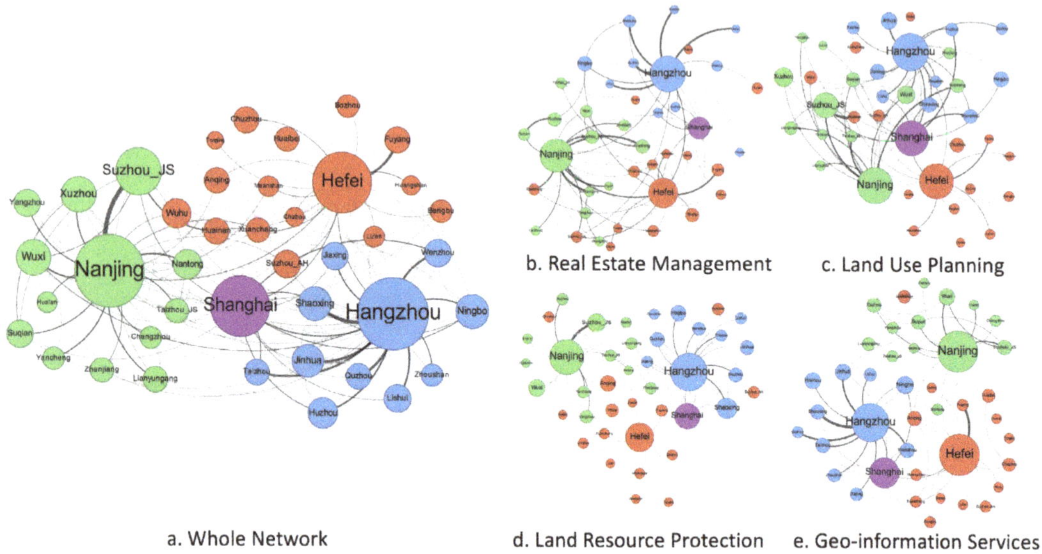

Figure 6. SLUP project procurement collaboration network in the YRD region.

Table 3. Network descriptive indicators.

Network	Nodes	Edges	Avg Weighted Degree	Density	Avg Clustering Coefficient	Centralization
Overall	41	180	1360.77	0.22	0.031	0.65
Land Use Planning	41	113	418.98	0.14	0.053	0.45
Land Resource Protection	41	97	359.53	0.12	0.0199	0.47
Geo-information Services	41	87	330.10	0.11	0.0347	0.43
Real Estate Management	41	85	252.16	0.10	0.074	0.48

In terms of average weighted degree, the four network types exhibit varying cooperation intensities. The Land Use Planning network ranks highest (418.98), reflecting frequent exchanges in this domain; Land Resource Protection network follows (359.53), indicating relatively close cooperation relationships; Geographic Information Services network shows a moderate average weighted degree (330.10); Real Estate Management network ranks lowest (252.16), suggesting relatively weaker cooperation in this field. Regarding structural characteristics, the four networks present different organizational patterns. By average clustering coefficient, the Real Estate Management network ranks highest (0.074), indicating this domain tends to form localized, clustered cooperation patterns; Land Use Planning network follows (0.053), also showing certain clustering features; Geographic Information Services network has a moderate average clustering coefficient (0.035); while Land Resource Protection network ranks lowest (0.020), suggesting more dispersed cooperation with fewer closed triangular cooperation structures. Examining centralization indices, the Real Estate Management network ranks highest (0.48), indicating cooperation in this domain highly depends on a few core cities; Land Resource Protection network follows (0.47), also showing high centralization; Land Use Planning network's centralization index is moderate (0.45), while Geographic Information Services network ranks lowest (0.43), suggesting a relatively balanced cooperation pattern with more equal participation among cities.

These differences in network indicators reflect the project characteristics and policy orientations across SLUP sub-domains. The Land Use Planning network's high connectivity and weighted degree reflect the importance of regional collaborative planning amid spatial planning system reforms. The Land Resource Protection network's high centralization and low clustering coefficient demonstrate both core city dominance and dispersed cooperation in this field. The Real Estate Management network's high clustering coefficient and centralization indicate both regional collaboration needs and dependence on core cities' technical capabilities. The Geographic Information Services network's relatively balanced centralization reflects the distinct territorial nature of such projects—since basic geographic data collection, processing, and application have clear territorial attributes, local procurement may be more efficient. These differentiated network structures reveal varying cooperation logics and resource integration patterns across different domains in SLUP.

4.2. Factors Influencing Partner Selection in Cooperation Networks

4.2.1. Proximity and Homogeneity

Multiple factors shape intercity cooperation networks, primarily categorized as proximity and homogeneity. Geographic proximity, emphasized in traditional regional science, suggests that cities closer to each other face lower cooperation costs and higher cooperation likelihood [59,60]. Recent studies indicate that interactions between geographic proximity and network relationships generate social capital, strengthen cooperation willingness, and promote collaboration [51]. However, some studies suggest that the importance of geographic proximity may gradually decline while homogeneity factors (such as industrial structure similarity) become increasingly influential [61].

The homogeneity hypothesis proposes that similar partners better understand and trust each other, facilitating cooperation. For instance, local governments more easily establish cooperation when they share similar interests, economic development levels, and institutional foundations [60,62]. Competing views argue that similar actors often compete with each other [63], and differences between cooperation partners can create complementary advantages [59]. Recent studies increasingly recognize that homogeneity and heterogeneity are not mutually exclusive but play different roles across various characteristics [63].

4.2.2. MRQAP Analysis

Based on the above discussion, we examine how proximity and homogeneity between cities influence their cooperation in SLUP.

First, proximity refers to geographic distance between cities. We use the shortest road distance between government buildings of 41 cities in the YRD region, calculated through the API provided by Amap (a leading Chinese map service provider), as a proxy for geographic distance ($Distance_{ij}$).

Second, homogeneity encompasses multiple dimensions, including scale homogeneity, institutional homogeneity, and cultural homogeneity [43]. Using data from China City Statistical Yearbooks, we obtained city-level information on population, GDP, tertiary industry proportion, administrative area, cultivated land area, general public budget expenditure, and general public budget revenue for the years 2018–2022. We calculated differences between the 41 cities in population (Pop_{ij}), GDP (GDP_{ij}), tertiary industry proportion ($Industry_{ij}$), administrative area ($Area_{ij}$), cultivated land area ($Farm_{ij}$), and fiscal pressure ($Fiscal_{ij}$) as proxies for scale homogeneity. Within China's multi-level government system, cities in the same province face similar policy environments. Therefore, we use whether cities belong to the same province as a proxy for institutional homogeneity ($Province_{ij}$), coded as 0 if two cities belong to the same province and 1 otherwise. Additionally, previous studies have confirmed that China's counterpart assistance mechanism significantly promotes cooperation between paired cities [64], so we control for counterpart assistance relationships between cities ($Assistance_{ij}$) [65]. We use city-level dialect similarity as a proxy for cultural homogeneity ($Culture_{ij}$) [66].

Finally, the dependent variable is the procurement amount of SLUP projects between cities, derived from the cooperation network constructed in the previous section. We process this as a directed weighted network. For example, if City A procures a SLUP project worth CNY 1 million from City B, the edge weight from City A to City B is 1. We further decompose the dependent variable into the overall cooperation network and four different categories of cooperation networks.

Table 4 presents the MRQAP analysis results across five models, each representing a different dependent variable: Model 1 for the overall cooperation network, Model 2 for Land Use Planning network, Model 3 for Land Resource Protection network, Model 4 for Geographic Information Services network, and Model 5 for Real Estate Management network. The analysis reveals complex mechanisms in the formation of intercity SLUP cooperation networks, with variables showing differentiated effects across different types of cooperation networks.

Table 4. MRQAP model results.

	Model 1	Model 2	Model 3	Model 4	Model 5
$Distance_{ij}$	−0.086 (0.034)	−0.089 (0.010)	−0.048 (0.011)	−0.078 (0.009)	−0.077 (0.006)
Scale homogeneity					
Pop_{ij}	0.171 * (0.029)	0.167 ** (0.007)	0.123 (0.010)	0.142 * (0.008)	0.134 (0.006)
GDP_{ij}	−0.012 (0.113)	−0.157 * (0.030)	0.070 (0.039)	−0.006 (0.032)	0.048 (0.022)
$Industry_{ij}$	−0.346 *** (1.187)	−0.263 *** (0.291)	−0.310 *** (0.416)	−0.256 *** (0.319)	−0.319 *** (0.227)
$Area_{ij}$	−0.005 (0.001)	−0.004 (0.000)	0.024 (0.000)	−0.040 (0.000)	−0.005 (0.000)
$Farm_{ij}$	−0.195 ** (0.033)	−0.161 ** (0.008)	−0.165 ** (0.011)	−0.149 ** (0.009)	−0.170 ** (0.006)
$Fiscal_{ij}$	0.114 ** (6.817)	0.043 (1.667)	0.091 * (2.319)	0.110 ** (1.979)	0.155 *** (1.386)
Institutional homogeneity					
$Province_{ij}$	−0.118 *** (7.491)	−0.062 * (2.225)	−0.100 *** (2.753)	−0.118 *** (2.302)	−0.113 *** (1.464)
$Assistance_{ij}$	0.009 (34.168)	0.015 (11.465)	0.003 (13.890)	0.007 (10.825)	0.008 (7.063)
Cultural homogeneity					
$Culture_{ij}$	−0.020 (12.522)	−0.004 (3.612)	−0.004 (4.248)	−0.041 (3.628)	−0.023 (2.377)
Intercept	0.000 *** (0.000)	0.000 *** (0.000)	0.000 *** (0.000)	0.000 *** (0.000)	0.000 *** (0.000)
R^2	0.102	0.094	0.057	0.069	0.086

* $p < 0.1$, ** $p < 0.05$, *** $p < 0.01$ (two-tailed tests).

Geographic distance ($Distance_{ij}$) is not significant in any of the five models, indicating that geographic distance between cities does not significantly affect the formation of SLUP cooperation networks. This suggests that in the context of digital technology development and improved transportation infrastructure, the inhibiting effect of geographic distance on intercity cooperation has weakened in SLUP networks. YRD cities can overcome spatial limitations to engage in SLUP cooperation. This finding further confirms the "space of flows" attribute of digital services [32]: the low-cost flow characteristics of data enable cities to transcend spatial boundaries and select partners based on technical needs rather than geographic proximity.

Scale homogeneity appears to facilitate greater cooperation between cities in SLUP. Industry structure differences ($Industry_{ij}$) show highly significant negative correlations (1% level) across all five models, emerging as the most consistent and influential factor. This strongly supports the scale homogeneity hypothesis, indicating that cities with similar industrial structures more readily form cooperative relationships. Such similarity enhances mutual trust and understanding while reducing coordination costs, thereby facilitating cooperation. Farmland area differences ($Farm_{ij}$) display significant negative correlations (5% level) across all models, suggesting that cities with similar agricultural resource endowments are more likely to engage in SLUP cooperation. This may stem from shared challenges in agricultural transformation and land protection, prompting these cities to seek collaborative solutions. GDP differences (GDP_{ij}) show significance (10% level, negative correlation) only in the Land Use Planning network, indicating that cities with similar economic development levels cooperate more easily in this domain, possibly due to facing comparable urbanization challenges and economic structural transformation issues. In the other four domains, economic development disparities have no significant impact on cooperation. Population differences (Pop_{ij}) exhibit heterogeneity, with significant positive correlations (10% level) in overall network and Geographic Information Services network, and stronger significance (5% level) in Land Use Planning network. This supports the scale complementarity effect between cities—larger cities often possess advanced planning concepts and technical resources, while smaller cities provide diverse implementation scenarios, creating mutually beneficial cooperation patterns. This effect is similarly observed in fiscal pressure differences ($Fiscal_{ij}$). Administrative area differences ($Area_{ij}$) remain insignificant across all models with coefficients near zero, indicating that variations in administrative division size have limited impact on SLUP cooperation.

Institutional environment similarity emerges as a significant facilitator of intercity cooperation. Provincial relationships ($Province_{ij}$) show significant negative correlations across all models—with overall network, Land Resource Protection, Geographic Information Services, and Real Estate Management networks significant at the 1% level, and Land Use Planning at the 10% level. This indicates closer cooperation between cities within the same province, aligning with institutional homogeneity theory. Cities sharing similar institutional environments face lower coordination costs and enjoy more convenient cooperation due to common policy frameworks and regulatory requirements. Counterpart assistance relationships ($Assistance_{ij}$) remain insignificant across all models, suggesting that traditional assistance mechanisms have limited impact on SLUP cooperation when controlling for other factors, possibly because cooperation in this field relies more on professional technical needs and market mechanisms rather than administrative directives. Cultural homogeneity ($Culture_{ij}$) shows no significant influence across all models, indicating that traditional cultural differences (such as dialect variations) have minimal impact on SLUP cooperation.

The MRQAP analysis reveals complex mechanisms underlying SLUP cooperation networks in the YRD region. Industry structure homogeneity, institutional environment similarity, and farmland area homogeneity emerge as key cooperation drivers, while population size and fiscal condition differences demonstrate complementary cooperation characteristics. The weakened influence of geographic distance and traditional cultural differences highlights new features of urban cooperation in the context of digitalization and regional integration. These findings provide empirical evidence for promoting coordinated regional development.

5. Discussion

Previous studies on SLUP primarily focused on technical applications, policy frameworks, and effectiveness assessment, with limited exploration from urban governance and regional cooperation network perspectives. This study addresses this research gap by systematically analyzing SLUP project characteristics, spatial distribution patterns, and influencing factors of intercity cooperation networks in the YRD region. Specifically, this study employs LDA topic model analysis to identify main SLUP project types, reveal spatial distribution characteristics, and explore cooperation network structures and influencing factors using social network analysis and MRQAP.

First, this study provides a in-depth analysis of SLUP implementation mechanisms from an urban governance perspective, emphasizing that SLUP extends beyond technical applications to integration within actual urban governance operations. Results show government procurement SLUP projects include four main types, reflecting distinct urban governance functions. Government policy preferences significantly influence project priorities, with Geographic Information Services and Land Use Planning identified as major investment areas. Additionally, Land Resource Protection projects show growth, reflecting policy orientations toward ecological protection and sustainable development. This finding suggests SLUP has evolved from a purely technical tool to an essential component of urban governance systems, supporting the view that the smart city is not technologically neutral but profoundly shaped by government policy preferences and strategic goals [2,15]. The variation in procurement proportions across SLUP types also reveals specific pathways of Chinese local governments in SLUP development, demonstrating trends from planning formulation to digital service delivery.

Second, this study reveals spatial inequality in SLUP development, closely echoing equity discussions in current smart city research. Findings show SLUP projects display a distinct "core-periphery" spatial pattern in the YRD region. Central cities like Shanghai, Suzhou, and Hangzhou significantly outpace peripheral cities in funding, with this distribution difference substantially influenced by economic development levels, administrative status, and innovation capacities. Central cities exhibit significantly lower cross-regional procurement rates than peripheral cities. This phenomenon indicates peripheral cities' heavy reliance on regional cooperation networks for technical support and services [28]. Such spatial inequality highlights equity issues in SLUP implementation, drawing attention to challenges faced by peripheral cities and emphasizing the importance of regional cooperation in narrowing technical capacity gaps.

Third, this study explores the procurement relationships between YRD region cities in the SLUP field and the influencing factors of relationship formation from a cooperation network perspective. Results indicate geographic distance's diminishing influence on intercity cooperation networks in the digital era, allowing cities to overcome traditional spatial constraints for technical cooperation. Additionally, industrial structure similarity significantly increases cooperation likelihood, supporting theoretical assumptions that scale homogeneity reduces cooperation costs and enhances mutual trust. The study also finds institutional environment similarity plays a key role in intercity cooperation, with cities in the same province more readily cooperating due to institutional and policy framework convergence. Conversely, traditional cultural similarities and counterpart assistance mechanisms have limited impact on cooperation network formation. These findings provide new perspectives and empirical evidence for theoretical research on intercity cooperation in the context of digitalization and regional integration [59,62].

Despite providing the first analysis of SLUP implementation mechanisms from an urban governance perspective, this study has limitations. It primarily relies on government procurement contract data without fully considering other types of government-implemented SLUP projects or informal SLUP initiatives, potentially overlooking critical factors. Future research could broaden data sources and incorporate interviews and surveys to examine government cooperation motivations and barriers more comprehensively. Additionally, while this study focuses on the YRD region, future research should pursue cross-regional and cross-national comparative analyses to assess how variations in policy environments and technological development across regions and countries influence SLUP implementation and network cooperation patterns, providing valuable insights for broader policy formulation.

Based on these findings, we propose several policy recommendations (Table 5). Governments should clearly define technical applications and requirements across different functional domains when formulating and implementing SLUP policies and projects, enhancing SLUP targeting and implementation efficiency. Regarding the high inequality in SLUP cooperation networks, governments should prioritize regional equity issues by establishing more robust technical sharing and cooperation mechanisms, strengthening collaboration between core and peripheral cities to significantly enhance regional SLUP capabilities and reduce technical and resource disparities. Additionally, considering the positive impact of institutional environment and industrial structure similarity on regional cooperation, governments should actively promote policy and institutional unification and standardization processes within regions to further reduce cross-regional cooperation barriers, optimize resource allocation efficiency, and achieve regional coordinated development goals.

Table 5. Policy recommendations for enhancing SLUP implementation and regional cooperation.

Policy Recommendation	Objective	Proposed Action
Define technical standards clearly	Improve SLUP targeting and implementation efficiency	Clearly define technical applications and requirements across different functional domains in SLUP policies and projects.
Enhance regional equity	Reduce technical and resource disparities among cities	Establish robust technical sharing and cooperation mechanisms between core and peripheral cities.
Promote institutional unification and standardization	Reduce cross-regional cooperation barriers; optimize resource allocation	Actively promote policy and institutional unification and standardization processes within regions.

6. Conclusions

This study aims to explore government procurement project characteristics and intercity cooperation network mechanisms in SLUP, focusing on three research questions: (1) What are the typological characteristics of SLUP government procurement projects in the YRD? (2) How are government SLUP projects spatially distributed across different cities in the region? (3) What are the structural characteristics of the intercity SLUP cooperation network, and which factors influence the formation of cooperative relationships? Using the YRD region as the research subject, the study analyzes 3689 SLUP government procurement projects identified from online government procurement contracts. The study first employs LDA topic modeling analysis to identify project types and describe their spatial distribution characteristics, then uses Social Network Analysis to construct and analyze intercity cooperation network structures, and finally applies MRQAP to examine how geographic distance, scale homogeneity, institutional environment homogeneity, counterpart assistance relationships, and cultural homogeneity influence intercity cooperation network formation.

The study finds that SLUP government procurement projects in the YRD region primarily comprise four categories: Real Estate Management, Land Resource Protection, Land Use Planning, and Geographic Information Services. Geographic Information Services account for the highest proportion of investment, followed by Land Use Planning, then Land Resource Protection, with Real Estate Management receiving the lowest. Regarding spatial distribution, regional investments display a distinct core–periphery structure, centered around economically developed central cities, with peripheral cities highly dependent on external procurement. Network structure analysis reveals a relatively sparse overall network with pronounced core–periphery characteristics, where a few core cities dominate resource and information allocation. Among factors influencing network formation, similarities in industrial structure and farmland area significantly promote city cooperation; institutional environment similarity also significantly increases cooperation probability, while geographic distance, traditional cultural similarity, and counterpart assistance relationships do not show influences. Theoretically, this study enriches regional-scale SLUP research by combining social network analysis and MRQAP methods to analyze SLUP cooperation network formation mechanisms for the first time, providing new analytical perspectives and empirical evidence for promoting regional coordinated development policies and practices.

Author Contributions: Conceptualization, G.L. and Y.X.; Methodology, G.L.; Software, G.L.; Validation, Y.X.; Formal analysis, G.L.; Investigation, G.L.; Resources, Y.X.; Data curation, G.L.; Writing—original draft, G.L.; Writing—review & editing, Y.X.; Supervision, Y.X.; Project administration, Y.X.; Funding acquisition, Y.X. All authors have read and agreed to the published version of the manuscript.

Funding: This research is supported by funding from the National Natural Science Foundation of China (42201206), Humanities and Social Science Project of the Ministry of Education of China (22YJC630176), and Zhejiang Federation of Humanities and Social Sciences (21YJRC05-2YB). This research is also supported by ZJU-CMZJ Joint Lab on Data Intelligence and Urban Future and China Institute of Urbanization Zhejiang University.

Data Availability Statement: The original contributions presented in the study are included in the article, further inquiries can be directed to the corresponding author.

Conflicts of Interest: The authors declare no conflicts of interest.

References

1. Cui, X.; Li, F.; de Vries, W.T. Smart Land Use Planning: New Theories, New Tools and New Practice. *Land* **2023**, *12*, 1315. [CrossRef]
2. Vali, A.; Comai, S.; Matteucci, M. Deep Learning for Land Use and Land Cover Classification Based on Hyperspectral and Multispectral Earth Observation Data: A Review. *Land* **2020**, *12*, 2495. [CrossRef]
3. Li, Z.; Chen, B.; Wu, S.; Su, M.; Chen, J.M.; Xu, B. Deep Learning for Urban Land Use Category Classification: A Review and Experimental Assessment. *Remote Sens. Environ.* **2024**, *311*, 114290. [CrossRef]
4. Wang, J.; Bretz, M.; Dewan, M.A.A.; Delavar, M.A. Machine Learning in Modelling Land-Use and Land Cover-Change (LULCC): Current Status, Challenges and Prospects. *Sci. Total Environ.* **2022**, *822*, 153559. [CrossRef]
5. Pham, Q.B.; Ali, S.A.; Parvin, F.; Van On, V.; Sidek, L.M.; Đurin, B.; Cetl, V.; Šamanović, S.; Minh, N.N. Multi-Spectral Remote Sensing and GIS-Based Analysis for Decadal Land Use Land Cover Changes and Future Prediction Using Random Forest Tree and Artificial Neural Network. *Adv. Space Res.* **2024**, *74*, 17–47. [CrossRef]
6. Peng, M.; Liu, Y.; Khan, A.; Ahmed, B.; Sarker, S.K.; Ghadi, Y.Y.; Bhatti, U.A.; Al-Razgan, M.; Ali, Y.A. Crop Monitoring Using Remote Sensing Land Use and Land Change Data: Comparative Analysis of Deep Learning Methods Using Pre-Trained CNN Models. *Big Data Res.* **2024**, *36*, 100448. [CrossRef]
7. Aslam, R.W.; Naz, I.; Quddoos, A.; Quddusi, M.R. Assessing Climatic Impacts on Land Use and Land Cover Dynamics in Peshawar, Khyber Pakhtunkhwa, Pakistan: A Remote Sensing and GIS Approach. *GeoJournal* **2024**, *89*, 202. [CrossRef]
8. Zhuo, Z.; Ye, J.; Wang, Y.; Chen, H.; Liang, B. Smart Cities, Smarter Land Use? Unveiling the Efficiency Gains from China's Digital Urban Transformation. *Ecol. Indic.* **2025**, *171*, 113151. [CrossRef]

9. Masoumi, Z.; van Genderen, J. Artificial Intelligence for Sustainable Development of Smart Cities and Urban Land-Use Management. *Geo-spat. Info. Sci.* **2024**, *27*, 1212–1236. [CrossRef]
10. Kalfas, D.; Kalogiannidis, S.; Chatzitheodoridis, F.; Toska, E. Urbanization and Land Use Planning for Achieving the Sustainable Development Goals (SDGs): A Case Study of Greece. *Urban Sci.* **2023**, *7*, 43. [CrossRef]
11. Chaturvedi, V.; de Vries, W.T. Machine Learning Algorithms for Urban Land Use Planning: A Review. *Urban Sci.* **2021**, *5*, 68. [CrossRef]
12. Wang, Z.; Sun, Q.; Zhang, X.; Hu, Z.; Chen, J.; Zhong, C.; Li, H. CUGUV: A Benchmark Dataset for Promoting Large-Scale Urban Village Mapping with Deep Learning Models. *Sci. Data* **2025**, *12*, 390. [CrossRef]
13. Batty, M. Digital Twins in City Planning. *Nat. Comput. Sci.* **2024**, *4*, 192–199. [CrossRef]
14. Ameyaw, P.D.; de Vries, W.T. Blockchain Technology Adaptation for Land Administration Services: The Importance of Socio-Cultural Elements. *Land Use Policy* **2023**, *125*, 106485. [CrossRef]
15. Marsal-Llacuna, M.-L.; López-Ibáñez, M.-B. Smart Urban Planning: Designing Urban Land Use from Urban Time Use. *J. Urban Technol.* **2014**, *21*, 39–56. [CrossRef]
16. Jia, C.; Feng, S.; Chu, H.; Huang, W. The Heterogeneous Effects of Urban Form on CO_2 Emissions: An Empirical Analysis of 255 Cities in China. *Land* **2023**, *12*, 981. [CrossRef]
17. Ameyaw, P.D.; de Vries, W.T. Toward Smart Land Management: Land Acquisition and the Associated Challenges in Ghana. A Look into a Blockchain Digital Land Registry for Prospects. *Land* **2021**, *10*, 239. [CrossRef]
18. Yu, T.; Huang, X.; Jia, S.; Cui, X. Unveiling the Spatio-Temporal Evolution and Key Drivers for Urban Green High-Quality Development: A Comparative Analysis of China's Five Major Urban Agglomerations. *Land* **2023**, *12*, 1962. [CrossRef]
19. Kamrowska-Załuska, D. Impact of AI-Based Tools and Urban Big Data Analytics on the Design and Planning of Cities. *Land* **2021**, *10*, 1209. [CrossRef]
20. Jiang, Y.; Yang, L.; Wei, X.; Zhang, X. The Impact of Government Digital Transformation on Land Use Efficiency: Evidence from China. *Land* **2024**, *13*, 2080. [CrossRef]
21. Wu, M.; Yan, B.; Huang, Y.; Sarker, M.N.I. Big Data-Driven Urban Management: Potential for Urban Sustainability. *Land* **2022**, *11*, 680. [CrossRef]
22. Lu, S.; Wang, H. How Political Connections Exploit Loopholes in Procurement Institutions for Government Contracts: Evidence from China. *Governance* **2022**, *36*, 1205–1224. [CrossRef]
23. Tang, W.; Wang, Y.; Wu, J. Local Favoritism in China's Public Procurement: Information Frictions or Incentive Distortion? *J. Urban Econ.* **2025**, *145*, 103716. [CrossRef]
24. Casady, C.B.; Petersen, O.H.; Brogaard, L. Public Procurement Failure: The Role of Transaction Costs and Government Capacity in Procurement Cancellations. *Public Manag. Rev.* **2023**, 1–28. [CrossRef]
25. Patrucco, A.S.; Kauppi, K.; Di Mauro, C.; Schotanus, F. Enhancing Strategic Public Procurement: A Public Service Logic Perspective. *Public Manag. Rev.* **2024**, 1–21. [CrossRef]
26. Konashevych, O. Constraints and Benefits of the Blockchain Use for Real Estate and Property Rights. *J. Prop. Plan. Environ. Law* **2020**, *12*, 109–127. [CrossRef]
27. Putri, R.A. Development of Digital Registration Information System in Kelurahan to Improve Administration Efficiency and Transparency. *Inf. Technol. Syst.* **2024**, *1*, 92–99. [CrossRef]
28. Liu, C.; Zhang, Z.; Zhang, S. Smart Initiatives for Land Resource Management: Perspectives and Practices from China. *J. Urban Technol.* **2023**, *30*, 3–21. [CrossRef]
29. Cao, C.; Dragićević, S.; Li, S. Land-Use Change Detection with Convolutional Neural Network Methods. *Environments* **2019**, *6*, 25. [CrossRef]
30. Jiménez-Espada, M.; Martínez García, F.M.; González-Escobar, R. Sustainability Indicators and GIS as Land-Use Planning Instrument Tools for Urban Model Assessment. *ISPRS Int. J. Geo-Inf.* **2023**, *12*, 42. [CrossRef]
31. Zhuo, Y.; Jing, X.; Wang, X.; Li, G.; Xu, Z.; Chen, Y.; Wang, X. The Rise and Fall of Land Use Mix: Review and Prospects. *Land* **2022**, *11*, 2198. [CrossRef]
32. Ejiaku, S. Technology Adoption: Issues and Challenges in Information Technology Adoption in Emerging Economies. *J. Int. Technol. Inf. Manag.* **2014**, *23*, 5. [CrossRef]
33. Arduini, D.; Belotti, F.; Denni, M.; Giungato, G.; Zanfei, A. Technology Adoption and Innovation in Public Services: The Case of E-Government in Italy. *Inf. Econ. Policy* **2010**, *22*, 257–275. [CrossRef]
34. Mu, R.; Wu, P.; Haershan, M. Pre-Contractual Relational Governance for Public–Private Partnerships: How Can Ex-Ante Relational Governance Help Formal Contracting in Smart City Outsourcing Projects? *Int. Rev. Adm. Sci.* **2023**, *89*, 112–128. [CrossRef]
35. Gil-Garcia, J.R.; Zhang, J.; Puron-Cid, G. Conceptualizing Smartness in Government: An Integrative and Multi-Dimensional View. *Gov. Inf. Q.* **2016**, *33*, 524–534. [CrossRef]
36. Zhang, R.; Lin, J.; Sun, D. The Role of Institutions and Markets in Shaping Intercity Investment Networks in China. *Cities* **2024**, *153*, 105221. [CrossRef]

37. Shi, S.; Wong, S.K.; Zheng, C. Network Capital and Urban Development: An Inter-Urban Capital Flow Network Analysis. *Reg. Stud.* **2022**, *56*, 406–419. [CrossRef]
38. Capello, R. The City Network Paradigm: Measuring Urban Network Externalities. *Urban Stud.* **2000**, *37*, 1925–1945. [CrossRef]
39. Castells, M. *The Rise of the Network Society*, 3rd ed.; John Wiley & Sons: Chichester, UK, 2010; pp. 407–453.
40. Taylor, P.; Derudder, B. *World City Network: A Global Urban Analysis*; Routledge: New York, NY, USA, 2015; pp. 7–27.
41. Coe, N.M.; Dicken, P.; Hess, M.; Yeung, H.W. Making Connections: Global Production Networks and World City Networks. *Glob. Netw.* **2010**, *10*, 138–149. [CrossRef]
42. Mignoni, J.; Bittencourt, B.A.; da Silva, S.B.; Zen, A.C. Orchestrators of Innovation Networks in the City Level: The Case of Pacto Alegre. *Innov. Manag. Rev.* **2023**, *20*, 194–210. [CrossRef]
43. Gebauer, A.; Nam, C.W.; Parsche, R. Regional Technology Policy and Factors Shaping Local Innovation Networks in Small German Cities. *Eur. Plan. Stud.* **2005**, *13*, 661–683. [CrossRef]
44. Leydesdorff, L.; Rafols, I. Local Emergence and Global Diffusion of Research Technologies: An Exploration of Patterns of Network Formation. *J. Am. Soc. Inf. Sci.* **2011**, *62*, 846–860. [CrossRef]
45. De Freitas, S.; Mayer, I.; Arnab, S.; Marshall, I. Industrial and Academic Collaboration: Hybrid Models for Research and Innovation Diffusion. *J. High. Educ. Policy Manag.* **2014**, *36*, 2–14. [CrossRef]
46. Shrestha, M.K.; Feiock, R.C. Toward a Multiplex Network Theory of Interlocal Service Contracting. *Public Adm. Rev.* **2021**, *81*, 911–924. [CrossRef]
47. Feiock, R.C.; Scholz, J.T. (Eds.) *Self-Organizing Federalism: Collaborative Mechanisms to Mitigate Institutional Collective Action Dilemmas*; Cambridge University Press: Cambridge, UK, 2009; pp. 3–33.
48. Andrew, S.A.; Short, J.E.; Jung, K.; Arlikatti, S. Intergovernmental Cooperation in the Provision of Public Safety: Monitoring Mechanisms Embedded in Interlocal Agreements. *Public Adm. Rev.* **2015**, *75*, 401–410. [CrossRef]
49. LeRoux, K.; Brandenburger, P.W.; Pandey, S.K. Interlocal Service Cooperation in U.S. Cities: A Social Network Explanation. *Public Adm. Rev.* **2010**, *70*, 268–278. [CrossRef]
50. Shen, R. Regional Governance and Multiplex Networks in Environmental Sustainability: An Exponential Random Graph Model Analysis in the Chinese Local Government Context. *Urban Aff. Rev.* **2024**, *60*, 571–613. [CrossRef]
51. Zhu, W.; Zhang, J.; Dai, J.; Wang, D.; Ma, C.; Xu, Y.; Chen, Y. Study on the Spatiotemporal Evolution Characteristics and Influencing Factors on Green Building Development of City Clusters in the Yangtze River Delta Region in China. *Sustainability* **2023**, *15*, 9457. [CrossRef]
52. Tao, Z.; Zhi, Z.; Shangkun, L. Digital Economy, Entrepreneurship, and High Quality Economic Development: Empirical Evidence from Urban China. *Front. Econ. China* **2022**, *17*, 393.
53. Central Government Procurement Network Home Page. Available online: http://www.ccgp.gov.cn (accessed on 13 May 2025).
54. Tianyancha Home Page. Available online: http://www.tianyancha.com (accessed on 13 May 2025).
55. Chauhan, U.; Shah, A. Topic Modeling Using Latent Dirichlet Allocation: A Survey. *ACM Comput. Surv.* **2022**, *54*, 1–35. [CrossRef]
56. Scott, J. *What Is Social Network Analysis?* Bloomsbury Academic: London, UK, 2012; pp. 31–57.
57. Krackhardt, D. Predicting with Networks: Nonparametric Multiple Regression Analysis of Dyadic Data. *Soc. Netw.* **1988**, *10*, 359–381. [CrossRef]
58. Feiock, R.C. Metropolitan Governance and Institutional Collective Action. *Urban Aff. Rev.* **2009**, *44*, 356–377. [CrossRef]
59. Chen, B.; Ma, J.; Feiock, R.; Suo, L. Factors Influencing Participation in Bilateral Interprovincial Agreements: Evidence from China's Pan Pearl River Delta. *Urban Aff. Rev.* **2019**, *55*, 923–949. [CrossRef]
60. Tavares, A.F.; Camões, P.J. New Forms of Local Governance: A Theoretical and Empirical Analysis of Municipal Corporations in Portugal. *Public Manag. Rev.* **2010**, *12*, 587–608. [CrossRef]
61. Feiock, R.C. The Institutional Collective Action Framework. *Policy Stud. J.* **2013**, *41*, 397–425. [CrossRef]
62. Feiock, R.C.; Lee, I.W.; Park, H.J. Administrators' and Elected Officials' Collaboration Networks: Selecting Partners to Reduce Risk in Economic Development. *Public Adm. Rev.* **2012**, *72*, S58–S68. [CrossRef]
63. Yi, H.; Suo, L.; Shen, R.; Zhang, J.; Ramaswami, A.; Feiock, R.C. Regional Governance and Institutional Collective Action for Environmental Sustainability. *Public Adm. Rev.* **2018**, *78*, 556–566. [CrossRef]
64. Yi, H.; Liu, W.; Ma, L. Designed Networks and the Emergence of Self-Organizing Interlocal Learning Network: Evidence from Chinese Cities. *Public Adm.* **2024**, *102*, 21–39. [CrossRef]
65. Xu, Y.; Zhu, Y.; Wu, Y.; Wang, X.; Zhang, W. The Population Flow under Regional Cooperation of "City-Helps-City": The Case of Mountain-Sea Project in Zhejiang. *Land* **2022**, *11*, 1816. [CrossRef]
66. Zhu, J.; Grigoriadis, T.N. Chinese Dialects, Culture & Economic Performance. *China Econ. Rev.* **2022**, *73*, 101783. [CrossRef]

land

MDPI

Article

Does Industrial Green Transformation Really Lead to High Land Use Efficiency? Evidence from China

Wenfang Pu [1,2], Mengba Liu [3] and Anlu Zhang [2,*]

1 School of Economy and Management, Zhejiang University of Water Resources and Electric Power, Qiantang District, Hangzhou 310018, China; wenfanglucky@webmail.hzau.edu.cn
2 College of Land Management, Huazhong Agricultural University, Wuhan 430070, China
3 College of City Construction, Jiangxi Normal University, Nanchang 330022, China; lmb_940312@jxnu.edu.cn
* Correspondence: zhanganlu@mail.hzau.edu.cn

Abstract: This research aimed to investigate whether transformation of the industrial sector in a region could improve industrial land use efficiency. Taking the urban agglomeration in the middle reaches of the Yangtze River in China as the research area, we compiled socioeconomic panel data from 2000 to 2020 in order to analyze the impact of the transformation of industrial sectors in an area on industrial land use efficiency from two dimensions: industrial structural optimization and industrial spatial layout. The research results show the following: (1) The rationalization and upgrading of the industrial sector, as well as the professional agglomeration of industry and diversified industrial agglomeration, have improved the efficiency of industrial land use. (2) The impact of industrial rationalization on industrial land use efficiency presents an inverted U-shaped curve, whereby the impact of industrial upgrading on industrial land use efficiency has a relatively small spatiotemporal heterogeneity. The spatiotemporal changes in the impact of industrial specialized agglomeration on industrial land use efficiency are relatively small, while the spatiotemporal changes in the impact of industrial diversified agglomeration on industrial land use efficiency are more obvious. (3) There is obvious spatial heterogeneity in the two dimensions industrial structural optimization and industrial spatial layout in the three sub-regions when improving industrial land use efficiency.

Keywords: industrial transformation; industrial land efficiency; spatial spillover effect; spatiotemporal heterogeneity; threshold effect

1. Introduction

Since the beginning of China's reform and opening-up in the late twentieth century, China has experienced rapid industrialization, moving from a large agricultural country to an industrialized country [1], and its industrial development has now entered the late stage of industrialization [2–5]. However, despite being a world industrial leader since entering the middle and late stages of industrialization [6], China's industrial development has not yet been fully transformed from its reliance on the "traditional" development mode, where the investment-driven and extensive development model is still prevalent. In particular, its industrial structure still has a large proportion of industries with low added value, high resource consumption, and serious pollution emissions and a factor input structure characterized by an over-reliance on general production factors, such as land and labor. There are many inconsistencies between China's industrial structure and land use [7–12]. In particular, the ratio of secondary and tertiary industries is uncoordinated, while the proportion of industrial land in areas is high. At the same time, during this era of

 https://doi.org/10.3390/land14051110

industrialization, the scale at which urban land has been transformed for industrial use has expanded rapidly, triggering a range of issues, such as the increasing industrialization of large amounts of cultivated land and ecological land, and a low industrial land use efficiency (ILUE). China's existing urban land has been only basically developed to date but has the potential for further development. However, due to some unreasonable utilization methods, a series of problems have arisen in land use in many regions, such as the rapid expansion of the scale of new land given over to industrial purposes, extensive industrial land use, and the inefficiency of construction land. Increasingly [13], problems are arising from mismatches between growth demands and the requirement for new construction land and the availability of land that is not subjected to cultivated land protection. Such protection aims to ensure that developments will not pose a risk to the ecological environment or lead to a series of associated issues, such as food security concerns.

To address these issues, the Chinese government has proposed important strategies to modify and optimize the economic structure of the country and promote industrial transformation. It is hoped that through industrial transformation in the industrial sector (ITI), the creeping industrialization of land could be reduced, ILUE could be improved, and sustainable development could be promoted. This research seeks to answer the following question: Can the industrial transformation promoted by the Chinese government play a role in enhancing ILUE?

To date, existing research has focused on exploring industrial transformation and land use efficiency, with the research considering three key aspects: (1) The first is research based on economic development, specifically, examining the connection between industrial transformation and ILUE in economic development and construction. For example, Gao et al. (2019) explored the changes in urban land use structure brought about by the promotion of regional economic integration through transformation of the industrial sector and market factors. They found that changing the land structure affects urban land use efficiency by changing the inputs and outputs of the land [14]. Scholars have drawn on the theory of smart growth and relied on high-quality social and economic development as the premise to construct an analysis framework for industrial land adjustment covering five dimensions: industrial transformation, urban planning, social and economic benefits, transportation convenience, and environmental protection [15–22]. The research results have shown that industrial transformation is an effective means to improve the internal structure of industrial land use, promote the intensification of industrial land, and improve the efficiency of industrial land utilization. Lu et al. (2020) analyzed the implications of land marketization on land use efficiency by considering changes in the industrial structure. They believed that land marketization could change the land use structure, thereby allowing for an optimization of resource allocation and improvements to land use efficiency. In parallel, from a regional perspective, in particular considering the western section to the central and eastern regions, they found that the role of the industrial structure in improving land use efficiency gradually weakened. (2) The second is research exploring the association between industrial transformation and land use efficiency at a more micro-scale. For example, Chen et al. (2018) focused on the manufacturing industry and studied the impact of manufacturing industrial transformation on ILUE. Their research results showed that industrial transformation could change the proportion of machinery manufacturing industries in the local land structure, and this process had a positive effect on improving the ILUE [23]. Chen et al. (2019) explored the ILUE of resource-oriented cities in China. They found that economic development, industrial transformation, and technological development all had significant effects on changing the land use types of resource-based cities and on improving the ILUE, while the internal labor structures of the enterprises and the ownership structures of the enterprises had serious negative outcomes on the

ILUE [24]. (3) The third is research analyzing the spatial and temporal differences in the effects of industrial restructuring on land use efficiency. For instance, Han et al. (2019) studied the impact of China's industrial transformation on land use efficiency and found that industrial transformation affects land use efficiency by changing the land structure, and this impact has regional heterogeneity. The impact was found to be more significant in large urban agglomerations, such as the Delta and the Pearl River Delta [25]. Similarly, Yin et al. (2019) explored how the transformation of leading industries guides urban land use. They found that the modernization and enhancement of leading industries in different cities have significantly different impacts on the ILUE, that changes in the leading industries can bring about changes in the land use intensity and land use patterns, and that these changes in land use structure have different impacts on the ILUE [26]. On a similar basis, Liu et al. (2021) explored the spatiotemporal differences in land use efficiency caused by transformation of the industrial structure. They found that from 2000 to 2015, China's land use patterns had changed through its promotion of industrial structural transformation, and this had improved the land use efficiency [27].

Undoubtedly, existing research has provided many interesting findings and inspiration for studying the impact of ITI on ILUE, but there are still some shortcomings in the research. For instance, in terms of theoretical logic, it is not uncommon for existing studies to explore the relationship between ITI and ILUE, but few consider land use as a way to establish the impact of land transformation brought about by ITI on ILUE from a comprehensive economic–environmental perspective. Also, existing research mostly analyzes the impact of ITI from one aspect: structural optimization of industrial transformation or the industrial spatial layout. Few studies have explored its impact from the two dimensions industrial structural optimization and industrial spatial layout. However, analyzing the impact of industrial structural optimization and industrial spatial layout brought about by ITI could allow for a more in-depth analysis of ITI, making the research results more valuable for practical reference. Lastly, ITI and ILUE have long-term and dynamic characteristics. The complexity of the ITI process can cause differences in the impact of ITI on ILUE at different stages in the development and transformation process. Yet, few existing studies have analyzed the spatiotemporal heterogeneity of ITI on ILUE from a dynamic perspective.

Based on the above discussion, this paper aimed to investigate whether ITI could improve the ILUE. Taking the urban agglomeration in the middle reaches of the Yangtze River (UMY) in China as the research area, we compiled socioeconomic panel data from 2000 to 2020 and then used them to analyze the effect of transformation of the industrial sector on ILUE from the two dimensions industrial structural optimization and industrial spatial layout.

2. Theoretical Analysis and Research Hypotheses

The ITI process inevitably causes land transformation, bringing about changes in the land use structure, methods, and intensity, which will then change the input–output that occurs on industrial land and affect the ILUE. This paper analyzes this impact from the changes in land use intensity and input level, land use structure and layout, land use subject behavior, and the comprehensive benefits of the changed land output caused by ITI.

Existing research has shown that the mechanism by which ITI affects land use transformation is by changing the proportion of land occupied by different industries in a city [28–30]. ITI leads to a transformation of the leading industries and their choice of location. Specifically, ITI changes the urban land structure as the city spreads out and leads to the reorganization of different land types within the city. ITI promotes agglomeration of the urban population and brings about a free flow and resetting of the production factors, which can lead to the reallocation of land resources in different industrial sectors. The expansion of urban land can lead to modifying the land use structure and patterns, which is one of the ways that ITI affects the ILUE. That is, different industrial developments will bring about changes in the regional land use proportions, which will in turn cause changes in the land use structure.

With the advancement of ITI, the proportion of high-polluting industries in the industrial sector, such as coal, metallurgy, wood, and chemicals, in the entire industrial system has gradually decreased, replaced by new industries based on new materials, medicine, digital information, and green energy [31–33]. In this process, areas with a low allocation and low efficiency of production land have dropped sharply, and been replaced by high-allocation emerging industrial land use. As the financial commitment to technology, innovation, and management in production processes has continued to increase, the ILUE has increased. The process of ITI can thus bring about the replacement of a city's leading industries [34], and this transformation of the leading industries will lead to changes in the ILUE. Leading industries are important as they lead the way in the industrial structure of economic systems. They not only play an important leading role in optimizing the industrial structure and industrial spatial layout in their location and neighboring areas but also drive the rise of emerging industries. From the perspective of the location selection process of the city's leading industries, the adjustment process of a city's industrial structure also involves transformation of the city's leading industry sectors. Leading industries have established their dominant position in the competition for urban land use space with their high-efficiency characteristics and tend to be distributed in urban centers with higher aggregation benefits [35–37]. The development of emerging leading industries prompts enterprises to increase investment in advanced production factors. Enterprises can improve their production efficiency by developing advanced production technologies and increasing their investment in scientific research. In this process, advanced production factors, such as technology, capital, and scientific management concepts, gradually replace natural production factors, such as land, which produces a factor substitution effect. In the production process, the land factor input is reduced, output is increased, and the ILUE is improved. At the same time, the spatial distribution of the urban land use structure will also be affected by the location selection of leading industries. Under the influence of the "retrospective effect" and "side effects" of leading industries, industries that are closely related to leading industries tend to gather around them to form an industrial complex, generating economies of scale and improving the regional ILUE.

Promoting the optimization of the land use spatial layout is another way for IT to change the ILUE [38]. Land rent in urban centers is typically expensive, as consumption levels and land costs are high. Due to the high rents and taxes, it is difficult for general industrial enterprises to obtain good benefits. Therefore, to reduce production expenses, industrial firms pursuing profit maximization tend to gradually migrate from the urban core to the city outskirts. This forms a driving force for industrial enterprises in the heart of the city to migrate to the periphery of the city.

Heavy industrial enterprises, such as metallurgy, chemicals, machinery, and metal smelting, which are highly dependent on market environments, such as production materials and transportation conditions, have mostly withdrawn from urban centers and gradually moved to urban suburbs where land costs are relatively low, promoting the agglomeration of industries on the edges of cities and creating scale effects, which can improve the ILUE to a certain extent. The relocation of traditional enterprises makes room for the development of emerging industries and high-efficiency industries. Emerging industries that occupy less land, with a high degree of land intensiveness, and that have high unit land output rates have the ability to pay higher land rents and tend to gradually cluster in the city center, where they can benefit from the good infrastructure and transportation conditions. These enterprises in the city center can generate higher product ancillary value, increase the land output, and improve the ILUE. Industries form two types of agglomeration distributions in urban centers and urban fringe areas, thereby achieving effective matches between the land elements and suitable enterprises. This process forms a "survival of the fittest" mechanism that enhances the allocation efficiency of the land production factors and enhances the ILUE.

In the ITI process in urban central areas, there are obvious differences in land use location requirements for different industrial developments. For instance, industries such as commerce, leisure, and entertainment have a high location requirement and need to be located in central areas with good transportation conditions and a large concentration of urban residents [39,40]. At the same time, these enterprises tend to have the characteristics of high profits and high output, which allows them to be able to pay high land rents and supports their choice to cluster in urban areas where land resources are tight. The proportion of land in the city center that covers large areas such as residences and warehousing has a low intensive land utilization and low output efficiency, and is gradually reducing as the urban land use structure and spatial layout change [41]. As a result, the urban building density and floor area ratio continue to increase, land use intensity continues to increase, and the ILUE is improving. Many investors in different industries in the city center want a central location with the most convenient transportation options and the highest economic benefits. There will be fierce competition for land in this location. On the one hand, the high land prices in the city center promote a simultaneous increase in investment intensity in the land by producers who aim to maximize their economic interests. On the other hand, the limited land resources in the city center and the demand for land for industrial development are pushing developers to turn to urban areas. The redevelopment and utilization of the existing land improves the ILUE.

The rapid expansion of tertiary industries during the ILUE process brings about a substantial increase in land use, while the increasing scarcity of urban land resources leads to an increase in land prices. Scarce land resources are allocated solely to efficient production companies at elevated prices through the market. According to land rent theory, the closer the land is to a city center, the higher the land rent and the land price that needs to be paid. In order to maintain survival and maximize profits, companies must find corresponding measures to reduce their production costs. Enterprises may gradually increase investment in non-land factors, such as manpower and technology to replace the original land investment, thereby increasing the ILUE. Value-added effects brought about by improvements to the urban infrastructure and rail transit in the city center to the surrounding land further deepens the factor substitution effect and accelerates the improvement in ILUE.

Another way that IT changes the ILUE is by promoting the market-oriented allocation of land resources. IT promotes the gradual expansion of the local and adjacent market demand scale [42], clearer production and division of labor in various industries, and the agglomeration of enterprises within a region, resulting in economies of scale. The production of different enterprises promotes economic investment within the region and changes the input intensity and utilization intensity of the land. The fierce competition among various types of enterprises in the process of pursuing profit maximization will also lead to a reduction in enterprise production costs, which can help expand the breadth and depth of the market. Expansion of the market scale will promote the flow of regional production factors, while more advanced investment factors will increase the intensity of land investment. Expansion of the market scale will also strengthen enterprise production capabilities, increase land output, and improve the ILUE.

From a spatial perspective, the spatial distribution of regional industries brought about by ITI essentially promotes the clustering of production factors, such as labor and capital, in a certain area. ITI promotes the agglomeration of different enterprises within a certain land space, and thereby creates a scale effect. The agglomeration of various related enterprises on different land gives them the ability to share production factors in the district and therefore promotes the flow of various factors of production to the agglomeration area [43]. The rapid accumulation of production inputs, including capital and labor, reduces production input, increases the economic output of the industrial land, and promotes improvement of the ITI.

Specifically, from the viewpoint of companies, having access to the same type of land in the city center brings about the agglomeration of similar enterprises in the same geographical space, resulting in industrial specialized clustering (KSL), which strengthens the advantages of the leading industries in the region. This continuous expansion improves the economic benefits of the land and promotes improvement of the ILUE. The agglomeration of similar industries brings about advanced production technology, and enterprises can be strengthened by cooperation, conducting relevant business training, sharing training and management costs, and reducing their production costs. The flow of production technology and technological innovation among enterprises in the same industry promotes cooperation between enterprises of the same type, promotes knowledge spillover, and is conducive to improving the overall regional production capacity.

Every enterprise can improve their resource utilization efficiency, reduce undesired outputs, and promote ILUE improvement. In the process of KSL, the agglomeration of similar enterprises on industrial land will cause competition. With the intention to dominate the competition and acquire more economic benefits, an abundance of homogeneous enterprises will compete against each other based on their production processes and products. On the one hand, enterprises will increase investment in research, improve production efficiency, and expand production scale through the application of more advanced production concepts. On the other hand, enterprises may promote innovation through the use of technology. These measures can reduce resource input in the production process, achieve greater economic outputs, and increase the ILUE. Additionally, enterprises from different industries may gather in urban centers, creating diversification agglomeration (DIV). This process promotes mutual learning and communication between enterprises, creating a learning effect. These can help reduce the risk of a single enterprise carrying out technology research, allow for optimizing technical resources, and increase the scope of technology application [44]. DIV can not only promote technological progress within various industries from spillover effects but also increase the benefits brought about by technological progress, increase the land output, and thereby improve the ILUE.

During the DIV process, the core technologies of various industries generate knowledge spillover benefits within and between enterprises, promoting productivity improvements in the entire region. Different enterprises gathered in the city center can share diverse production inputs, like production financial resources and technological innovation, promoting the concentration of different production factors in the region, especially the construction of shared infrastructure. Enterprises can reduce transportation costs by sharing infrastructure construction dividends, which can increase the economic benefits of the industrial land, and promote improvements in the ILUE [45]. Enterprises in various fields form advanced technology and knowledge diffusion effects through exchange and cooperation with each other, which can also promote the emergence of advanced production concepts. Diversified industries can also help build a solid economic structure. The development of various industries will drive the land economic output of the entire region and enhance the ILUE.

From the perspective of the labor force, on the one hand, the KSL process is conducive to promoting the accumulation of productive capital by labor at the same level, especially core technical talents, forming a labor matching effect. Within the KSL area, for enterprises that rely on technology to carry out production, KSL reduces learning and production costs through cross-industry production. The agglomeration of professional talent is also conducive to reasonably matching enterprises and the talent they need, thereby strengthening the regional talent market structure. Enterprises can save costs by being able to efficiently find the labor they need, which also reduces the production cost of recruiting labor. The agglomeration of the same types of skilled labor brought about by KSL can reduce the capital investment in land production, improve production efficiency, and increase the ILUE. On the other hand, DIV promotes a concentration of skilled labor with expertise in various production technologies in the same area, and creates a labor "reservoir" function to meet the employment needs of enterprises with various production technologies to the greatest extent. Meanwhile, the labor force has more opportunities to choose job options to work in professionally related enterprises, which reduces the production costs of training workers new to the sector and increases the economic effects from them being able to get up to speed quicker, which increases firm and land output. If a firm in an agglomeration area faces bankruptcy, its internal labor force would not need to look for jobs across other regions or move cities as their skills could be transferable to other nearby industries. The "reservoir" of labor brought about by DIV therefore effectively reduces unemployment risks for the labor force. During the agglomeration process, the labor force can also learn different skills and gain new knowledge and technology expertise to improve their employment security. A good employment environment can also absorb more high-quality labor, improve production efficiency, and enhance the ILUE.

Overall, land transformation brought about by ITI causes changes in the allocation relationship of capital, labor, and other resources on the land, which leads to changes in the structure of the input resources. According to the C-D production function [41], at different stages of the production technology level and economies of scale, changes in the land input structure will lead to an increase or decrease in land economic output, thus affecting the land input–output structure, that is, the ILUE (Figure 1).

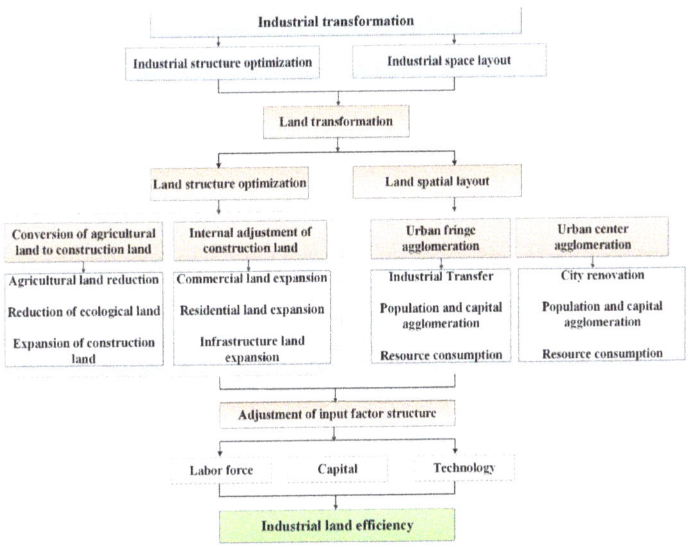

Figure 1. Theoretical framework for the impact of industrial transformation on industrial land use efficiency.

Based on the above analysis, this paper puts forward a research hypothesis:
ITI affects the input–output structure of land, which then affects the ILUE.

3. Study Area, Data, and Methodology

3.1. Study Area

For this study and as discussed in this paper, the urban agglomeration in the middle reaches of the Yangtze River (UMY) in China, including the three provinces of Hunan, Hubei, and Jiangxi, was chosen as a representative area for the research. UMY is a large urban agglomeration mainly formed by Wuhan urban agglomeration and urban agglomeration around Changsha, Zhuzhou, and Lake Poyang. Due to the lack of some data in Tianmen City after 2010, Tianmen City was not included in the study area. Excluding Tianmen City, the study area of this paper was made up of 30 prefecture-level cities in the three provinces of Hunan, Hubei, and Jiangxi in the UMY (Figure 2).

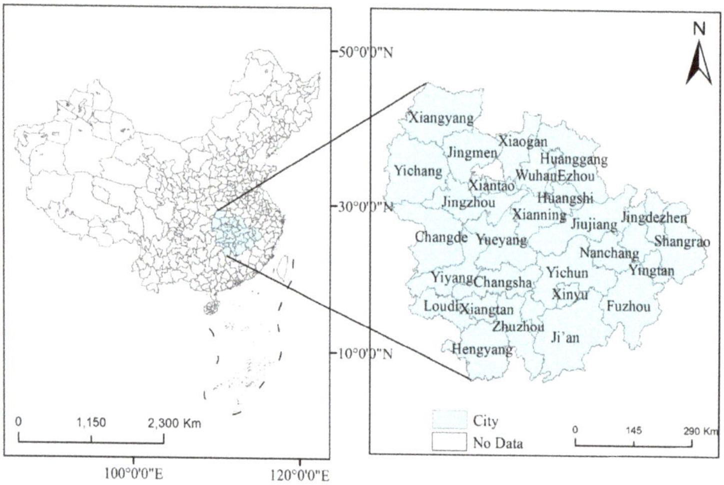

Figure 2. Location of study area.

The UMY is a new growth pole of China's economy, a pioneer area for new urbanization in Central and Western China, a demonstration area for inland opening-up and cooperation, and a guiding area for the construction of a "two-type" society. ITI in this region has a typical demonstration role to play in China.

3.2. Data Sources

The ITI data in this paper come from the statistical year book of each city, while the ILUE calculation data come from the China Land Market Network, https://www.landchina.com (accessed on 27 July 2021). The statistical year book of each city, the "China City Statistical Year book", and the national economic and social development of each city gazette. The data for the control variables were sourced from the China Urban Statistical Year book and the National Economic and Social Development Bulletin of each city. Because the Qianjiang Municipal Bureau of Statistics did not collect data on the output value of industrial sectors from 2000 to 2010, the output value of each industrial sector in Qianjiang City was replaced by the main business income of each industrial sector. Missing values in some years were interpolated.

3.3. Methods

3.3.1. Measurement of the ILUE: Data Envelopment Analysis by the DEA-SBM Model

In this research, we chose the SBM model, including undesired outputs, built by Zhao et al. (2014) [43] based on tone. The formula is as follows:

$$\text{ILUE} = \min \frac{1-\frac{1}{N}\sum_{n=1}^{N}\frac{n_x^s}{k'n_{t'}^x}}{1+\frac{1}{M+I}\left(\sum_{m=1}^{M}\frac{m_y^s}{k'm_{t'}^y}+\sum_{i=1}^{I}\frac{i_b^s}{k'i_{t'}^b}\right)}$$

$$\begin{cases} \sum\limits_{t=1}^{T}\sum\limits_{k=1}^{K} k_t^z k n_t^x + n_x^s = k'n_{t'}^x, n = 1,\ldots,N, \sum\limits_{t=1}^{T}\sum\limits_{k=1}^{K} k_t^z k m_t^y - m_y^s = k'm_{t'}^y, m = 1,\ldots,M \\ \sum\limits_{t=1}^{T}\sum\limits_{k=1}^{K} k_t^z k i_t^b + i_b^s = k'i_{t'}^b, i = 1,\ldots,I \\ k_t^z \geq 0, n_x^s \geq 0, m_y^s \geq 0, i_b^s \geq 0, k = 1,\ldots,K \end{cases} \tag{1}$$

In the formula, ILUE is the ILUE value to be calculated; N, M, and I are the number of inputs, desired outputs, and undesired outputs, respectively; $\left(n_x^s, m_y^s, i_b^s\right)$ represents the input–output relaxation vector; $\left(k'n_{t'}^x, k'm_{t'}^y, k'i_{t'}^b\right)$ is k''s input–output value of a production unit in period t'; and $k_{/}^z$ represents the weight of the decision-making unit. The objective function ρ strictly decreases monotonically with respect to n_x^s,m_y^s,i_b^s.

3.3.2. Measurement of the ILUE

Based on existing research [40], the input–output indicators for ILUE measurement were chosen as follows (Table 1):

Table 1. Calculation index system of industrial land use efficiency.

Indicator Selection	Classification Indicator	Single Indicator	Unit
Input variable	Land	Industrial land area	Hectare
	Labor force	Secondary industry employment	Thousands of people
	Capital	Industrial fixed asset investment	Billion RMB
Desired output	Economic output	Industrial output	Billion RMB
Undesired output	Environmental pollution	Carbon emission	Million tons

Input: industrial land area, number of employees in the secondary industry, and industrial fixed asset investment. Output: the desired as well as undesired outputs. The desired output value is the total industrial output value, and the undesired output value here is the industrial carbon emissions. The carbon emissions data used in this paper come from the China Carbon Accounting Database (CEADs) (https://www.ceads.net.cn/) created by Professor Guan Dabo's team at Tsinghua University.

3.3.3. Measurement of Industrial Industry Rationalization (ITL)

This paper refers to Gan et al. (2011) [46] and uses Theil's coefficient to calculate the ITL:

$$\text{ITL} = \sum_{i=1}^{n}\left(\frac{V_i}{V}\right)\ln\left(\frac{\frac{V_i}{P_i}}{\frac{V}{P}}\right) \tag{2}$$

In Formula (2), V_i represents the industrial added value of i industrial sector; P_i represents the number of laborers engaged in i industrial sector; V represents the gross production value of 33 industrial sectors; P represents the sum of the labor force of

33 industrial sectors; and n represents the number of industrial sectors. If ITL were to gradually approach zero, the industrial industry would reach an economic equilibrium state. Therefore, the smaller the ITL value, the more optimized the industrial structure of the industrial industry is.

3.3.4. Measurement of Industrial Sector Upgrading (IW)

Existing studies mostly employ the quotient of the tertiary industry's production value over the secondary industry's production value as an indicator to measure how advanced the industrial structure is [47]. On the basis of this method, this study innovatively used high-tech industries to account for the output value of industrial industries to represent the IW. Through data matching and deletion, the selected high-tech industrial industries included instrumentation, medicine, electronic and communication equipment, computers and office equipment, aerospace equipment, and machinery. The IW calculation formula was as follows:

$$\mathrm{IW} = \mathrm{IH}/\mathrm{IO} \tag{3}$$

In this formula, IO represents the production value of the industrial sector, and IH represents the output worth of the high-tech industry. The larger the value of IW, the greater the proportion of high-tech industry in the industrial sector and the higher the level of IW.

3.3.5. Measurement of Industrial Sector Industry Specialization Agglomeration (IKSL)

$$\mathrm{Iksl} = \sum\nolimits_{i=1}^{I} \left| \frac{L_{i,r}}{L_r} - \frac{L_i}{L} \right| \tag{4}$$

In the formula above, i represents an industry sector, r represents the city, and L represents the amount of people. The bigger the Iksl index, the higher the Iksl level.

3.3.6. Measurement of Industrial Sector Industry Diversification Agglomeration (IDIV)

Drawing on the research of Zhang et al. (2019) [48], this paper uses the Herfindahl index to judge the extent of industrial diversification and agglomeration, using the formula below:

$$\mathrm{IDiv} = 1/\sum\nolimits_{i=1}^{I} (L_{i,r}/L_r)^2 \tag{5}$$

The definitions of the items in the equation are the same as those in the Iksl formula above. The larger the IDiv index, the more dispersed the industrial and the higher the IDiv.

3.3.7. Industrial Sector Transformation on the ILUE Using the Spatial Durbin Model

Effect of Industrial Structure Optimization on the ILUE

$$\begin{aligned} \mathrm{ILUE} = {} & \rho \mathrm{AlnILUE} + \partial_1 \mathrm{ITL} + {+\partial_2 \mathrm{IW}} + \partial_3 \mathrm{lnGDP} \ + \varphi_4 \mathrm{CZB} + \varphi_5 \mathrm{InGZ} + \varphi_6 \mathrm{SCHL} + \varphi_7 \mathrm{KJB} + \varphi_1 \mathrm{AITL} + \varphi_2 \mathrm{AIW} \\ & + \partial_3 \mathrm{AlnGDP} \ + \varphi_4 \mathrm{ACZB} + \varphi_5 \mathrm{AInGZ} + \varphi_6 \mathrm{ASCHL} + \varphi_7 \mathrm{AKJB} + \gamma l_n + \varepsilon \end{aligned} \tag{6}$$

This paper uses the spatial Durbin model to explore the effect of ITI on the ILUE. In Formula (6), ρ is the spatial autocorrelation coefficient, A represents the spatial weight matrix, and l_n represents $n \times 1$ vector. In addition, the control variables lnGDP, KJB, CZB, InGZ, SCHL that affect the ILUE were added. Also, ∂, φ, and γ are vectors of their respective regression coefficients, while ε is the error term.

Impact of the Industrial Space Layout on the ILUE

$$\begin{aligned} \mathrm{ILUE} = {} & \rho \mathrm{AlnILUE} + \partial_1 \mathrm{InIKsl} + {+\partial_2 \mathrm{InIDiv}} + \partial_3 \mathrm{lnGDP} \ + \varphi_4 \mathrm{CZB} + \varphi_5 \mathrm{InG} + \varphi_6 \mathrm{SCHL} + \varphi_7 \mathrm{KJB} + \varphi_1 \mathrm{AIKsl} + \varphi_2 \mathrm{AInIDiv} \\ & + \partial_3 \mathrm{AlnGDP} \ + \varphi_4 \mathrm{ACZB} + \varphi_5 \mathrm{AInGZ} + \varphi_6 \mathrm{ASCHL} + \varphi_7 \mathrm{AKJB} + \gamma l_n + \varepsilon \end{aligned} \tag{7}$$

The definitions of the variables are consistent with those in Formula (6).

3.3.8. Industrial Transformation Effect on the Spatial and Temporal Heterogeneity of the ILUE

Geographically weighted regression (GWR) is a local linear regression method based on modeling spatially relationships. This regression method can generate a regression model describing local relationships by exploring each part of the region, thereby allowing for an accurate analysis of localized variable relationships and spatial heterogeneity. The geographically time-weighted regression (GTWR) model used in this paper adds time effects to the GWR model as an extension of the GWR model [49,50]. This paper uses this model to explore the spatiotemporal heterogeneity of carbon emissions caused by industrial transformation at different times and in different spatial dimensions (Table 2).

Table 2. Descriptive statistics of all variables used.

Variable	Meaning	Mean	Std	Min	Max
ILUE	Industrial land efficiency	0.688	0.266	0.215	1
ITL	Industrial industry rationalization	0.097	0.083	0.004	0.549
IW	Industrial industry upgrading	0.147	0.073	0.007	0.522
IKSL	Industrial industry specialization agglomeration	0.767	0.215	0.091	1.441
InIDIV	Industrial industry diversified agglomeration	2.219	0.435	0.747	2.924
InGDP	The level of economic development	15.888	1.072	13.198	18.905
KJB	Technology investment	0.005	0.007	0.00001	0.057
InGZ	Urban residents' wages	10.106	0.781	7.999	11.586
SCHL	Land marketization level	0.701	0.332	0.063	1
CZB	Government management	0.142	0.060	0.032	0.394

4. Results

4.1. Spatiotemporal Distribution of the ITL

From 2000 to 2020, the ITL of the UMY dropped from 0.250 in 2000 to 0.033 in 2020. The ITL coefficient became significantly smaller and the degree of ITL became higher (Figure 3).

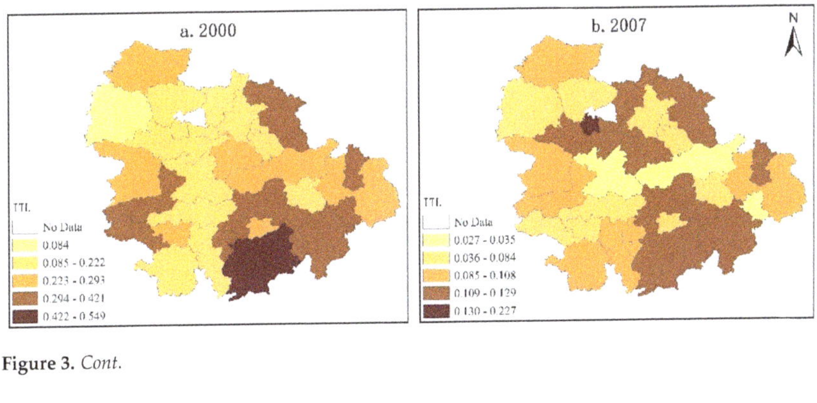

Figure 3. *Cont.*

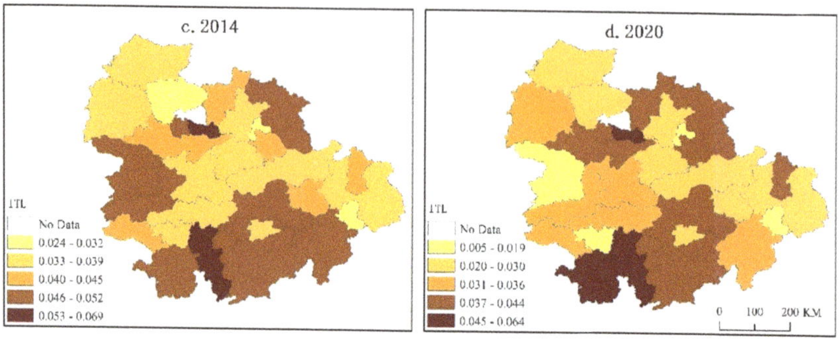

Figure 3. Spatiotemporal distribution of the industrial rationalization of industrial industries in UMY.

4.2. Spatiotemporal Distribution of the IW

IW increased from 0.121 in 2000 to 0.184 in 2020. The IW coefficient also increased, indicating that the proportion of high-tech industries in the industry gradually increased, high-tech industries gradually expanded, and the IW level increased (Figure 4).

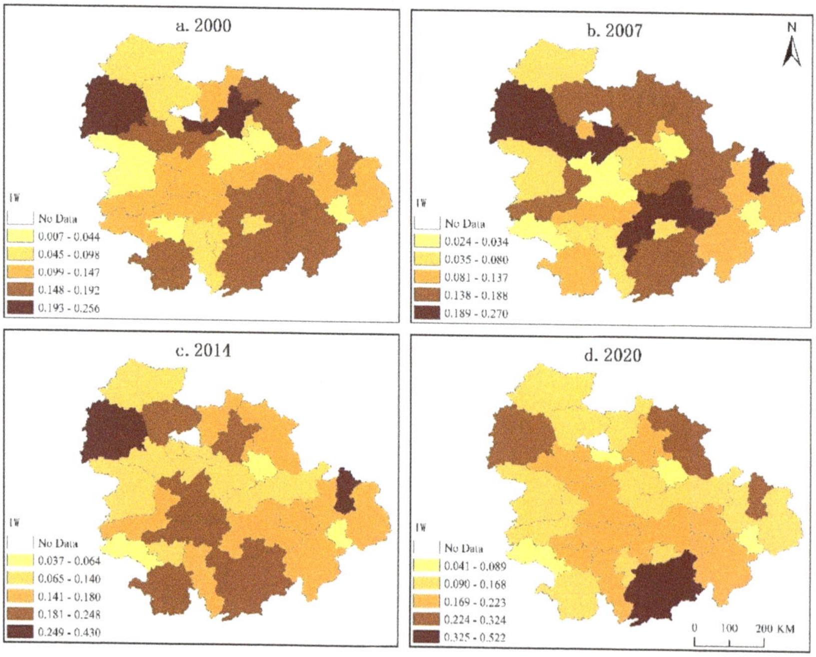

Figure 4. Spatiotemporal distribution of industrial upgrading in UMY.

4.3. Spatiotemporal Distribution of the IKSL and IDIV

The IKSL of the UMY dropped from 0.797 in 2000 to 0.735 in 2020, and the level of industrial professional agglomeration declined (Figure 5). The IDIV level increased from 8.570 in 2000 to 10.766 in 2020 (Figure 6).

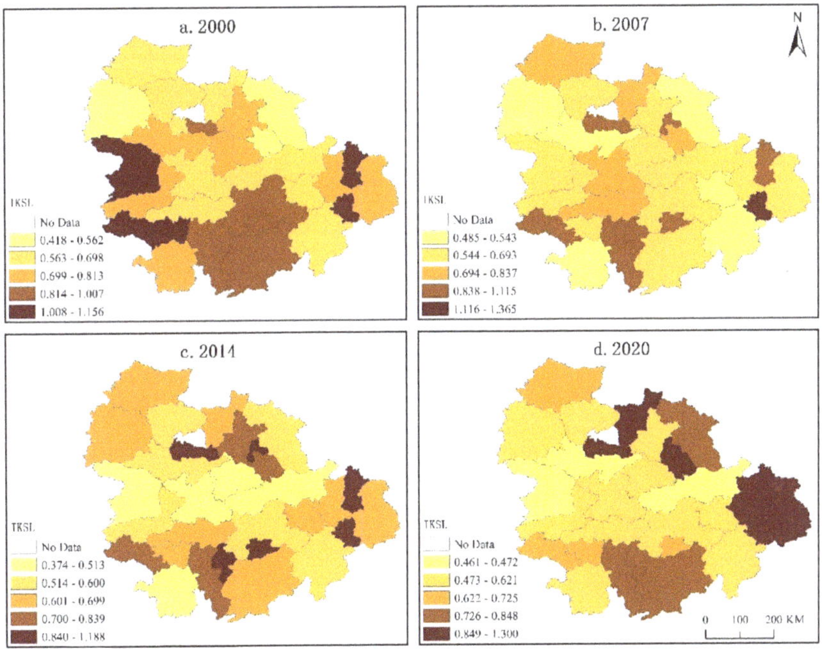

Figure 5. Spatiotemporal distribution map of industrial specialization agglomeration in UMY.

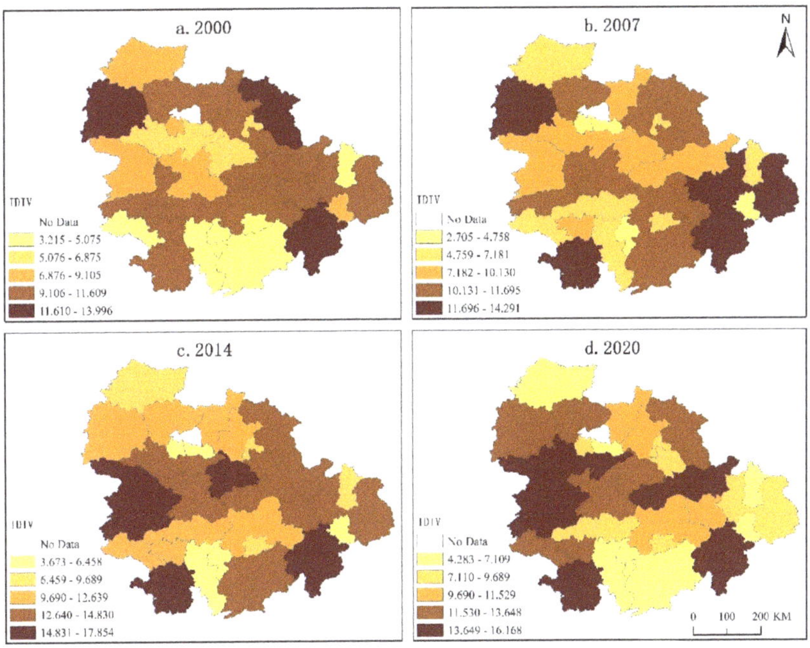

Figure 6. Spatiotemporal distribution of industrial diversification in UMY.

4.4. Spatiotemporal Distribution of the ILUE

The ILUE of the urban agglomeration in the UMY increased from 0.635 in 2000 to 0.779 in 2020. During this time, the lowest ILUE value was 0.601 in 2010, while the highest

ILUE value appeared in 2020 as 0.779. Generally speaking, from 2000 to 2020, the ILUE of the UMY showed an upward trend year by year (Figure 7).

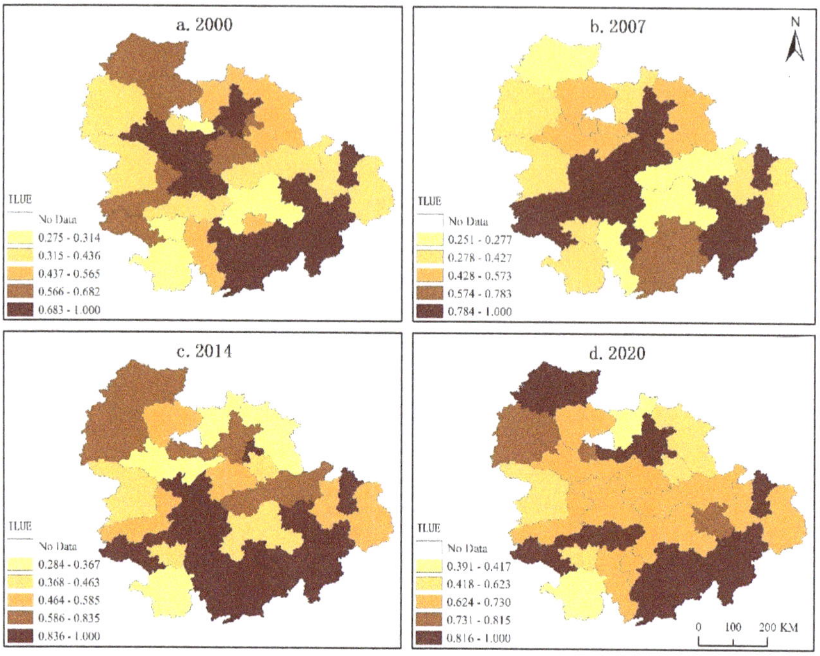

Figure 7. Spatiotemporal distribution of industrial land efficiency in UMY.

4.5. Spatial Spillover Effects of Industrial Transformation on the ILUE

4.5.1. Spatial Spillover Effect of Industrial Structure Optimization on the ILUE

Both IKSL and IDIV in the UMY played a role in improving the ILUE. Table 3 reports the impact of industrial structure optimization on the ILUE under three types of spatial weight matrices. The spatial autoregression coefficients of the three models were 0.228, 0.251, and 0.156, respectively. Notably, all Spearman's rho showed positive numbers and all passed the 1% significance level test. Overall, the model results show that under the three weights, ITL played a role in improving the ILUE. Under the adjacency distance weight matrix, ITL increased by 1% and ILUE increased by 0.370%, while under the economic distance weight matrix, ITL increased by 1% and ILUE increased by 0.898% and under the economic geography nested weight matrix, ITL increased by 1% and ILUE increased by 0.193%. This is because ITL reduces enterprise production costs and environmental losses by promoting a sharing of knowledge, technology, and labor within industrial enterprises, which improves the ILUE. Meanwhile, ITL can promote the extent of the industrial chain, promote the integration of urban advantageous resources, and improve the ILUE. Pollution-intensive industries, as represented by the energy industry and chemical industry, will eventually achieve internal upgrades through high-tech and clean transformation. On the one hand, pollution and energy consumption will be reduced, and on the other, the undesired land output will be reduced. The transformation and upgrading of industrial sectors rely on high-tech industries and new technologies, with technological innovation the main driving force for development. Advanced technology increases the intensity of land use, increases the land economic benefits while promoting ecological and environmental benefits, and is conducive to the rational use of land to increase the land economic output and improve the ILUE.

Table 3. Spatial spillover effects of industrial structure optimization on ILUE.

ILUE	Adjacency Distance			Economic Distance			Economic Geography Nesting		
rho	0.228 ***			0.251 ***			0.156 ***		
	(4.090)			(3.120)			(4.42)		
	direct effect	indirect effect	total effect	direct effect	indirect effect	total effect	direct effect	indirect effect	total effect
ITL	0.099	0.271	−0.370	0.066	0.832 **	−0.898 ***	0.114	0.079	−0.193
	(0.660)	(1.010)	(1.380)	(0.450)	(2.580)	(2.780)	(0.78)	(0.42)	(0.95)
IW	−0.155	−0.128	0.283	−0.157	−0.247	0.405	0.269 *	−0.324	0.055
	(−1.020)	(−0.310)	(−0.620)	(−1.060)	(−0.370)	(−0.590)	(−1.80)	(1.52)	(0.20)
lnGDP	0.136 ***	−0.194 **	0.058	0.134 ***	−0.014	0.120	0.122 ***	−0.215 ***	0.093
	(3.320)	(−2.040)	(−0.590)	(3.080)	(−0.100)	(0.880)	(3.21)	(−4.25)	(−1.55)
KJB	−0.814	1.518	0.704	−1.398	2.772	1.373	0.431	−0.650	0.219
	(−0.710)	(1.030)	(0.500)	(−1.090)	(1.290)	(0.790)	(0.39)	(−0.53)	(−0.19)
CZB	0.266	−0.227	−0.039	0.325	−1.094	−0.770	0.282	−0.604 *	−0.322
	(1.010)	(−0.370)	(0.060)	(1.250)	(−1.020)	(−0.700)	(1.11)	(−1.81)	(−0.79)
lnGZ	0.088	0.065	0.153	0.027	0.057	0.084	0.010	0.200 ***	0.211 ***
	(1.430)	(0.630)	(1.520)	(0.440)	(0.450)	(0.710)	(0.19)	(3.37)	(3.31)
SCHL	0.034	−0.083	−0.049	0.058	−0.205	−0.147	0.097 **	−0.207 ***	−0.111 *
	(0.690)	(−0.960)	(−0.580)	(1.270)	(−1.560)	(−1.070)	(2.05)	(−3.48)	(−1.68)
R^2	0.012			0.007			0.015		

Note: * statistical significance at the $p < 0.10$ level, ** $p < 0.05$ level, *** $p < 0.01$ level.

Looking at IW, the model results show that under the three weights, IW played a role in improving the ILUE. Under the adjacency distance weight, IW increased by 1% and ILUE increased by 0.283%, while under the economic distance weight, IW increased by 1% and ILUE increased by 0.405% and under the economic geography nested weight, IW increased by 1% and ILUE increased by 0.055%. The main way that IW affects the ILUE is by first optimizing the internal land resource utilization structure and industrial layout of industry. IW can accelerate the elimination of polluting industries, such as coal, textiles, and electricity, while also accelerating the development of emerging industries, like pharmaceuticals, electronic communication equipment, computers, and supplies. In this process, the land area used for production plants and other related facilities of industrial enterprises with high inputs and low outputs is gradually decreased, while the land used by high-tech enterprises with low inputs and high outputs is increased, and the output of the industrial land is thereby increased. Another way in which IW affects the ILUE is through its driving role in optimizing the peripheral land use layout of industrial enterprises. In order to reduce costs, companies must find locations suitable for their development and adjust their industrial layout. On the one hand, heavy industrial industries, such as machinery and metals, are gradually being squeezed out of the city center and are moving to the suburban districts. The original idle and extensive land resources in the suburbs have now been developed and utilized, and ILUE has improved. On the other hand, high-tech industries that can pay high land rents have gathered in urban centers, forming new industrial agglomeration groups, which has not only expanded the scale of industrial production but also increased the intensive utilization of industrial land, significantly improving ILUE. Among the control variables, the stage of economic development, investments in science and technology, and the wage level of urban residents all increase the ILUE, while the level of government management and land marketization decrease the ILUE.

4.5.2. Spatial Spillover Effect of the Industrial Spatial Layout on the ILUE

The two dimensions of the industrial layout in the UMY, IKSL and IDIV, have played a role in improving the ILUE. Table 4 reports the impact of the industrial spatial layout on the ILUE under three types of spatial weight matrices. The spatial autoregression coefficients of the three models were 0.240, 0.257, and 0.157, respectively, while Spearman's rho was positive and passed the 1% significance level, indicating the existence of spatial spillover effects. Looking at IKSL first, all three weight matrices showed that IKSL increased the ILUE. With the adjacency distance weight matrix, a 1% increase in IKSL increased the ILUE by 0.259%. In the economic distance weight matrix, every 1% increase in IKSL increased the ILUE by 0.479%. With the economic geography nested weight matrix, for every 1% increase in IKSL, ILUE increased by 0.188%. On the one hand, IKSL promoted the agglomeration of the same industries within the market and sharing of production materials, such as capital, technology, and labor, within the industry, generating spillover effects and improving the ILUE. On the other hand, the IKSL process was conducive to promoting the accumulation of productive capital by labor at the same level, especially core technical talents, and forming a labor matching effect. The labor pool formed by IKSL reduces the unemployment risk of the labor force to a certain extent. Industrial enterprises can also access a suitable labor force to carry out production, improve productivity, and enhance the ILUE. Marshall's externality theory believes that the input–output correlation between producer services and manufacturing is strong, and labor sharing occurs during the agglomeration process. This process reduces production costs, increases land output, and improves the ILUE. Looking at IDIV, IDIV also improved the ILUE under the three weights. In the adjacency distance weight, every 1% increase in IDIV increased the ILUE by 0.017%, while with the economic distance weight, every 1% increase in IDIV increased the ILUE by 0.176% and under the economic geography nested weight, every 1% increase in IDIV increased the ILUE by 0.003%. On the one hand, under the agglomeration effect of IDIV, the operating costs of industrial enterprises were reduced, which helped to enhance corporate competitiveness, improve corporate productivity, and promoted an improvement in the ILUE. Additionally, IDIV will also promote interactions between industrial enterprises and generate a synergistic effect of "the whole is greater than its parts", which helps improve overall competitiveness and enhances the ILUE.

Table 4. Spatial spillover effect of industrial spatial layout on ILUE.

ILUE	Adjacency Distance				Economic Distance		Economic Geography Nesting		
Rho	0.240 ***				0.257 ***		0.157 ***		
	(4.410)				(3.240)		(4.450)		
	direct effect	indirect effect	total effect	direct effect	indirect effect	total effect	direct effect	indirect effect	total effect
IKsl	−0.084	0.343 ***	0.259 **	−0.054	0.532 ***	0.479 ***	−0.077	0.264 ***	0.188 **
	(−1.370)	(2.960)	(2.160)	(−0.900)	(3.020)	(2.600)	(−1.320)	(3.640)	(2.090)
InIDiv	−0.095 ***	0.112	0.017	−0.082 **	0.257 **	0.176	0.074**	−0.071 *	0.003
	(−2.680)	(1.430)	(0.200)	(−2.330)	(2.340)	(1.560)	(−2.150)	(1.650)	(−0.050)
InGDP	0.123 ***	−0.178 *	−0.055	0.122 ***	−0.109	0.013	0.112***	−0.191 ***	−0.079
	(3.100)	(−1.900)	(−0.570)	(2.860)	(−0.810)	(0.100)	(3.040)	(−3.960)	(−1.400)
KJB	−0.510	2.093	1.583	−1.100	4.061 *	2.961	0.580	−0.426	0.154
	(−0.450)	(1.410)	(1.130)	(−0.870)	(1.740)	(1.560)	(0.530)	(−0.350)	(0.130)
CZB	0.326	−0.509	−0.183	0:375	−1.386	−1.011	0.372	−0.789 **	−0.417
	(1.290)	(−0.860)	(−0.280)	(1.480)	(−1.260)	(−0.890)	(1.530)	(−2.390)	(−1.020)
InGZ	0.091	0.044	0.135	0.042	0.082	0.124	0.036	0.163 ***	0.199 ***
	(1.570)	(0.430)	(1.330)	(0.710)	(0.640)	(1.030)	(0.710)	(2.830)	(3.140)
SCHL	0.034	−0.072	−0.038	0.070	−0.201	−0.131	0.074 *	−0.196 ***	−0.121 *
	(0.710)	(−0.840)	(−0.460)	(1.530)	(−1.560)	(−0.970)	(1.650)	(−3.310)	(−1.940)
R^2	0.012			0.017			0.023		

Note: * statistical significance at the $p < 0.10$ level, ** $p < 0.05$ level, *** $p < 0.01$ level.

4.5.3. Robustness Check

In order to test the robustness of the results, we conducted a robustness test on the model by changing the sample period and matrix, reducing the research time by 4 years, that is, changing the research time period to 2000–2016, and using the inverse distance weight matrix to verify the research robustness of the results. Tables 5 and 6 present the results of the robustness test. The results for the industrial structure optimization and industrial spatial layout on the ILUE were basically the same as in the original model, which shows that the estimation results of the original model had good stability.

Table 5. Robustness test of the impact of industrial structure optimization in industrial industries on industrial land use efficiency.

Explained Variable ILUE			
Explanatory variable	Direct effect	Indirect effect	Total effect
ITL	0.069 *** (3.020)	0.145 ** (−2.550)	−0.214 * (−1.910)
IW	0.164 (−0.890)	0.200 (−0.390)	0.364 (−0.64)
InGDP	0.135 *** (2.950)	−0.114 (−1.040)	0.021 (0.180)
KJB	−0.601 (−0.490)	2.183 (1.390)	1.582 (1.020)
CZB	0.086 (0.250)	−0.234 (−0.320)	−0.148 (−0.180)
InGZ	0.057 (0.790)	−0.008 (−0.070)	0.048 (0.380)
SCHL	0.011 (0.210)	−0.419 (−0.440)	−0.030 (−0.330)
Rho	0.222 *** (3.660)		
R^2	0.034		

Note: * statistical significance at the $p < 0.10$ level, ** $p < 0.05$ level, *** $p < 0.01$ level.

Table 6. Robustness test of the impact of industrial spatial layout of industrial industries on industrial land use efficiency.

Explained variable ILUE			
Explanatory variable	Direct effect	Indirect effect	Total effect
IKsl	0.111 (−1.630)	0.377 *** (2.950)	0.266 ** (2.030)
InIDiv	−0.102 ** (−2.520)	0.169 ** (1.850)	0.067 *** (3.020)
InGDP	0.130 *** (2.910)	−0.103 (−0.920)	0.028 (0.240)
KJB	−0.194 (−0.160)	2.422 (1.550)	2.228 (1.500)
CZB	0.195 (0.600)	−0.502 (−0.700)	−0.307 (−0.380)
InGZ	0.063 (0.940)	−0.036 (−0.300)	0.027 (0.210)

Table 6. *Cont.*

Explained variable ILUE			
SCHL	0.010 (0.180)	−0.031 (−0.340)	−0.022 (−0.240)
Rho	0.230 *** (3.860)		
R^2	0.053		

Note: ** statistical significance at the $p < 0.05$ level, *** $p < 0.01$ level.

4.6. Industrial Transformation Effects on the Temporal and Spatial Heterogeneity of the ILUE

The impact of ITL on the ILUE showed an inverted U-shaped curve in the box (Figure 8). From 2000 to 2014, the length of the box showed an increasing trend year by year, with a small change. The box was longest in 2014, indicating that the spatiotemporal heterogeneity of the impact of ITL on the ILUE was the largest in 2014. Since 2015, the box length has been weakening year by year, indicating that the spatiotemporal heterogeneity of the impact of ITL on the ILUE has been gradually reducing since 2015. This shows that after 2015, the role of ITL in improving the ILUE gradually became stable. The spatiotemporal heterogeneity of the effect of IW on the ILUE changed slightly, whereby, from 2000 to 2005, the box length increased and the spatiotemporal heterogeneity increased. From 2006 to 2015, the length of the box gradually decreased, and the spatiotemporal heterogeneity gradually weakened. After 2016, the box length gradually increased, and the spatiotemporal heterogeneity increased again. The spatiotemporal changes in the impact of IKSL on ILUE were relatively small, and the box curve remained always relatively stable. The impact of IDIV on the ILUE showed a large change in length and obvious spatiotemporal changes. From 2000 to 2008, the box curve increased from small to large. During this period, the impact of IDIV on the ILUE gradually increased each year. From 2009 to 2015, the impact of IDIV on the ILUE gradually decreased each year. After 2015, the impact of IDIV on the ILUE gradually increased each year.

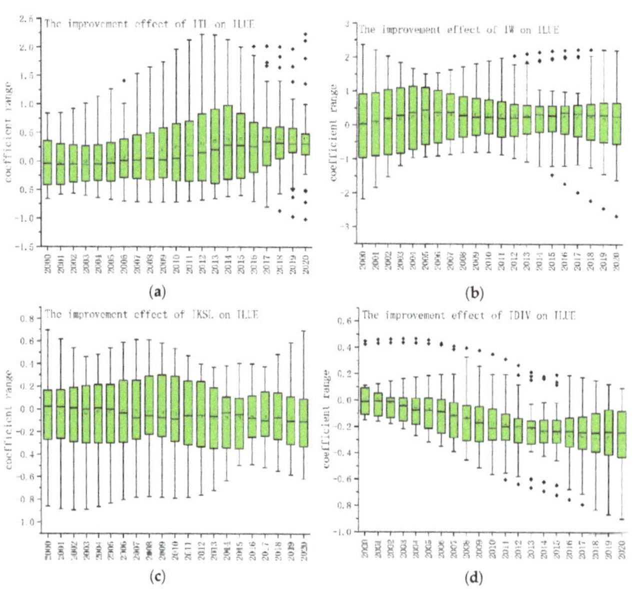

(a) (b) (c) (d)

Figure 8. The impact of industrial transformation on ILUE of space–time heterogeneity.

From 2000 to 2010, the effect of IKSL in Hubei Province on improving the ILUE decreased year by year, then increased from 2010 to 2015, and finally gradually decreased from 2015 to 2020 (Figure 9). The effect of IKSL in Jiangxi Province on improving the ILUE showed an inverted U-shaped curve, increasing year by year from 2000 to 2010 but then decreasing year by year from 2010 to 2020. Compared with Hubei Province and Jiangxi Province, the effect of IKSL on improving the ILUE in Hunan Province varied greatly, and the curve was not smooth, with highs and lows observed during the study period. From 2000 to 2005, the effect of IDIV on ILUE in Hubei Province was linear and relatively stable. After 2005, it showed a downward trend year by year. The effect of IDIV on ILUE in Jiangxi Province was relatively stable, with a slight increase from 2000 to 2005, a slight decrease from 2005 to 2015, and a stable trend from 2015 to 2020. For IKSL, the effect of IDIV on the ILUE in Hunan Province varied greatly, and the curve was not smooth. It had highs and lows during the study period and generally showed a downward trend.

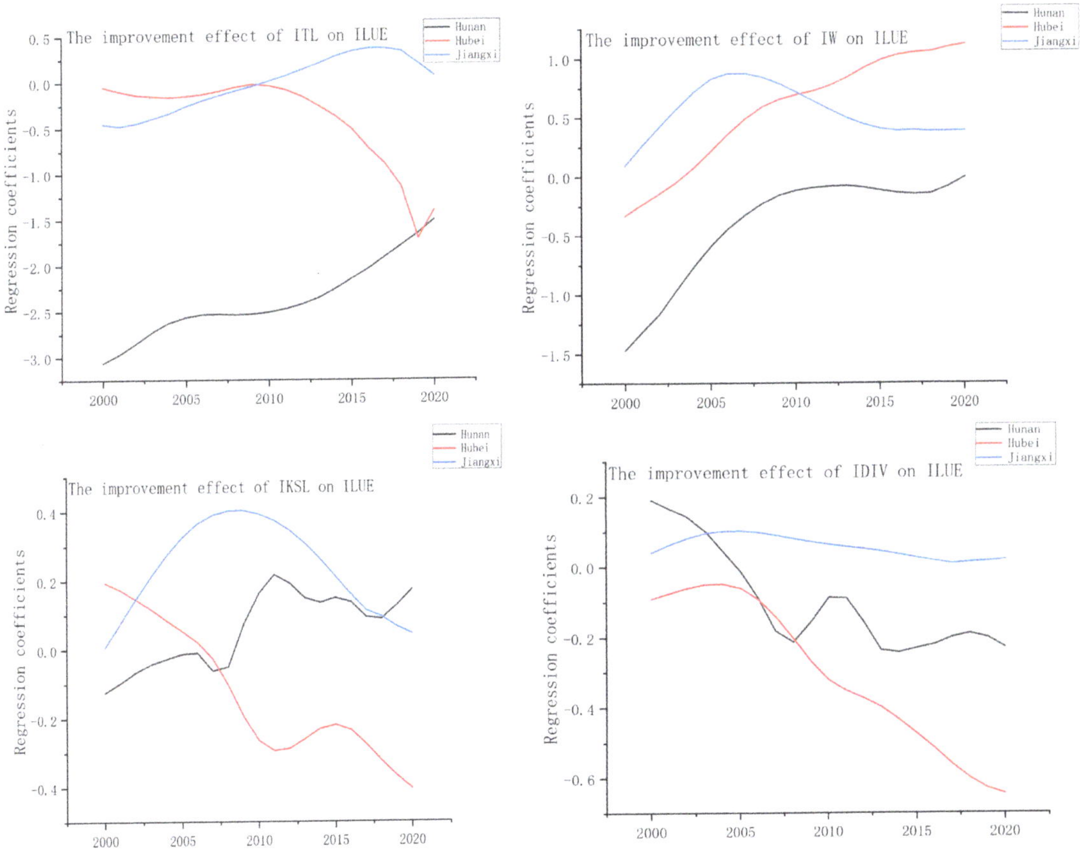

Figure 9. Spatial heterogeneity of the effect of industrial spatial layout on ILUE in industrial sectors.

5. Discussion

This study used land as a carrier to establish a complete logical system of the impact of land transformation brought about by ITI on ILUE in an integrated economic–environmental system.

An attempt was made to establish a theoretical analysis framework for the effect of ITI on the ILUE. We incorporated IT into a CO_2 emissions analysis to build a logical analysis framework for CO_2 emissions caused by land transformation in the IT process and then analyzed the socioeconomic effect of CO_2 emissions caused by land transformation driven by ITI.

We detailed our analysis of how the two dimensions industrial structural optimization and industrial layout brought about by land use transformation in the ITI process affects the ILUE. Specifically, we explored the impact of ITI from the two dimensions structural optimization and spatial layout, taking into account both industrial structural optimization and spatial layout, with four industrial transformation measurement index systems established to empirically analyze the carbon reduction effect of IT at different scales.

We also used a local variable parameter model that considers spatiotemporal heterogeneity to explore the influence of economic activities on the ILUE, which allows for new analytical perspectives on the changes in different dimensions of time and space in the research. We also sought to analyze the dynamic process of ITI's effect on ILUE from a dynamic perspective, and the results provide a significance reference for governmental decision-making and policy regulation to promote ITI and achieve high land use efficiency goals based on different economic development stages.

However, this study has some limitations to note. The industry can be subdivided into different industries that are labor-intensive, capital-intensive, and technology-intensive. Further research could further subdivide the industry to explore the impact of its internal transformation on the ILUE. In addition, this study still has some unfinished research on the impact mechanism of IT on changes in the ILUE, which could inform and motivate subsequent research. Furthermore, the theoretical analysis framework constructed in this paper is not absolutely complete, and further research is needed to analyze the interactive effects of IT and ILUE and expand the theoretical framework to more deeply explore the relationship between IT and ILUE.

6. Conclusions

This research took the UMY as a representative area for investigation, compiled socioeconomic panel data from 2000 to 2020, and analyzed the impact of ITI in 33 industrial sectors on ILUE from the two dimensions industrial structural optimization and industrial spatial layout. The research results of this paper can be summarized as follows:

Firstly, from 2000 to 2020, the ITL of the UMY dropped from 0.250 in 2000 to 0.033, and the ITL level became higher. IW increased from 0.121 in 2000 to 0.184 in 2020, and IW increased. ILUE increased from 0.635 in 2000 to 0.779 in 2020, and ILUE showed an upward trend year by year.

Secondly, the industrial transformation had a spatial impact on the ILUE. ITL and IW as well as IKSL and IDIV all improved the ILUE.

Thirdly, ITL's impact box on the ILUE was indicated by an inverted U-shaped curve, while the spatiotemporal heterogeneity of the impact of IW on ILUE changed only a little. The spatiotemporal changes of IKSL's impact on the ILUE were relatively small, while the spatiotemporal changes of IDIV's impact on the ILUE were more obvious.

Fourthly, by analyzing the regression coefficients for Hunan, Hubei, and Jiangxi provinces, it was found that there was an obvious spatial heterogeneity in the ILUE in the two dimensions structural optimization and spatial layout of industries in the three regions.

Based on the research results of this paper, we put forward the following recommendations:

It is important to strengthen regional cooperation and promote joint governance. The results of this paper indicate that there is a spatial correlation between IT and ILUE. This means that industrial structural adjustments or changes in industrial spatial layout between adjacent areas will likely have spatial spillover effects on the surrounding areas, affecting the ILUE in neighboring areas. This shows that it is not feasible for each city in the region to develop unilaterally in isolation from its surrounding environment, and the economic activities of cities in a region influence each other. Therefore, in the process of future industrial transformation, cooperation between regions should be promoted, and a cross-regional collaborative development mechanism should be established to improve the ILUE through regional cooperation. Also, the spatial spillover effect of IT should be fully utilized to achieve overall regional economic development. Local governments should actively promote healthy economic competition among regions while promoting regional cooperation. By establishing a benign regional cooperation mechanism, we can reduce the potential harm caused by vicious competition between regions on the market and economic efficiency and eliminate wastage of resources caused by barriers to the flow of factors of production between areas. Strengthening the flow of capital, labor, technology, energy, data, and other factor resources between cities in the region and optimizing the efficiency of production factor allocation are critical factors for regions to improve the ILUE. At the same time, it is also important to promote the comparative advantages of cities of different sizes and levels; form a development pattern of functional complementarity, industrial interaction, and knowledge and technology interoperability; enhance the spatial spillover effect of industrial transformation on improving ILUE; and promote IT to improve the ILUE. Cities at various levels can establish regional internal governance based on their different resource endowments and location characteristics, while strengthening active cooperation between cities at various levels. A regional community of interests could be built by promoting cross-regional governance. During the period of regional cooperation, it is necessary to actively carry out technical cooperation and exchange, promote the complementarity of advantages between regions, and focus on the linkage effects between cities. Differentiated development plans should be formulated based on the industrial structure, and economic development needs of different areas. Cross-regional industrial collaborative layouts should be promoted through regional cooperation, while facilitating industry linkage development, accelerating communication among local governments, industries, and enterprises in the industry chain and using the comparative advantages of each industry chain, while relying on funds, knowledge, and information through other methods, Through the above measures, we can achieve the sharing of resources and technology, use developed regions to drive underdeveloped regions, improve the ILUE of the entire region.

Focus on the long term, and promote IT according to local conditions. The results of this paper show that the distribution of IT and ILUE in the UMY shows obvious stage characteristics and regional non-equilibrium characteristics, while empirical tests found that IT has a spatiotemporal heterogeneity in improving the ILUE in the region, which means that in the process of promoting IT, long-term and differentiated industrial policies need to be formulated to promote regional IT. On the one hand, IT is dynamic and transformation requires a certain process. Whether it is industrial rationalization and industrial upgrading or the agglomeration of industrial specialization and industrial diversification, it is a gradual dynamic evolution process. To prevent the emergence of problems, such as uncoordinated industrial structural development and industrial oversupply, a long-term concept must be implemented throughout the entire process of IT, improving the ILUE, and a long-term mechanism should be established to collaboratively promote transformation and improve the ILUE. On the other hand, governments should consider the development positioning of each region, pay attention to regional differences, and formulate industrial policies tailored to actual local conditions according to the current industrial structure and energy structure. Each region should promote IT according to its own actual situation, and the IT model should be adapted to the actual situation. Regions with good economic development tend to have abundant technology and capital, while they may have a relative shortage of resources. Regions with lower economic development levels tend to lack advanced technology and capital but have relatively abundant natural resources. These regional differences are all related to differences in formulation and implementation and provide a realistic basis for the policy of globalized IT and for further promoting IT by formulating industrial policies tailored to local conditions. For regions with developed economies and high levels of industrial structure and agglomeration, the role of IT in traditional departments in improving ILUE in developed regions has been very limited. In the future, industrial policies should focus on driving the development of new industries and in promoting the development of ILUE in developed regions. At the same time, IT also provides IT experiences to other regions. For regions with backward economic development and a low industrial structure, supportive industrial policies can be appropriately tilted toward underdeveloped cities and IT can be promoted in underdeveloped cities.

Author Contributions: W.P.: writing—original draft preparation, and review and editing; M.L.: software; A.Z.: conceptualization, review and editing, and supervision. All authors have read and agreed to the published version of the manuscript.

Funding: This research was supported by the Major Project of National Social Science Foundation of China (18ZDA054), the National Natural Science Foundation of China (72273048, 42401324), The Ministry of Education of Humanities and Social Science project (22YJA790065, 24YJCZH175), The Fundamental Research Fund Project of Central Universities (2662022GGYJ004), Zhejiang Province Social Science Planning Special Project "Research and Interpretation of the Spirit of the 20th National Congress of the Communist Party of China and the Second Plenary Session of the 15th Provincial Party Committee" (202327051), and the 2025 Zhejiang Province Soft Science Research Project (2025C25027).

Data Availability Statement: The data used in this research are available from the authors upon reasonable request.

Conflicts of Interest: The authors declare neither conflicts of interest nor competing interests.

References

1. Pu, W.; Zhang, A.; Zhang, Z.; Qin, S.; Xia, Q. Can urban land market reform mitigate industrial emissions? Environmental evidence from 257 prefecture-level cities in China. *Environ. Res.* **2023**, *236*, 116707. [CrossRef] [PubMed]
2. Tian, C.; Luan, W.; Wang, X.; Jin, X. Land price, government debt, and land-use efficiency in the coastal cities of China under the constraint of land scarcity. *J. Environ. Manag.* **2024**, *371*, 123144. [CrossRef]
3. Geng, Y.; Li, X.; Chen, J. Integration of land use resilience and efficiency in China: Analysis of spatial patterns, differential impacts on SDGs, and adaptive management strategies. *Appl. Geogr.* **2025**, *175*, 103490. [CrossRef]
4. Wu, X.; Zhong, S.; Chen, G.; Wu, C.; Wu, C.; Han, J.; Qian, Z. Influence of land use intensity on urban carbon efficiency under a carbon neutrality target: Evidence from the Yangtze River Delta urban agglomeration, China. *Environ. Impact Assess. Rev.* **2025**, *110*, 107689. [CrossRef]
5. Qi, Y.; Lin, R.; Zhu, D. Impact of rising industrial land prices on land-use efficiency in China: A study of underpriced land price. *Land Use Policy* **2025**, *151*, 107490. [CrossRef]
6. Jaber, S. Land use efficiency and governance disparities: Unveiling the nexus in the Arab world. *Environ. Dev.* **2025**, *55*, 101169. [CrossRef]
7. Zhou, D.; Hu, Y.; Sun, Q.; Xie, D. Land resource mismatch and energy efficiency: Evidence from 243 cities in China. *Energy Policy* **2023**, *183*, 113800. [CrossRef]
8. Zhang, H.; Zheng, J.; Hunjra, A.; Zhao, S.; Bouri, E. How does urban land use efficiency improve resource and environment carrying capacity? *Socio-Econ. Plan. Sci.* **2024**, *91*, 101760. [CrossRef]
9. Li, B.; Wang, Z.; Xu, F. Exploring the effects of market-oriented reforms on industrial land use eco-efficiency in China: Evidence from a spatial and non-linear analysis. *Environ. Impact Assess. Rev.* **2023**, *102*, 107211. [CrossRef]
10. Ma, L.; Xu, W.; Zhang, W.; Ma, Y. Effect and mechanism of environmental regulation improving the urban land use eco-efficiency: Evidence from China. *Ecol. Indic.* **2024**, *159*, 111602. [CrossRef]
11. Zhang, N.; Sun, F.; Hu, Y. Carbon emission efficiency of land use in urban agglomerations of Yangtze River Economic Belt, China: Based on three-stage SBM-DEA model. *Ecol. Indic.* **2024**, *160*, 111922. [CrossRef]
12. Yang, Y.; Xue, R.; Zhang, X.; Cheng, Y.; Shan, Y. Can the marketization of urban land transfer improve energy efficiency? *J. Environ. Manag.* **2023**, *329*, 117126. [CrossRef] [PubMed]

13. Lei, J.; Xie, Y.; Chen, Y.; Zhong, T.; Lin, Y.; Wang, M. The Transformation of Peri-Urban Agriculture and Its Implications for Urban–Rural Integration Under the Influence of Digital Technology. *Land* **2025**, *14*, 375. [CrossRef]
14. Gao, X.; Zhang, A.; Sun, Z. How regional economic integration influence on urban land use efficiency? A case study of Wuhan metropolitan area, China. *Land Use Policy* **2019**, *90*, 104329. [CrossRef]
15. Kvartiuk, V.; Bukin, E.; Herzfeld, T. "For whoever has will be given more"? Land rental decisions and technical efficiency in Ukraine. *Land Use Policy* **2024**, *146*, 107336. [CrossRef]
16. Su, B.; Shen, X.; Wang, Q.; Zhang, Q.; Niu, J.; Yin, Q.; Chen, Y.; Zhou, S. The Evolution and Performance Response of Industrial Land Use Development in China's Development Zone: The Case of Suzhou Industrial Park. *Land* **2024**, *13*, 2182. [CrossRef]
17. Wang, F.; Zhang, H.; Zhou, J. Impact of Green Finance on Chinese Urban Land Green Use Efficiency: An Empirical Study Based on a Quasinatural Experiment. *Land* **2025**, *14*, 332. [CrossRef]
18. Gao, M.; Shao, Z.; Zhang, L.; Qiao, Z.; Yang, Y.; Zhao, L. Coupling and Coordination Relationship Between Carbon Emissions from Land Use and High-Quality Economic Development in Inner Mongolia, China. *Land* **2025**, *14*, 354. [CrossRef]
19. Wu, Y.; Luo, M. Study on Spatial-Temporal Evolution Law of Green Land Use Efficiency in Resource-Based Cities. *Land* **2025**, *14*, 360. [CrossRef]
20. Zhang, X.; Yang, M.; Guo, R.; Li, Y.; Zhong, F. Land Use Transition and Regional Development Patterns Under Shared Socioeconomic Pathways: Evidence from Prefecture-Level Cities in China. *Land* **2025**, *14*, 454. [CrossRef]
21. Zhang, H.; Song, Y.; Zhang, M.; Duan, Y. Land use efficiency and energy transition in Chinese cities: A cluster-frontier super-efficiency SBM-based analytical approach. *Energy* **2024**, *304*, 132049. [CrossRef]
22. Lu, X.; Jiang, X.; Gong, M. How land transfer marketization influence on green total factor productivity from the approach of industrial structure? Evidence from China. *Land Use Policy* **2020**, *95*, 104610. [CrossRef]
23. Chen, W.; Shen, Y.; Wang, Y.; Wu, Q. The effect of industrial relocation on industrial land use efficiency in China: A spatial econometrics approach. *J. Clean. Prod.* **2018**, *205*, 525–535. [CrossRef]
24. Chen, W.; Chen, W.; Ning, S.; Liu, E.; Zhou, X.; Wang, Y.; Zhao, M. Exploring the industrial land use efficiency of China's resource-based cities. *Cities* **2019**, *2019*, 215–223. [CrossRef]
25. Han, W.; Zhang, Y.; Cai, J.; Ma, E. Does Urban Industrial Agglomeration Lead to the Improvement of Land Use Efficiency in China? An Empirical Study from a Spatial Perspective. *Sustainability* **2019**, *11*, 986. [CrossRef]
26. Yin, G.; Lin, Z.; Jiang, X.; Qiu, M.; Sun, J. How do the industrial land use intensity and dominant industries guide the urban land use? Evidences from 19 industrial land categories in ten cities of China. *Sustain. Cities Soc.* **2019**, *53*, 101978. [CrossRef]
27. Liu, J.; Hou, X.; Wang, Z.; Shen, Y. Study the effect of industrial structure optimization on urban land-use efficiency in China. *Land Use Policy* **2021**, *105*, 105390. [CrossRef]
28. Flynn, H.; Canals, L.; Keller, E.; King, H.; Sim, S.; Hasting, A.; Wang, S.; Smith, P. Quantifying global greenhouse gas emissions from land-use change for crop production. *Glob. Change Biol.* **2012**, *18*, 1622–1635. [CrossRef]
29. Xu, L.; Tan, J. Financial development, industrial structure and natural resource utilization efficiency in China. *Resour. Policy* **2020**, *66*, 101642. [CrossRef]
30. Koroso, N.; Lengoiboni, M.; Zevenbergen, J. Urbanization and urban land use efficiency: Evidence from regional and Addis Ababa satellite cities, Ethiopia. *Habitat Int.* **2021**, *117*, 102437. [CrossRef]
31. Xie, H.; Wang, W.; Yang, Z.; Choi, Y. Measuring the sustainable performance of industrial land utilization in major industrial zones of China. *Technol. Forecast. Soc. Change* **2016**, *112*, 207–219. [CrossRef]
32. Guastella, G.; Pareglio, S.; Sckokai, P. A spatial econometric analysis of land use efficiency in large and small municipalities. *Land Use Policy* **2017**, *63*, 288–297. [CrossRef]
33. Mao, W.; Wang, W.; Sun, H.; Yao, P.; Wang, X.; Luo, D. Urban industrial transformation patterns under natural resource dependence: A rule mining technique. *Energy Policy* **2021**, *156*, 112383. [CrossRef]
34. Jiang, W.; Luo, S.; Zhou, G. Financial development, OFDI spillovers and upgrading of industrial structure. *Technol. Forecast. Soc. Change* **2020**, *155*, 119974. [CrossRef]
35. Matsumoto, H. International urban systems and air passenger and cargo flows: Some calculations. *J. Air Transp. Manag.* **2004**, *10*, 239–247. [CrossRef]
36. Balassa, B.; Toutjesdijk, A. Economic integration among developing countries. *J. Common Mark. Stud.* **1975**, *14*, 37–55. [CrossRef]
37. Auer, P. Protected mobility for employment and decent work: Labour market security in a globalized world. *J. Ind. Relat.* **2006**, *48*, 21–40. [CrossRef]
38. Hall, P.; Pain, K. *The Polycentric Metropolis: Learning from Mega-City Regions in Europe*; Routledge: Abingdon-on-Thames, UK, 2006.
39. Tian, Y.; Jiang, G.; Ma, W.; Wu, S.; Tian, Y.; Zhou, T. Understanding the relationship between population–land–industry element inputs and function outputs of rural settlement: An efficiency-based perspective. *Habitat Int.* **2025**, *159*, 103370. [CrossRef]
40. Chen, W.; Ning, S.; Chen, W.; Liu, E.; Wang, Y.; Zhao, M. Spatial-temporal characteristics of industrial land green efficiency in China: Evidence from prefecture-level cities. *Ecol. Indic.* **2020**, *113*, 106256. [CrossRef]
41. Cobb, C.; Douglas, P. A theory of production. *Am. Econ. Rev.* **1928**, *18*, 139–165.

42. Pu, W.; Zhang, A. Can China's Market-Oriented Reform Improve the Efficiency of Industrial Land Use? A Panel Data Empirical Analysis at Prefecture Level From 2007–2019. *Front. Environ. Sci.* **2022**, *10*, 884958. [CrossRef]
43. Zhao, L.; Sun, C.; Zheng, D. Measurement of inter-provincial water resources utilization efficiency and spatial spillover effect in China. *Acta Geogr. Sin.* **2014**, *69*, 121–133. [CrossRef]
44. Chen, S.; Jefferson, G.; Zhang, J. Structural change, productivity growth and industrial transformation in China. *China Econ. Rev.* **2011**, *22*, 133–150. [CrossRef]
45. Huang, Z.; He, C.; Zhu, S. Do China's economic development zones improve land use efficiency? The effects of selection, factor accumulation and agglomeration. *Landsc. Urban Plan.* **2017**, *162*, 145–156. [CrossRef]
46. Gan, C.; Zheng, R.; Yu, D. The impact of China's industrial structure changes on economic growth and fluctuations. *Econ. Res.* **2011**, *46*, 4–16+31.
47. Bai, Z.; Han, L.; Liu, H.; Li, L.; Jiang, X. Assessment of coordinated development between urban land use efficiency and ecological carrying capacity: Case study of the cities in Inner Mongolia. *Ecol. Indic.* **2023**, *155*, 110933. [CrossRef]
48. Zhang, W.; Wu, Q.; Wang, B.; Huang, J. Multidimensional study on the impact of industrial specialization and diversified agglomeration on urban land use efficiency. *Chin. Popul. Resour. Environ.* **2019**, *29*, 100–110.
49. Xiang, S.; Zhou, M.; Huang, L.; Shan, L.; Wang, K. Assessing the dynamic land utilization efficiency and relevant driving mechanism in in-situ urbanized rural areas: A case study of 1979 administrative villages in Hangzhou. *Environ. Impact Assess. Rev.* **2023**, *101*, 107111. [CrossRef]
50. Wang, H.; Liu, Y.; Sun, L.; Ning, X.; Li, G. Assessment of Chinese urban land-use efficiency (SDG11.3.1) utilizing high-precision urban built-up area data. *Geogr. Sustain.* **2025**, *6*, 100210. [CrossRef]

land

MDPI

Article

Cross-Provincial City-Regionalism in China: Evidence from Smart Planning and Integrated Governance of the Yangtze River Delta

Tianren Ge [1], Yang Yu [2,*], Xiaohua Zhong [3] and Yongli Jiao [4,*]

1 Department of Political Science, School of Political Science & International Relations, Tongji University, Shanghai 200092, China; getianren@tongji.edu.cn
2 Department of Urban Planning and Management, School of Public Administration and Policy, Renmin University of China, Beijing 100872, China
3 Department of Sociology, School of Political Science & International Relations, Tongji University, Shanghai 200092, China; xhzhong@tongji.edu.cn
4 Department of Teaching and Research, China Executive Leadership Academy Pudong (CELAP), Shanghai 201204, China
* Correspondence: yuyang319@ruc.edu.cn (Y.Y.); yljiao@celap.org.cn (Y.J.)

Abstract: Taking the Demonstration Zone of Green and Integrated Ecological Development of the Yangtze River Delta as a case study, we find that city-regional development of the Yangtze River Delta has advanced to the fifth stage, so-called cross-provincial city-regional integrated development. The ongoing reform experiment in China presents a new model of city-regional development, which distinguishes itself from previous approaches used in both China and Europe/America. We propose a theoretical framework of cross-provincial city-regionalism from the two dimensions of smart planning and integrated governance. Based on the new framework, this article reveals how the top-down intervention of the central government has helped local governments break down the administrative barriers at the provincial level and stimulated them to participate in the cross-provincial coordinated development from the bottom up. The new framework alters the assumption and institutional logic of the traditional city-regionalism theory and extends its explanatory scope.

Keywords: cross-provincial city-regionalism; smart planning; integrated governance; Yangtze River Delta

1. Introduction

Global capital is intensifying regional competition worldwide, leading to increased attention to the city-regionalism theory. This theory emphasizes that neoliberal globalization and its relaxation of regulations have led to intense competition between entrepreneurial governments, further triggering the restructuring of city-regional governance: from urban growth machines at the city level to regional governance alliances at the city-regional level [1]. However, the theory has the shortcoming of relatively neglecting the willingness of local governments to cooperate from the bottom up and the differences in city-regional restructuring methods under different national institutional backgrounds [2]. The traditional city-regionalism theory overemphasizing the economic perspective cannot fully explain the new changes in city-regional development in China.

Taking the Demonstration Zone of Green and Integrated Ecological Development of the Yangtze River Delta as a research case, we find that a new model of cross-provincial city-regional development supported by the central government has gradually emerged in the last decade. The traditional city-regionalism theory focuses too much on the economic logic of China's city-regional governance restructuring [3,4]. However, with the intensification of global regional competition, the roles and goals of the state have undergone transformation, leading to significant differences in the new city-regional restructuring process compared to before. As the regional economy develops into a new stage, the bottom-up cooperation motivation of local governments has been growing, and the new stage of cross-provincial city-regional governance practices is showing a new institutional logic.

Therefore, this article aims to construct a new theoretical explanation for the new phenomenon of cross-provincial city-regional development in the Yangtze River Delta since 2018. Specifically, the two research questions are as follows: (1) How can the cross-provincial city-regional integrated development become possible? (2) Compared with the traditional regional development models, what are the characteristics of the new city-regional development model? To answer the research questions, we attempt to propose a new theoretical framework of cross-provincial city-regionalism different from the traditional city-regionalism theory.

The article consists of six parts: aside from the introduction in the first part, the second part reviews the traditional city-regionalism theory and its new changes; the third part constructs a theoretical framework of cross-provincial city-regionalism; the fourth part reviews the history of the research case and analyses its new institutional logic; the fifth part explores the technological and institutional innovations; and the sixth part is the conclusion and discussion, proposing prospects for future research.

2. Literature Review

City-regionalism studies have evolved over several decades, integrating theories from urban studies, political economy, planning, and sociology. Research on city-regionalism theory and its transformation in the context of economic globalization mainly originates from the political economic geography in the West based on the experience in Europe and North America. With China's deep involvement in the globalization process and the rise of Chinese city-regions since the 21st century, research on Chinese city-regionalism theory has attracted much attention, gradually becoming a hot topic in academia. Therefore, we will first review the Western city-regionalism theory, and then examine the research on Chinese city-regionalism theory.

2.1. Western City-Regionalism Theory

From the perspective of the political economic geography in the West, modern cities are shaped by capitalism. Cities are seen as a material manifestation of capital territorialization, the result of capital impacting on space [5]. However, from the second industrial revolution in the 19th century to the third industrial revolution in the 1950s and 1960s, as the territorially organized nation-states began to intervene in urban development and governance, they gradually built up a Fordist mode of production centered on cities, known as "organized capitalism" [6]. However, entering the 1970s, the breakthrough growth of productivity generated by technological revolution and the deepening contradictions between capitalist production relations led to the emergence of neoliberal globalization and the spatial restructuring of city-regional areas. With the de-territorialization and reterritorialization of global capital, the governance model based on the scale of cities faced challenges [7]. Brenner explains this phenomenon as "the denationalization of territoriality". He points out that the essence of this phenomenon is that the increasingly powerful global capital no longer conforms to the existing national governance spatial framework, but reconstructs "territoriality" in line with the logic of global capital expansion [8].

In response to the expansion of urban economy to regional scale, political economic geography has sparked a wave of "regional turn" research, initiating a theoretical transition from city to city-region [9]. According to the new political economic geography, city-regional governance is gradually seen by the state as a new governance strategy to resolve excessive competition between cities, involving the restructuring of state governance models at different geographical scales [10]. In other words, it is the relaxation of regulatory policies advocated by neoliberalism and the new crisis of governance generated by urban competition driven by entrepreneurial governments at the traditional city scale that has led to the expansion of state governance space to the regional scale. Thus, compared to the spatial production represented by urban renewal, "city-region" has become a new form of territorial organization. Multi-city-regional governance alliances have replaced the urban growth machine as the dominant governance structure driving city-regional economic growth [11]. Since the late 1990s, city-regionalism has gradually become a research hotspot in the new political economic geography in the West, focusing on the rescaling of territoriality in the globalization era and the measurement of economic networks, transportation networks, and governance structures in city-regional areas [12–17]. Contemporary research, such as Scott and Storper's re-examination of 21st century urban theory, Peter Hall and Kathy Pain's analysis of polycentric metropolitan areas, Allen Scott's and Michael Storper's studies on innovation systems, enriches the theoretical framework of city-regionalism. This research prioritizes sustainable development, smart cities, inclusive urbanization, and urbanization in the Global North, providing new perspectives and methods for understanding complex city-regional dynamics [18–21].

Although the Western academic community generally acknowledges the explanatory power of the city-regionalism theory for the transformation of state governance structures in the post-Keynesian era, there have been increasing criticisms of the theory after 2010 [22]. These criticisms mainly focus on the theory's excessive emphasis on the decisive role of global capital in determining state governance structures, while neglecting the potential impact of bottom-up local factors, especially the differences between different national institutional backgrounds [23]. As Wu Fulong points out, although the rescaling is common, the forms of rescaling under different institutional backgrounds are significantly different [3]. In addition, some scholars criticize that the previous literature often focused too much on the theoretical study of changes in state governance structures at the global scale, lacking solid empirical research and insufficient exploration of the driving mechanisms leading to changes in state governance structures at the local scale [1].

2.2. Chinese City-Regionalism Theory

In the early 1980s, China joined the process of economic globalization through reform and opening up and embarked on large-scale rapid urbanization. The reterritorialization of global capital has reshaped the basic pattern of Chinese cities. As China's economic system transitioned from centrally planned to market-oriented, Chinese cities followed a path similar to that of Western cities. However, unlike Western countries, the socialist state has maintained a strong influence on spatial restructuring during the market transition. Thus, different from the rescaling, horizontal coordinated governance of European and American counterparts, Chinese city-regions have experienced a process of continuous scaling-up, vertical administrative integration.

Early research on Chinese city-regional governance started with the analysis of governance failures due to vicious competition among local governments. The market-oriented reforms in China have led to the decentralization of economic management power from the central government to local authorities, giving rise to the emergence of entrepreneurial government and the formation of local interest groups based on administrative boundaries [24–26]. This phenomenon has been summarized as "local corporatism" by Jean Oi, referring to the cooperation between local governments and businesses within their jurisdictions, forming diverse commercial common interests [27]. The political promotion tournament of local economic growth has strengthened the vertical "central–local" relationship while the horizontal intergovernmental relations have been divided by the interests of administrative jurisdictions, leading to further economic competition between local governments [28].

To address the excessive economic competition between local governments, China initiated the reform process of city-regional integration governance. Some scholars have divided this process into four stages of development: the regional office stage, the city-managing county stage, the abolishment of counties and establishment of districts stage, and the strengthening of core cities within provinces stage [29]. In the early 1980s, the regional office system, as an alternative to the regional revolutionary committee in Mao's era, first appeared in history. This institution, as an agency dispatched by the provincial government, was mainly responsible for coordinating affairs and assignments between cities and counties. However, for developing core cities, the regional office system was gradually replaced by the city-managing county system after the mid-1980s. Despite the shift of financial resources towards cities, the competitive landscape between cities and counties did not fundamentally change. In the late 1990s, the vigorous development of township enterprises drove the economic growth of counties. In developed coastal provinces such as Zhejiang, Jiangsu, and Guangdong, the competition between strong industrial counties and their affiliated prefecture-level cities became increasingly intense. As a result, some strong industrial counties were successfully upgraded to county-level cities, and even prefecture-level cities, such as Zhoushan City in Zhejiang Province and Dongguan City in Guangdong Province. Therefore, in the late 1990s, the abolishment of counties and establishment of districts became a new governance tool to solve the problem of negative competition between cities and counties. Some county-level cities were directly reclassified as districts under the jurisdiction of cities, such as Wujin District under the jurisdiction of Changzhou City in Jiangsu Province. While enhancing the overall planning capabilities of prefecture-level cities, the abolishment of counties and establishment of districts also intensified the competition for capital between prefecture-level cities. After 2000, the competition between prefecture-level cities resulted in the continuous accumulation of market resources towards the core cities within provinces, highlighting the polarization of city-regions within provinces. According to Shen's classification, territorial restructuring can be divided into two types: vertical restructuring and horizontal restructuring [30]. In fact, the above four stages of development are all forms of vertical restructuring. To achieve the economic goal of strengthening capital accumulation, the fundamental role of China's state intervention in the restructuring of city-regional relations cannot be overlooked. The policy of city-regional integrated governance has undergone continuous advancement from counties to cities and then to provinces [31,32]. In this context, the city-regionalism theory has received increasing attention from the Chinese academic community, becoming the mainstream theory of research on the regional development and governance in China [33–35]. Recent studies mainly focus on some specific aspects of city-regional development, such as industrial clustering, regional planning, land use management, inter-regional disparities and sustainability, to investigate the city-region's transition from rapid growth to high-quality development [36–39].

However, the framework of analysis of the Chinese city-regionalism theory makes it difficult to provide a sufficient explanation for the new phenomenon of cross-provincial integrated governance in recent years. This issue has become more prominent with the acceleration of regional economic integration. Although the existing literature has deepened our understanding of the rescaling city-regional governance model, it also has certain limitations. Most studies are confined to theoretical frameworks based on Western experiences, focusing on how global capital shapes the development pattern of city-regions in the process of reterritorialization. However, against the Chinese institutional background, the formation of city-regions is not simply an economic clustering process, but also the result of political choices and policy implementation. Therefore, Western city-regionalism theory from a purely economic perspective lacks sufficient explanatory power for the institutional logic behind the formation of Chinese city-regions, while traditional Chinese city-regionalism theory focusing on vertical administrative division adjustments also fails to explain the fifth stage of regional development in China. The new practice of cross-provincial city-regional integrated development deserves further empirical research to enrich the existing city-regionalism theory.

3. Theoretical Framework of Cross-Provincial City-Regionalism

Based on the traditional Chinese city-regionalism, this section attempts to propose a new theoretical framework to explain the fifth stage of city-regional integrated development in China. We name the new theoretical framework "cross-provincial city-regionalism", distinguishing it from traditional city-regionalism, which does not completely deny the logic of the traditional city-regionalism, but extends it in the new development stage. Cross-provincial city-regionalism aims to provide an analytical perspective for the empirical study by focusing on the two research questions mentioned above. The first question can be interpreted as what insurmountable obstacles traditional city-regionalism has encountered in addressing the development of cross-provincial city-regions, thus leading to failure. The second question can be interpreted as what innovations cross-provincial city-regionalism has made to effectively overcome obstacles to the development of cross-provincial city-regions.

3.1. The Obstacles Faced by Traditional City-Regionalism in Cross-Provincial City-Regional Development

At the stage of cross-provincial regional development, the biggest obstacle facing traditional city-regionalism lies in the intergovernmental relations between the central and local governments. Since the reform and opening up, the central–local relationship in China has gone through a process of change from a highly centralized system to a relatively decentralized one. Local governments have gained more autonomy, but at the same time, the problem of "fragmentation between departments and regions" has emerged, that is, there are difficulties in division and coordination among departments and regions. Xu Chenggang summarizes this characteristic of the central–local relations in China as "the fragmented authoritarian system of regional decentralization" [40].

The "fragmentation" feature is manifested as the existence of segmentation between vertical sectors and horizontal blocks. "Vertical sectors" refer to the top-down vertically managed departments, and "horizontal blocks" refer to local governments at all levels. When different vertically managed departments carry out work locally, there may be certain conflicts with the overall planning and coordination of local governments. For example, when some vertically managed environmental protection departments implement environmental policies, there may be contradictions with the short-term goals of local governments in pursuing economic development [41,42]. Under the intense pressure of regional competition, the fragmented administrative system leads local governments to formulate land use plans and economic development plans that may be contradictory to each other and implement competing policies within their respective administrative regions. When vicious competition occurs, there is a lack of necessary coordination mechanisms among local governments at the same level [43,44]. Generally speaking, the fragmentation of the administrative system has created dual barriers at both the planning technical level and the governance institutional level.

The smooth progress of the previous four stages of governmental reform can mainly be attributed to the high level of authority that Chinese provincial governments have in solving the problem of fragmentation of local administrations within provinces. Under the Chinese-style local government structure, provincial governments, as the highest level of local government, often resolve the vertical "city–city" competition within provinces through vertical administrative division adjustments, resulting in the establishment of the city-regional integrated governance model centered around super-large cities within the province. However, when it comes to the problem of cross-provincial integrated governance, it has reached the power boundary of the local state.

3.2. New Institutional Logic of Cross-Provincial City-Regionalism for High-Quality Green Development

Since the new central leadership took power in China in 2012, a new trend of combining vertical and horizontal territorial restructuring beyond the provincial power boundary has emerged especially in the developed coastal areas, and the development of cross-provincial regional integration attempts to solve the horizontal cooperation issues among provinces. This not only changes the economic logic of traditional city-regionalism, but also proposes the new institutional logic of promoting green development. Green development refers to a sustainable approach to urban planning, construction, and land use that aims to minimize environmental impact while promoting economic growth and social equity.

The main connotation of cross-provincial city-regionalism is that the central government strongly intervenes to promote the cross-provincial city-regional smart planning technology innovation and integrated governance reform to overcome the dual barriers. At the planning technical level, a unified planning management information platform has been established. By leveraging digitization and networking, the smart planning aims to achieve the intelligent management of the entire life cycle of territorial and spatial planning, and explore multi-level construction methods, paths and policy mechanisms for high-quality land use at the cross-provincial scale. At the governance institutional level, the new structure of cross-provincial governance has been constructed through integrating vertical and horizontal territorial restructuring and bridging multi-level local governments rather than just resolving the vertical "city–city" competition within provinces through vertical administrative division adjustments. In addition, market forces and social forces are also incorporated into the governance structure to form a developer alliance, consisting of local governments, private developers, and social organizations (Figure 1).

Looking further, the success of cross-provincial city-regionalism is attributed to the strong intervention of the central government which breaks the existing fragmented administrative system and integrates the vertical sectors by restructuring the horizontal blocks, thus achieving intelligent planning and integrated governance of cross-provincial urban areas. The reason why the central government needs to intervene and establish new administrative institutions is that it is impossible to break through provincial restrictions merely through local government reforms. On the one hand, provincial governments are already the highest level of local governments and cannot exercise jurisdiction over local affairs outside their own provinces. On the other hand, the difficulty of cross-provincial collaborative governance is extremely high. It is impossible to effectively achieve the goals unless a powerful superior government intervenes. Previously, various relatively soft integrated governance methods, such as non-governmental integrated collaborative governance and network governance, were adopted, but the effects were rather limited. The failure of integrated governance was caused by the inability to avoid vicious competition between horizontal blocks. After the reform of cross-provincial city-regionalism, the relationship between horizontal blocks has now transformed into a cooperative one instead of a competitive one by activating intrinsic dynamics for local governments. Moreover, the newly established blocks are detached from the original administrative subordination relationships. Therefore, the institutional basis for the vicious competition between the original horizontal blocks no longer exists.

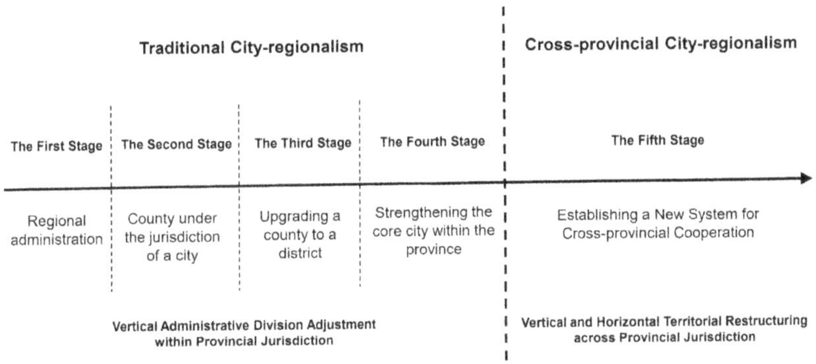

Figure 1. The traditional city-regionalism vs. cross-provincial city-regionalism. Source: Prepared by the authors.

3.3. Technological and Organizational Innovations of Cross-Provincial City-Regionalism Based on the New Institutional Logic

Cross-provincial city-regionalism has a new institutional logic compared to traditional city-regionalism. This is partly due to the changes in the relationship between the central and local governments caused by the strengthening of the central government's power over the last decade. With support from the central government, the cross-provincial administrative structure has been established to address various challenges in cross-provincial regional cooperation. Additionally, the green development goal set by the central government requires an increased level of economic integration beyond the provincial level, prompting local governments to enhance cooperation in order to reduce governance costs. As a result, the institutional logic that underpins the traditional city-regionalism has undergone significant changes. This also means that the theoretical framework of cross-provincial city-regionalism possesses a new institutional logic characterized by the central government's strong intervention and continuously improving technical regulatory capabilities.

The central government establishes a new model of smart planning and integrated governance by digital technology and institutional innovation (Figure 2). On the one hand, the unified planning management information platform based on the GIS database and AI technology facilitates the implementation of smart planning across provinces. The Demonstration Zone has built the first smart brain for cross-provincial planning in China. Its main function is to assist decision-making through a unified standard for data and institutions, cross-provincial planning implementation and annual inspection of planning implementation. On the other hand, some independent regional integrated governance administrative bodies have been founded to coordinate various regional governance affairs in the newly designated Demonstration Zone. Different from the scaling-up vertical administrative integration of traditional city-regionalism, the new governance structure under the intervention of the central government reflects the characteristics of vertical and horizontal territorial restructuring, which not only integrates the vertical administrative structure, but also rebuilds the horizontal administrative mechanisms. The vertical administrative structure mainly includes the leading group for integrated governance at the national and strategic level, the Demonstration Zone council at the decision-making and coordination level, the executive committee at the policy implementation level and the developer alliance at the project management level, in which the central government, local governments, private enterprises and social organizations get involved, forming a new pattern of multi-level cooperative governance with distinctive features. The horizontal administrative mechanisms, including the personal co-employment mechanism, the co-management mechanism for project planning and construction and the fiscal and tax sharing mechanism, provide the institutional foundation for cross-provincial collaborative governance.

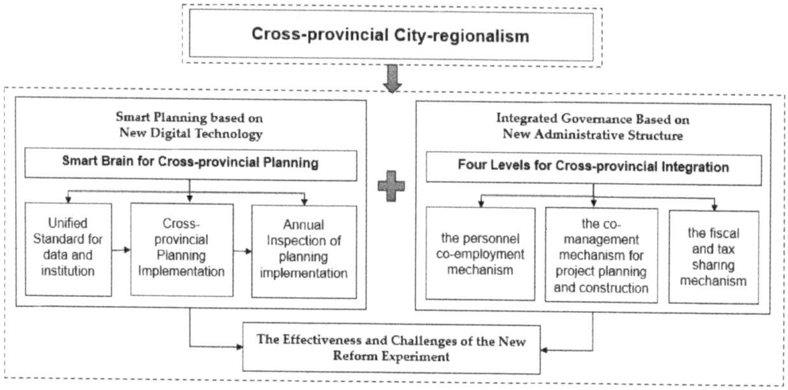

Figure 2. Theoretical framework of cross-provincial city-regionalism.

4. The New Institutional Reform Experiment Driven by Top-Down and Bottom-Up Forces

This article has selected the Demonstration Zone of Green and Integrated Ecological Development of the Yangtze River Delta as a research case for three main reasons. Firstly, the Demonstration Zone aims to conduct policy experiments extensively and deeply, especially in terms of applicability. Currently, the central leadership's Office for Promoting the Integrated Development of the Yangtze River Delta has requested the promotion and proliferation of the institutional innovation experiences gained in the Demonstration Zone. This indicates that the success of the pilot area has been recognized by the central government. Secondly, the theoretical value of this case study lies in its representativeness. As a typical policy experiment in the context of China's city-regional governance reform, its experience can provide a valuable reference for future reforms in other city-regions in China, such as the Beijing–Tianjin–Hebei Region and the Greater Bay Area. The Demonstration Zone reflects the latest trend of China's city-regional evolution. Thirdly, although there is much research on the regional development of the Yangtze River Delta region, its fifth stage, or cross-provincial integrated development, has received less attention. We believe that the case study can enrich the existing body of the literature on Chinese city-regionalism.

With the intensifying global city-regional competition and China's economy entering a new stage of development, the Central Committee of the Communist Party of China and the State Council issued the Outline of the Development Plan for the Yangtze River Delta and approved the establishment of the Demonstration Zone of Green and Integrated Ecological Development of the Yangtze River Delta in the border area of Shanghai, Jiangsu Province and Zhejiang Province in 2019 (Figure 3). The Demonstration Zone includes Qingpu District of Shanghai, Wujiang District of Suzhou in Jiangsu Province, and Jiashan County of Jiaxing in Zhejiang Province. The total area is approximately 2413 square kilometers. The top-down intervention of the central government helps local governments break down the administrative barriers at the provincial level and stimulates them to participate in the cross-provincial integrated development from the bottom up.

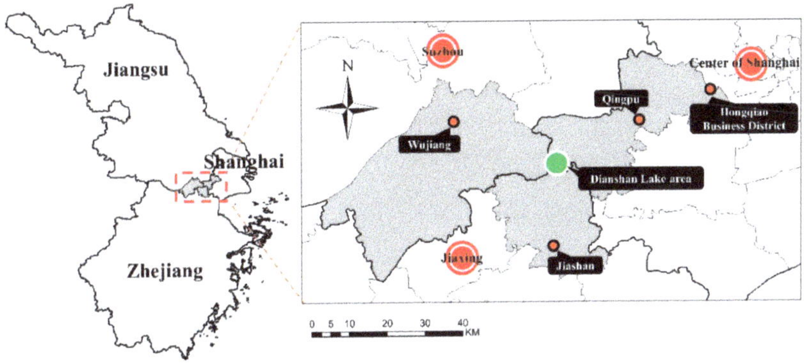

Figure 3. Location of the Demonstration Zone of the Green and Integrated Ecological Development in the Yangtze River Delta. Source: Prepared by the authors.

Our field survey started in September 2021 and ended in October 2024, conducting 12 on-site interviews and 2 online interviews with 22 interviewees, and collecting numerous first-hand documents through participatory observation. All the interviews were recorded and documented in both audio and written forms for empirical study. The interviewees mainly included officials from the central government, key personnel of the Demonstration Zone's executive committee, department heads, and policy research personnel, managers of other experimental areas, officials from local governments, experts, and scholars (Table 1).

Table 1. List of the interviewees.

No.	Interviewee	Date
01	Official from the Executive Committee of the Yangtze River Delta Demonstration Zone	29 September 2021
02	Head of a branch organization of the Executive Committee of the Yangtze River Delta Demonstration Zone	29 September 2021
03	Group Leader of the Management Team of the Executive Committee of the Yangtze River Delta Demonstration Zone	28 March 2022
04	Economic Officer, Regional Economics Department, National Development and Reform Commission	28 April 2022
05	Official from the Regional Economics Department, National Development and Reform Commission	28 April 2022
06	Official from Shanghai Development and Reform Commission	1 April 2022
07	Official from Suzhou Municipal Government, Jiangsu Province	1 April 2022
08	Leader from China Academy of Urban Planning and Design	1 April 2022
09	Head of the Shanghai Branch, China Academy of Urban Planning and Design	1 April 2022
10	Professor, Center for Urban Governance Research, Fudan University	15 April 2022
11	Professor, School of Government, Peking University	15 April 2022
12	Professor, School of Public Administration and Policy, Renmin University of China	13 August 2022
13	Leader of Beijing Tongzhou District Sub-Center	12 November 2022

Table 1. *Cont.*

No.	Interviewee	Date
14	Official from the Collaborative Development Center of Hebei Langfang Sanhe City, Beijing–Tianjin–Hebei Region	2 December 2022
15	Official from Tongzhou District Development and Reform Commission, Beijing	4 January 2023
16	Professor, School of International and Public Affairs, Shanghai Jiao Tong University	4 January 2023
17	Architect from State Engineering Survey and Design	2 September 2023
18	Academician of Chinese Academy of Sciences	2 September 2023
19	Official from the Shanghai Municipal Bureau of Planning and Natural Resources	2 November 2023
20	Professor, East China Normal University, Institute of Urban Development	24 November 2023
21	Official from Qingpu District Bureau of Planning and Natural Resources, Shanghai	20 March 2024
22	Official from the Bureau of Natural Resources and Planning, Wujiang District, Suzhou City, Jiangsu Province	21 October 2024
23	Head of a branch organization of the Executive Committee of the Yangtze River Delta Demonstration Zone	21 October 2024

4.1. *Top-Down Promotion from the Central Government*

Globalization has promoted the development of industrial clusters and the complexity of industrial chains, leading to the global economic competition expanding from global cities to global city-regions. To gain advantage in global regional competition, Chinese city-regions need to strengthen integrated governance. China's city-regions including the Beijing–Tianjin–Hebei region, the Yangtze River Delta, and the Pearl River Delta are undergoing this process. In the past four decades, rapid economic growth has not only exacerbated regional disparities, but also led to serious environmental problems. Since Xi Jinping came to power in 2012, the central government has been committed to promoting green and sustainable development while facilitating industrial transformation and upgrading. Therefore, changing the mode of economic development has become the new strategic goal of the state. But the regional attribute of the ecological environment requires local governments to strengthen cooperation to solve environmental problems and promote coordinated sustainable development. Due to the long-term accumulation of regional disparities and territorial divisions, achieving cross-provincial integrated city-regional governance faces technological and institutional challenges, requiring strong intervention and effective leadership from the central government.

For this purpose, the central government has high expectations for smart planning and integrated governance of the Yangtze River Delta. In November 2018, it elevated the coordinated development reform of the Yangtze River Delta to a national strategy. This demonstrates the determination of the central government to forcefully resolve the technological and institutional challenges. As the most economically developed city-region in China, the Yangtze River Delta has a solid foundation to undertake the significant mission of participating in global competition. The city-region covers an area of 358,000 square kilometers, with an economic output accounting for about a quarter of the national total, and it has two comprehensive national scientific centers, approximately a quarter of the top universities, best laboratories, and national research centers in the country. The annual

R&D expenditure and the number of effective invention patents account for about one-third of the national total. Because of the strong economic strength of each province and city in the Yangtze River Delta, the competition and conflict of interests in the city-region are the most prominent in China. For this reason, with the strong promotion of the central government, the Demonstration Zone across provincial boundaries was designated as a pilot zone for policy experimentation. As a senior official from the central government told us, "The designation of the demonstration zone is of great significance to promote the green and integrated development of the Yangtze River Delta, enhance its innovation and competitiveness, and improve economic agglomeration, regional connectivity and policy coordination efficiency, which is important to the high-quality development of China" (Interview No. 04).

The regional disparity in the Demonstration Zone is relatively significant, with serious governance deficits in infrastructure, ecological environment, industrial development, and public services, making the achievement of coordinated development in the city-region particularly challenging. From 2013 to 2022, despite the overall economic growth trend, there were obvious differences in economic volume between Qingpu, Wujiang, and Jiashan (Figure 4). There are also significant differences among the three in the indicators of per capita GDP, per capita disposable income, public budget revenue and expenditure (Figures 5 and 6). The above indicators reflect that, although the two districts and one county in the Demonstration Zone are geographically adjacent, there are relatively large regional disparities in economic development.

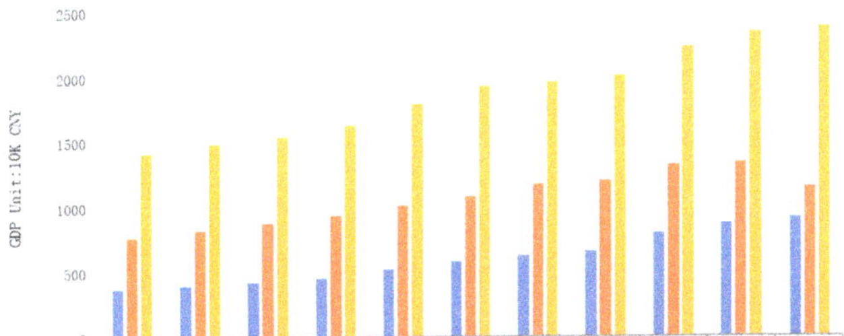

	2013	2014	2015	2016	2017	2018	2019	2020	2021	2022	2023
Jiashan	374. 3	402. 4	425. 54	457	520. 95	582. 6	626. 81	655. 8	789. 26	863. 48	908. 11
Qingpu	771. 9	827. 4	878. 2	939. 7	1009. 2	1074. 3	1166. 25	1194. 01	1317. 25	1334. 45	1140. 08
Wujiang	1415. 47	1486. 51	1540. 09	1628. 33	1788. 98	1925. 03	1958. 16	2002. 83	2224. 53	2331. 97	2377. 28

Years

Figure 4. Comparison of GDP in total for two districts and one county from 2013 to 2023. Source: Statistical Bulletin on National Economic and Social Development.

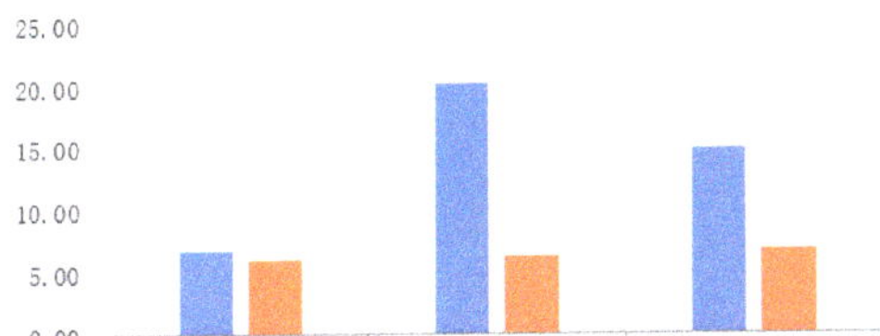

	Qingpu District	Jiashan County	Wujiang District
Per Capita GDP (ten thousand Yuan)	6. 82	20. 16	14. 89
Per Capita Disposable Income (ten thousand yuan)	5. 99	6. 23	6. 79

Figure 5. Comparison of GDP per capita and disposable income per capita (2022). Source: Statistical Bulletin on National Economic and Social Development.

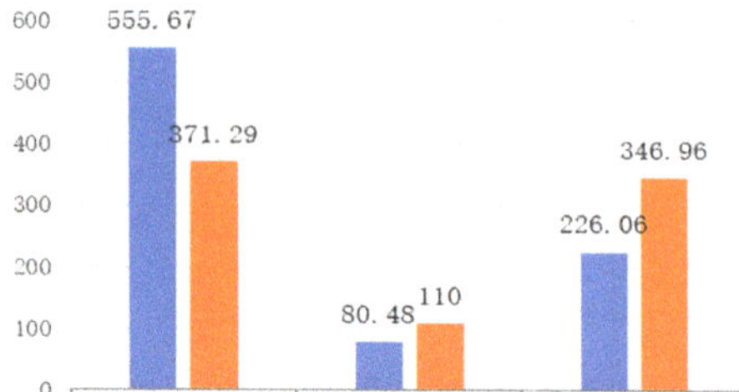

	Qingpu District	Jiashan County	Wujiang District
Revenue in the general public budget	555. 67	80. 48	226. 06
expenditure of general public budget	371. 29	110	346. 96

Figure 6. Comparison of general public budget revenues and expenditures (2022). Source: Statistical Bulletin on National Economic and Social Development.

4.2. Bottom-Up Cooperation from Local Governments

Despite continuous adjustments to the administrative boundaries of the provinces in the Yangtze River Delta in the past, the region has a dep basis of cooperation in terms of historical and cultural heritage, shared markets, and industrial clusters, gradually forming an urban spatial area radiating outward from Shanghai as a metropolitan area. Since modern times, Shanghai has become a pivotal node in the global city network and has earned its status. However, Shanghai has always been aware that this gateway hub status relies on strong support from the Yangtze River Delta's economic hinterland. Therefore, the development of Shanghai and its surrounding areas is integrated and inseparable.

Since the 1980s, the Yangtze River Delta region has spontaneously formed various regional cooperation models. The first is the Adjacent Area Model. For instance, Kunshan, adjacent to Shanghai, actively integrates with Shanghai for investment promotion, and the level of integration between Kunshan and Shanghai is already very high. The second is the Leapfrog Model. In terms of integrated technological innovation, the G60 expressway links the technology industries of major cities such as Shanghai, Suzhou, Hangzhou, and Hefei. The third is the Enclave Economic Model. The Shanghai Economic Development Zone has established many branch zones in the cities of Jiangsu Province and Zhejiang Province, forming a base for close industrial cooperation.

From above, we can clearly see that the Yangtze River Delta with a solid foundation of economic ties and cultural similarities has exhibited an augmented intrinsic motivation among the local governments to promote regional development. Due to the deep cooperation basis, close exchanges of personnel, funds, and information have driven the economical integrated development of the Yangtze River Delta with strong endogenous dynamics, which has become even more prominent after the central government proposed a new goal of green ecology. The establishment of the Demonstration Zone has provided an experimental field for local governments to actively explore the new model of cross-provincial integrated development, focusing on solving some specific problems like transportation bottlenecks, administrative barriers to data flow, and obstacles to cross-provincial enterprise registration. As a senior official from the Demonstration Zone's executive committee told us in the interview, "The key premise for the implementation of policy experiments in the demonstration zone is the good foundation laid in the previous decades. The officials from the three regions identify with each other and work together harmoniously to solve many difficult problems that would be hard to be solved only by themselves" (Interview No. 03).

5. Technological and Institutional Innovations of the New Reform Experiments

5.1. Smart Planning Based on New Digital Technology

Establishing a dynamic monitoring, evaluation, early warning, and regulatory mechanism for land spatial planning is a strategic deployment made by the central government. In recent years, in order to address issues such as an excessive number of planning types, overlapping and conflicting planning content, complex approval processes with long cycles, and frequent modifications of plans by local governments, China has restructured the spatial planning system. It has replaced the previous urban–rural planning and land use planning with territorial spatial planning. The Yangtze River Delta region, especially the Demonstration Zone, serves as a testing ground for promoting and practicing this new planning system. To meet the needs of enterprises and the industrial sector to break through administrative boundaries and layout industrial chains and supply chains, satisfy the horizontal cooperation needs of local governments, and avoid cross-regional disorderly competition and construction, under the guidance of the Ministry of Natural Resources, the Demonstration Zone has formulated the Overall Territorial Spatial Plan of the Demonstration Zone of Green and Integrated Ecological Development of the Yangtze River Delta (2021–2035), which is the first legal territorial spatial plan jointly formulated across provinces and submitted to the State Council for approval in China. Therefore, this is an innovative formulation and implementation of a new type of plan in the Yangtze River Delta region and a policy response at the national level to local demands. Under normal circumstances, plans at the county and district levels only need to be submitted by the city to the province for approval. However, this cross-regional plan has received the approval of the central government, thus enhancing the authority of the plan. The joint formulation method has connected the basic spatio-temporal data, enabling discussions on a unified platform and resolving conflicts in project and infrastructure layout among regions. In terms of the application of new technologies, the Demonstration Zone has built a cross-provincial smart brain in combination with plan formulation and implementation. By leveraging new technologies such as big data and artificial intelligence, it has integrated the four systems of plan formulation and approval, implementation and supervision, legal and policy, and technical standards onto a unified platform and connected them horizontally.

From the perspective of the spatial structure of the Overall Territorial Spatial Plan, the Demonstration Zone is located in the triangular area composed of three core cities of Shanghai, Suzhou and Jiaxing and four development corridors, and its coordinated development effect has a key impact on the integrated regional development of the Yangtze River Delta (Figure 7). To borrow the words of an official in the interview, "The demonstration zone places ecological green development as the top priority, reflecting the core value orientation of the new era's development transformation. It uses new thinking and new approaches to drive regional environmental, economic and social development" (Interview No. 19). The Demonstration Zone primarily achieves unified regional dynamic spatial planning management through building China's first cross-provincial smart brain with the following three characteristics (Figure 8).

Firstly, the system unifies regional planning standards and data standards and builds a data hub platform for cross-provincial spatiotemporal data collection and sharing. Relying on the data foundation of the Demonstration Zone's smart brain platform, a standard specification for basic geographic information data and a standard for collecting public credit information have been established. Techniques such as remote sensing surveying, IoT perception, and AI recognition have been employed to analyze monitoring data, achieving dynamic planning evaluation of land use. The executive committee has promoted data interconnectivity and sharing among the two districts and one county through various data collection methods, such as offline data aggregation, map service registration, and functional component registration.

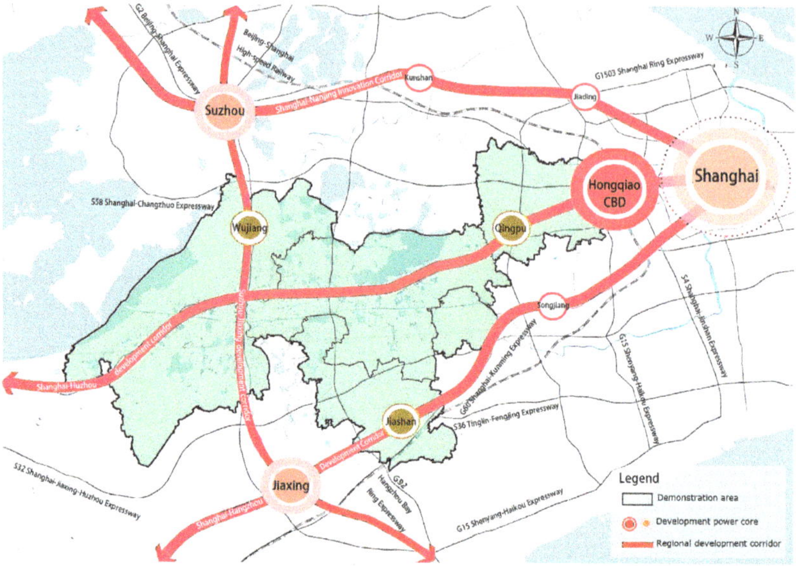

Figure 7. The spatial structure of the Overall Territorial Spatial Plan. Source: Prepared by the authors.

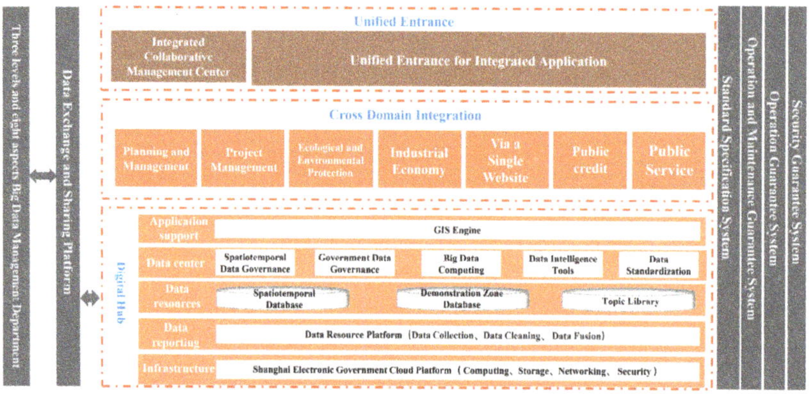

Figure 8. The technical architecture of the first cross-provincial smart brain in China. Source: Prepared by the authors.

Secondly, based on the data hub platform, a cross-provincial land spatial planning system has been constructed that provides digital technical support for various application scenarios, such as spatial connection intensity analysis based on population movement data and spatial pattern analysis based on nightscape lighting data. Relying on this system, the goal of managing the whole region with a blueprint has been achieved. Through digital technological innovation, the executive committee has explored the solutions to issues like cross-provincial digital infrastructure connectivity, data sharing, and land use management, achieving joint approval of detailed planning across provinces and joint review of urban development boundary adjustments.

Thirdly, the algorithm model of artificial intelligence is applied in dynamic monitoring of land spatial planning implementation. Also, the monitoring model has been connected with the China Spatial Planning Online Monitoring Network. This not only enhances the consistency between planning proposals and actual implementation, but also increases the efficiency and effectiveness of regional environmental monitoring, such as carbon emissions. As a planning expert pointed out, "The intelligent planning monitoring system represents a new starting point for China's spatial planning, which has achieved the national leading effect of planning implementation" (Interview No. 16).

5.2. Integrated Governance Based on New Administrative Structure

In addition to the technological innovation of smart planning, institutional innovation also promotes the green development of cross-provincial city-regions. The integrated governance of the Yangtze River Delta used to be achieved through vertical administrative boundary adjustments in the past decades. However, with the increasing demand for cross-provincial integrated development, the Demonstration Zone needs to build a new administrative structure to ensure the long-term sustainability of integrated governance. To this end, the Demonstration Zone proposes the idea of "not breaking administrative subordination, but breaking administrative boundaries" and focuses on institutional innovation through establishing a cross-provincial decision-making and execution system that integrates administrative resources and promoting horizontal and vertical intergovernmental cooperation.

Following the principle of collaborative governance, the decision-making and execution system incorporates different levels of government, market and social forces into the system and creatively constructs a four-level institutional framework (Figure 9). At the national and strategic level, the leading group for integrated governance was established to implement the deployment and decisions from the central government. This is a prevalent approach in China for resolving local administrative conflicts in regional development.

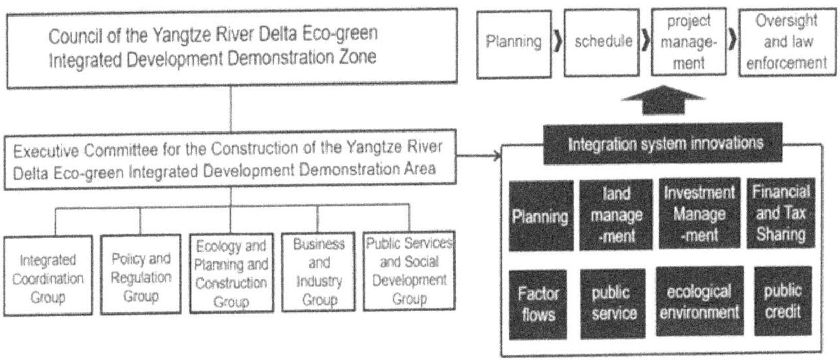

Figure 9. The institutional framework of the decision-making and execution system. Source: Prepared by the authors.

At the decision-making and coordination level, the Demonstration Zone council takes charge of making decisions and coordinating all the local public affairs in the city-region. This is mainly to achieve the integration of "province-to-province" relationships within the city-region and maximize the participation of market and social forces. The operation of the council adopts a rotation system, with senior provincial officials from Shanghai, Jiangsu and Zhejiang serving as the chairpersons of the council, holding regular council meetings and regional cooperation joint meetings to study and determine regional development plans and coordinate the promotion of major projects. The composition of the council reflects the characteristics of multi-level governance, breaking away from China's traditional administrative hierarchy. Provincial, municipal and county departments are all equal council members. Moreover, it also includes representatives from enterprises, social organizations, universities, think tanks, financial institutes, and research institutes as the council members.

At the policy implementation level, the executive committee is responsible for implementing the decisions made by the council and addressing cross-provincial administrative conflicts. It has five sub-groups: the comprehensive coordination group, the policy and regulation group, the ecological and planning construction group, the business and industry group, and the public service and social development group. Since its establishment, the executive committee has carried out a large number of institutional innovations in eight areas, including planning formulation, land management, investment management, fiscal revenue sharing, factor flow, public services, ecological environment and public credit, and has optimized a set of institutional processes from planning formulation, plan arrangement, and project management to supervision and law enforcement. As a scholar summarized, "the core organizational structure of 'the Council-Executive Committee' has indeed played a key role, and it has important reference significance for the integrated development and governance of the Beijing–Tianjin–Hebei region and other parts of the country" (Interview No. 12).

At the project management level, the Demonstration Zone has established the developer alliance consisting of consulting agencies, enterprises and industry associations, universities and research institutes, financial companies, and media to further strengthen the collaborative governance and solve specific issues in project management. As an advisory agency, the developer alliance provides timely feedback and advice to the executive committee in the process of land development and project operation. In addition, it also acts as a bridge for communication between public authority and private developers. Regarding the construction of the four-level decision-making and execution system that absorbs market and social forces, an official from the Beijing–Tianjin–Hebei integration pilot zone compared it with the Demonstration Zone of the Yangtze River Delta: "Compared to Beijing–Tianjin–Hebei, the market and social forces in the Yangtze River Delta are very strong, which is an advantage for the integration of the Yangtze River Delta. The wisdom of the demonstration zone lies in incorporating these non-governmental forces into the governance structure through institutional innovation, allowing them to play a positive and promoting role in advancing cross-provincial city-regional integrated development" (Interview No. 14).

In addition, the Demonstration Zone has also innovated three major governance mechanisms. The first is the personnel co-employment mechanism, which combines the modern enterprise personnel management system with the party cadre management system and selects outstanding cadres from the two provinces and one municipality to rebuild a new civil servant team. All the officials are co-employed by the executive committee rather than by each province separately, and their salaries are linked to the fiscal growth and governance performance of the Demonstration Zone. They do not distinguish between Shanghai cadres, Zhejiang cadres, and Jiangsu cadres, and there are greater salary incentives here, all working for a unified goal. The second is the co-management mechanism for project planning and construction. It includes the first online approval and supervision platform for cross-provincial investment projects in China. This platform is directly connected to the national information center platform, making it convenient for the projects to be directly included in the project management database of the National Development and Reform Commission. The third is the fiscal and tax sharing mechanism. On the basis of artificial intelligence technology, it is used for the tax calculation generated in the Demonstration Zone, cross-provincial revenue distribution, tax filing, law enforcement, and fiscal supervision.

This city-regional governance innovation possesses distinctive value. It does not adopt the historical approach of large-scale government mergers in Europe and America, nor does it rely on intergovernmental cooperation agreements or governance alliances lacking hard constraints. It also does not follow China's previous approach of resolving local administrative fragmentation through vertical administrative territorial adjustments. Instead, it adopts a new form of integrated governance: the provincial governments supported by the central government establishes a joint agency to be responsible for the management of cross-provincial city-regional development. From a global perspective, this represents the most significant institutional innovation in the new stage of cross-provincial integrated governance in the Yangtze River Delta.

5.3. The Effectiveness and Challenges of the New Reform Experiment

Since the establishment of the Demonstration Zone in 2019, the new reform experiment has achieved preliminary positive results, which has improved top-level design, formulated regional spatial plans, unified regional standards for ecological environment protection, and established unified mechanisms for personnel co-employment, project co-management, and fiscal sharing. After more than three years of dedicated efforts, the Demonstration Zone has successfully implemented 154 institutional innovations, out of which 48 have been effectively replicated and widely promoted across the nation, and 180 projects have already been implemented (Interview No. 21). Consequently, it has significantly enhanced its coordination and efficiency in terms of ecological co-governance, shared responsibility, and outcome sharing.

It is worth noting that the long-standing issue of unified regional management has been systematically resolved. The executive committee has issued many unified plans, standards and norms for regional development, ecological environment protection, air pollution, and water environment governance. For instance, the Guidelines for the Planning and Construction of the Pilot Area for the Demonstration Zone in the Yangtze River Delta is the first set of cross-provincial planning and construction standards in China. The Comprehensive Plan for the Territory Space for the Demonstration Zone of Green and Integrated Development in the Yangtze River Delta (2019–2035) is another example, as the first territorial spatial plan jointly formulated by cross-provincial cooperation. In October 2020, the Demonstration Zone issued a list of 7 unified standards, including the Technical Specification for Environmental Air Quality Forecast, becoming the first batch of unified regional standards in the Yangtze River Delta region. The two provinces and one municipality jointly carried out a unified monitoring network system for air, water, emergency response, and pollution sources, established a joint law enforcement team, and clarified unified law enforcement standards. The Demonstration Zone has developed specialized cooperative governance schemes for key transboundary water bodies, establishing standardized implementation criteria for ecological restoration and enhancement projects targeting issues such as eutrophication, cyanobacterial blooms, and cross-border management.

From the perspective of green integrated development, the ecological protection quality in the Demonstration Zone has been significantly improved. For example, the proportion of surface clean water bodies has increased substantially from 75% to 98.1%, and the water environment quality has already met the 2025 planning target ahead of schedule (Interview No. 16). In addition, substantial progress has been made in promoting transportation integration in the Demonstration Zone. An official in charge of transportation affairs in the Demonstration Zone said, "Nine inter-provincial dead-end roads have been opened to traffic. The westward extension project of Shanghai Metro Line 17 has been completed and put into operation, and the Shanghai-Suzhou-Huzhou High-speed Railway was officially opened on December 26th" (Interview No. 23). From the perspective of economic growth, the combined GDP of the two districts and one county in 2023 reached 472.5 billion yuan, and the total industrial output value of industrial enterprises above designated size reached 872.9 billion yuan. Compared with 2019, the average annual growth rates were 5.94% and 8.19%, respectively. In particular, the intensity of R&D investment increased to 4.2%, which is higher than the average level of the three provinces and one municipality in the Yangtze River Delta and leads the national average level (Interview No. 22). From the perspective of people's well-being, the reform and trials in the Demonstration Zone have brought tangible happiness to the people. As an official of the Executive Committee of the Demonstration Zone put it, "At present, 531 designated medical institutions and 926 pharmacies have realized cross-provincial medical insurance settlement. There are over 3500 cross-provincial online service items, and the cumulative number of processed cases has reached 116,000" (Interview No. 2).

However, the Demonstration Zone still faces the dual challenges of efficient factor flow and deep institutional integration in the face of intensified global city-regional competition. Compared with leading city-regions globally, the Yangtze River Delta region still has shortcomings in terms of economic aggregate, modern industrial system, innovation capability, and interconnected infrastructure. Although a series of major projects have been initiated in recent years, such as high-speed railways and intercity metro lines, some provincial or municipal border areas still need to be connected. In addition, due to the fierce competition between China and the United States, the resilience and security level of the industrial and supply chains in the Yangtze River Delta region have also been impacted. Hence, it is imperative to engage a diverse array of market participants and localities in the developmental process while employing market-oriented approaches to tackle underlying issues. The most important challenge for the future integrated green development of the Yangtze River Delta is how to leverage city-regional integration to achieve free and efficient flow of factors. This requires further promotion of the deep integration of relevant institutions. The long-standing administrative barriers in the region are unlikely to disappear soon. There is still significant regional disparity in terms of economic strength, fiscal capacity, infrastructures and public services. Although the Demonstration Zone has explored fiscal sharing, the scale and proportion are relatively small, leaving considerable room for improvement. Due to the involvement of multi-level local governments, such as provinces, cities, districts, counties, and townships, within the Demonstration Zone, the intertwined administrative relationships are relatively complex. Although the Demonstration Zone has undertaken many technological and institutional innovations, local governments at the district and county levels remain the most important administrative entities in terms of actual control over the development of various regions. Therefore, further promotion of interests and actions between different regions still requires the continuous operation, promotion, and improvement of new technologies, institutions and mechanisms.

6. Conclusions and Discussion

Based on the review of western city-regionalism theory, this article proposes a new theoretical framework for cross-provincial city-regionalism in China's new development stage, through reinterpreting the institutional logic of the new reform experiment based on smart planning and integrated governance. It is revealed that intensified global regional competition has prompted the central government to shift policy objectives and intervene strongly, thereby activating intrinsic dynamics for local governments in cross-provincial city-regional development and integration governance. Consequently, there has been a transition from adjusting vertical administrative boundaries at the provincial level to integrating vertical and horizontal territorial restructuring beyond the provincial level.

This article also reveals that the new administrative structure has promoted smart planning by digital technology through the establishment of the integrated governance system in the cross-provincial city-region. With the strong support of the central government, the Demonstration Zone has undergone comprehensive and robust governance restructuring in the personnel co-employment mechanism, project co-management mechanism, and fiscal sharing mechanism. Some institutional innovation achievements have already been promoted and replicated. The Demonstration Zone, despite still facing some challenges, undoubtedly represents a rare occurrence in China's previous regional integration governance reform and holds significant theoretical value.

From a theoretical perspective, the framework of cross-provincial city-regionalism extends traditional city-regionalism theory, explains the phenomenon of smart planning and integrated governance in the fifth stage of regional development, and provides a new supplement to western city-regionalism theory by incorporating new institutional logic and national vertical integration. On the one hand, the new framework breaks the economic logic of the existing analytical framework and focuses more on explaining the new institutional logic of green development. Under the authoritarian system in the Chinese style, the local governments can change their logic of action due to the central government's shift in policy objectives, transitioning from growth-oriented competition to cooperative development and ultimately transforming into a new regional governance structure that transcends the power boundaries of provincial governments. On the other hand, the new framework is evidently different from the framework of Western city-regionalism, which only focuses on the reterritorialization of global capital or excessively emphasizes bottom-up horizontal cooperative or integrated governance.

The theoretical value of the case is that China's new reform experiment offers a unique governance model, integrating vertical administrative levels and horizontal cooperative governance structures, distinct from the previous models used in China and Europe/America. However, for other countries, the new theoretical framework of China's cross-provincial city-regionalism only has partial technical applicability and institutional limitations. Regarding the former, the Chinese government has successfully promoted the formulation of overall plans for city-regions by building a smart brain with shared data and used digital technologies to dynamically monitor the implementation of city-regional plans. This has strongly promoted the green integrated development of urban regions. This is quite instructive for other countries. As for the latter, China adopts a unitary state structure form with the characteristics of a highly centralized system. The central government can directly lead local governments and directly promote the implementation of the new model of cross-provincial city-regional governance. It is difficult for other countries to imitate this if the central government lacks sufficient institutional authority. China's practice also shows that various "soft" forms of city-regional governance, due to the lack of strong institutional guarantees, may ultimately fail to truly and effectively solve the problems of city-regional collaborative governance. The reform experiment of smart planning and integrated governance in the Yangtze River Delta is still ongoing. Despite achieving preliminary results, it also faces many challenges. Therefore, the cross-provincial city-regionalism framework based on the current practice is only a preliminary theoretical thinking. Future practices may still need further refinement.

Author Contributions: Conceptualization, T.G., Y.Y. and Y.J.; field work and data collection, T.G. and Y.J.; writing—original draft preparation, T.G. and Y.Y.; writing—review and editing, T.G. and Y.Y.; visualization, T.G., Y.Y. and X.Z.; funding acquisition, T.G. and X.Z. All authors have read and agreed to the published version of the manuscript.

Funding: This research was funded by National Social Science Fund of China, grant numbers 22BSH022 and 24BSH151; Tongji University Xinrui Young Talents Grant for Social Sciences and Humanities, grant number 2024XR03.

Data Availability Statement: The original contributions presented in the study are included in the article material, further inquiries can be directed to the corresponding authors.

Conflicts of Interest: The authors declare no conflicts of interest.

References

1. Jonas, A.E.; Moisio, S. City Regionalism as Geopolitical Processes: A New Framework for Analysis. *Prog. Hum. Geogr.* **2018**, *42*, 350–370. [CrossRef]
2. Herrschel, T. *Cities, State and Globalization: City-Regional Governance in Europe and North America*; Routledge: London, UK, 2016.
3. Wu, F. China's Emergent City-Region Governance: A New Form of State Spatial Selectivity through State-Orchestrated Rescaling. *Int. J. Urban Reg. Res.* **2016**, *40*, 1134–1151. [CrossRef]
4. Li, Y.; Wu, F. Understanding City-Regionalism in China: Regional Cooperation in the Yangtze River Delta. *Reg. Stud.* **2018**, *52*, 313–324. [CrossRef]
5. Harvey, D. *The Urbanization of Capital: Studies in the History and Theory of Capitalist Urbanization*; John Hopkins University Press: Baltimore, MD, USA, 1985.
6. Lefebvre, H. *The Production of Space*; Wiley-Blackwell Publishing: Oxford, UK, 1992.
7. Brenner, N. Urban Governance and the Production of New State Spaces in Western Europe, 1960–2000. *Rev. Int. Political Econ.* **2004**, *11*, 447–488. [CrossRef]
8. Brenner, N. Globalization as Reterritorialization: The Re-Scaling of Urban Governance in the European Union. *Urban Stud.* **1999**, *36*, 431–451. [CrossRef]
9. Hall, P. Looking Backward, Looking Forward: The City Region of the Mid-21St Century. *Reg. Stud.* **2009**, *43*, 803–817. [CrossRef]
10. Brenner, N. *New State Space: Urban Governance and the Rescaling of Statehood*; Oxford University Press: New York, NY, USA, 2004.
11. Wachsmuth, D. Competitive Multi-City Regionalism: Growth Politics Beyond the Growth Machine. *Reg. Stud.* **2017**, *51*, 643–653. [CrossRef]
12. Keating, M. The Invention of Regions: Political Restructuring and Territorial Government in Western Europe. *Environ. Plan. C Gov. Policy* **1997**, *15*, 383–398. [CrossRef]
13. Lovering, J. Theory Led by Policy: The Inadequacies of the 'New Regionalism' (Illustrated from the Case of Wales). *Int. J. Urban Reg. Res.* **1999**, *23*, 379–395. [CrossRef]
14. Macleod, G. New Regionalism Reconsidered: Globalization and the Remaking of Political Economic Space. *Int. J. Urban Reg. Res.* **2001**, *25*, 804–829. [CrossRef]
15. Macleod, G.; Jones, M. Territorial, Scalar, Networked, Connected: In What Sense a 'Regional World'? *Reg. Stud.* **2007**, *41*, 1177–1191. [CrossRef]
16. Dicken, P.; Kelly, P.F.; Olds, K.; Yeung, H.W.-C. Chains and Networks, Territories and Scales: Towards a Relational Framework for Analyzing the Global Economy. *Glob. Netw.* **2001**, *1*, 89–112. [CrossRef]
17. Dixon, T.J.; Karuri-Sebina, G.; Ravetz, J.; Tewdwr-Jones, M. Re-Imagining the Future: City-Region Oversight and Visioning in an Era of Fragmented Governance. *Reg. Stud.* **2023**, *57*, 609–616. [CrossRef]
18. Scott, A.; Storper, M. The Nature of Cities: The Scope and Limits of Urban Theory. *Int. J. Urban Reg. Res.* **2014**, *39*, 1–15. [CrossRef]
19. Hall, P.; Pain, K. *The Polycentric Metropolis: Learning from Mega-City Regions in Europe*; Routledge: London, UK, 2006.
20. Scott, A. *Global City-Regions Trends, Theory, Policy*; Oxford University Press: Oxford, UK, 2001.
21. Storper, M. *Keys to the City: How Economics, Institutions, Social Interaction, and Politics Shape Development*; Princeton University Press: Princeton, NJ, USA, 2013.
22. Macleavy, J.; Harrison, J. New State Spatiality's: Perspectives on State, Space, and Scalar Geographies. *Antipode* **2010**, *42*, 1037–1046. [CrossRef]
23. Brenner, N. Open Questions on State Rescaling. *Camb. J. Reg. Econ. Soc.* **2009**, *2*, 123–139. [CrossRef]
24. Hubbard, M. Bureaucrats and Markets in China: The Rise and Fall of Entrepreneurial Local Government. *Governance* **1995**, *8*, 335–353. [CrossRef]
25. Zhang, T. Urban Development and a Socialist Pro-Growth Coalition in Shanghai. *Urban Aff. Rev.* **2002**, *37*, 475–499. [CrossRef]
26. Wu, F. China's Changing Urban Governance in the Transition Towards a More Market-Oriented Economy. *Urban Stud.* **2002**, *39*, 1071–1093. [CrossRef]
27. Oi, J.C. Fiscal Reform and the Economic Foundations of Local State Corporatism in China. *World Politics* **1992**, *45*, 99–126. [CrossRef]

28. Li, H.; Zhou, L. Political Turnover and Economic Performance: The Incentive Role of Personnel Control in China. *J. Public Econ.* **2004**, *89*, 1743–1762.
29. Zhang, J.; Wu, F. China's Changing Economic Governance: Administrative Annexation and the Reorganization of Local Governments in the Yangtze River Delta. *Reg. Stud.* **2006**, *40*, 3–21. [CrossRef]
30. Shen, J. Scale, State, and City: Urban Transformation in Post-Reform China. *Habitat Int.* **2007**, *31*, 303–316. [CrossRef]
31. Lin, G.C.S. Scaling-Up Regional Development in Globalizing China: Local Capital Accumulation, Land-Centered Politics, and Reproduction of Space. *Reg. Stud.* **2009**, *43*, 429–447. [CrossRef]
32. Ye, L. State-Led Metropolitan Governance in China: Making Integrated City Regions. *Cities* **2014**, *41*, 200–208. [CrossRef]
33. Luo, X. Why City-Region Planning Does Not Work Well in China: The Case of Suzhou–Wuxi–Changzhou. *Cities* **2008**, *25*, 207–217. [CrossRef]
34. Luo, X.; Shen, J. A Study on Inter-City Cooperation in the Yangtze River Delta Region, China. *Habitat Int.* **2008**, *33*, 52–62. [CrossRef]
35. Chen, X. A Tale of Two Regions in China: Rapid Economic Development and Slow Industrial Upgrading in the Pearl River and the Yangtze River Delta. *Int. J. Comp. Sociol.* **2007**, *48*, 167–201. [CrossRef]
36. Kang, J.; Xu, W.; Yu, L.; Ning, Y. Localization, Urbanization and Globalization: Dynamic Manufacturing Specialization in the YRD Mega-city Conglomeration. *Cities* **2020**, *99*, 102641. [CrossRef]
37. Chen, W.; Yuan, F.; Li, Y. Governing Cities through Regions: Evolution of Regional Plans for the Yangtze River Delta Mega city-region. *Trans. Plan. Urban Res.* **2023**, *2*, 71–84. [CrossRef]
38. Yi, P.; Shi, R.; Li, W.; Dong, Q. Evaluation of the Coordination-difference-driven Sustainability of 12 Urban Agglomerations in China based on the Dynamic Probability Weighting Method. *Sustain. Cities Soc.* **2024**, *116*, 105904. [CrossRef]
39. Tian, Y.; Mao, Q. The Effect of Regional Integration on Urban Sprawl in Urban Agglomeration Areas: A Case Study of the Yangtze River Delta, China. *Habitat Int.* **2022**, *130*, 102695. [CrossRef]
40. Xu, C. The Fundamental Institutions of China's Reforms and Development. *J. Econ. Lit.* **2011**, *49*, 1076–1151. [CrossRef]
41. Landry, P. *Decentralized Authoritarianism in China: The Communist Party's Control of Local Elites in the Post-Mao Era*; Cambridge University Press: New York, NY, USA, 2008.
42. Chen, H.; Feng, L.; Sun, X. Beyond Central-local Relations: The Introduction of a New Perspective on China's Environmental Governance Model. *Humanit. Soc. Sci. Commun.* **2024**, *11*, 701. [CrossRef]
43. Ma, W.; Jiang, G.; Chen, Y.; Qu, Y.; Zhou, T.; Li, W. How Feasible is Regional Integration for Reconciling Land Use Conflicts across the Urban–rural Interface? Evidence from Beijing–Tianjin–Hebei Metropolitan Region in China. *Land Use Policy* **2020**, *92*, 104433. [CrossRef]
44. Zheng, F.; Shen, Q.; Jian, B.; Zheng, J. Regional Governance, Local Fragmentation, and Administrative Division Adjustment: Spatial Integration in Changzhou. *China Rev.* **2010**, *10*, 95–128.

land

MDPI

Article

The Identification of Land Use Conflicts and Policy Implications for Donghai County Based on the "Production–Living–Ecological" Functions

Jianying Xiao [1], Jinjin Dai [1,*], Longqian Chen [1] and Yan Song [2]

1 Research Center of Digital Rural Service, School of Public Management, China University of Mining and Technology, Xuzhou 221116, China; xiaojianying@cumt.edu.cn (J.X.); chenlq@cumt.edu.cn (L.C.)
2 Department of City and Regional Planning, The University of North Carolina at Chapel Hill, Chapel Hill, NC 27599-3140, USA; ys@email.unc.edu
* Correspondence: daijinjin2022@163.com

Abstract: The rapid development of urbanization has continuously encroached on people's living space and ecological space, leading to an imbalance in territorial spatial functions. Identifying potential land use conflicts and optimizing land use structure are conducive to carrying out territorial spatial planning rationally. In this paper, we adopt the suitability assessment method to evaluate the suitability of land for production, living, and ecological functions and then use the land use conflict identification matrix to identify land use conflicts in Donghai County and make relevant suggestions according to the intensity of land use conflicts. The results of this study show the following: (1) the areas of suitable land use zones, strong conflict zones, medium conflict zones, and weak conflict zones in Donghai County are, respectively, 58.83%, 10.62%, 26.31%, and 4.24%. (2) The spatial distribution differences in the different conflict zones could determine the pertinence of conflict mitigation and spatial planning. In the process of the urbanization of Donghai County, ecological environmental protection is still the top priority. (3) It is necessary to economically and intensively use construction land, improving its fine management level. Land use efficiency should be maximized, and the spatial distribution of national territory should be reasonably optimized while strengthening the guiding role of planning. This study addresses land conflicts from the perspective of spatial planning rather than economic behavior. It also provides significant insight into land use layout at the county level, which is exactly what China is exploring in the new era.

Keywords: production–life–ecological functions; land use conflicts; suitability assessment

1. Introduction

Since the middle of the last century, human beings have intensified the development of land resources in order to meet the rapid growth of the population and the expansion of social demand. This has led to enormous pressure on resource allocation and aggravated the functional conflict of land use. The contradiction between limited land resources and increasing demand and the problem of land use conflicts have seriously affected the coordinated development of regions. This has become a common problem faced by different countries in the process of industrialization and urbanization [1]. In China, with urbanization entering into a rapid development process, the problem of land use conflict has seriously affected the coordinated development of regions. Urbanization and industrialization have led to rapid economic development. While significant attention has been directed toward the speed of development, the issues arising from the careless use of resources have often been overlooked. The rapid pace of production has encroached upon both living and ecological spaces, resulting in an imbalance in the functions of geographical space. Due to the land resource characteristics of territoriality, finiteness, and scarcity, the problem of land use function conflict has become more and more prominent [2,3]. The report of the 20th National Congress of the Communist Party of China (CPC) points out that it is necessary to build a regional economic layout and territorial space system with complementary advantages and high-quality development. It has become an important task to identify land use conflicts, rationally carry out land space planning, mitigate land use function conflicts, achieve the coordinated development of land with multiple objectives, and promote the high-quality and sustainable use of land resources.

Since the 1970s, the problem of land use conflict has attracted scholars' attention at home and abroad. It has become a research focus and hotspot [4]. "Agenda 21" outlined conflict as a key topic in land use planning. Different scholars have carried out extensive research on land use conflict from their own perspectives [5,6]. The consensus about the definition is that land use conflict is a kind of conflict caused by the scarcity of land resources, the multiplicity of functions, and different stakeholders' diverse needs [7].

The content of research on land use conflict has included the root causes of conflicts [8–10], land use conflict management [11,12], land use conflict simulation [13,14], land use conflict coordination and solutions [15,16], land use conflict identification and intensity diagnosis [17], land use conflict in different countries and global sustainable development [18], etc. In fact, the prevention and mitigation of land use conflicts and the formulation of reasonable territorial spatial planning must be based on the identification of land use conflicts [19]. While relevant research about land use spatial conflict is inadequate, spatial identification has become the most important issue in the land use conflict research field [20].

Land use conflict identification methods include qualitative and quantitative analyses. Qualitative methods of land use conflict identification include the participatory survey method, participatory mapping method [21,22], game theory analysis method, etc. [23,24]. Quantitative methods include the PASIR model [25], PSR model [26], multi-objective planning method [27], landscape ecological risk method [28,29], suitability evaluation method, etc. [30,31]. Among them, suitability evaluation is usually used in spatial conflict identification [32]. Within research on land suitability and competitiveness evaluation, the combination of land use conflict types and intensity has great application value. A variety of methods have been used for land use suitability assessment, including overlay mapping [33], the multi-criteria evaluation method [34], the weighted superposition method of GIS and RS [35], and so on. Among them, the methods of determining suitability weight include the analytic hierarchy process, the expert scoring method combined with the analytic hierarchy process, the network analysis method [36–38], and so on. Referring to the relevant literature, we selected the entropy weight method in this paper to evaluate the suitability of land use function [39].

Regarding survey regions, previous spatial research on land conflicts has been more concentrated in economically developed regions at the macro level, such as at the national [40], province [41], city [42], urban agglomeration [43], and river basin [44] levels. In China, the General Offices of the CPC Central Committee and the State Council issued "opinions on promoting urbanization construction with county seat as an important carrier" in 2022. The county seat has been listed as an important carrier of future urbanization construction. It has become an important task to study how to coordinate the production, life, ecology, and security needs of the county and jointly promote the development of new urbanization in the implementation of the national strategy. Land use function conflicts are important factors limiting the sustainable development of urbanization [45]. Based on the new trend of territorial spatial planning and economic and social sustainable development considerations, studying land use conflicts from the perspective of "production–living–ecological" functions becomes a practical requirement [37,46]. Existing studies of land use conflicts with spatial analysis have paid less attention to the county level, especially the township level. Research on County areas could improve the precision of evaluation results, which is of greater significance in guiding the land use planning of provincial and grassroots governments.

This study identifies the types and intensity of land use conflicts in Donghai County from the perspective of the suitability of "production-living-ecological" functions (PLEFs).

It analyzes the zones of land use conflicts within Donghai County and its townships. The main contributions of this research include: (1) identifying and diagnosing the intensity of land use conflicts in Donghai County, which could provide case references for coordinating ecological environmental protection with social development, thereby offering a scientific basis for decision-making by local and regional land use spatial planning administrators. (2) Using a land use conflict identification matrix, this study examines the land use conflict situation in each township from a county-wide perspective. The exploration of land use conflict issues at both the county and township scales provides valuable support materials for grassroots planning and governance. (3) Through the evaluation of the appropriateness of PLEFs, the research findings will contribute to systematic planning and layout, enhance the protection of land resources, and promote sustainable societal development.

2. Theoretical Analysis Framework

2.1. *Land Use Conflicts Based on PLEFs*

This study analyzes land conflicts from the perspective of spatial configuration. The same land can serve multiple functions simultaneously. The functional suitability of land use refers to the degree to which a particular land area is appropriate for agricultural and industrial production, urban construction, ecological protection, and other human activities. The scarcity of land resources, the multifunctional suitability of land use, and the diversity of human needs contribute to conflicts during the land use process [37]. From the perspective of "production-living-ecological" functions, we define land use conflict as the contradiction arising from the multiplicity of land functions, the competitiveness of land resources, and the diversity of social needs during production, economic construction, ecological protection, and other land use processes [29,43]. This definition encapsulates the spatial dynamics involved in land development and utilization, which are caused by conflicts among multiple land functions. When two or more functions exhibit an equal degree of suitability within the same land area, subsequent land use processes may give rise to corresponding conflicts due to different needs. Based on the evaluation of production, living, and ecological functions, this study identifies zones of land with high or medium suitability for two or three functions as potential areas for land use conflicts. To assess the intensity of these potential conflicts, a land use conflict matrix will be constructed, categorizing spatial land into strong conflict zones, medium conflict zones, and weak conflict zones.

2.2. *Analysis Framework of Land Use Conflict Identification*

The natural supply of land resources is limited. The development of urban and rural economic construction has led to an increasing demand for land resources, resulting in an increasingly tense relationship between humans and land. The scarcity of land resources has given rise to competition for land use, which is a primary cause of land use conflicts. With the proposal and widespread application of sustainable development theory, the study of land use functions has gradually expanded beyond agriculture to encompass economic, social, and ecological fields [30]. Land possesses multifunctionality, serving productive, subsistence, and ecological purposes [45]. This multiplicity of land functions is a significant factor contributing to the emergence of land use conflicts. In the use process, different stakeholders have varying needs, leading to competition in selecting land use methods and primary functions, which in turn results in conflicts. The conflicts between the production, living, and ecological functions of land reflect the competing demands of economic development, social construction, and ecological environment protection. Natural attributes represent the inherent qualities of the land, while location conditions arise from the interplay of various regional land attributes influenced by human economic activities. Lands in different areas exhibit varying functional suitability due to differences in their natural properties and location conditions. The suitability of production, living, and ecological functions is considered goal oriented. An indicator system is constructed to evaluate the suitability of regional land functions by selecting indicators from both natural and locational perspectives [32].

Based on the analysis of theoretical concepts and the existing literature [39,42,47], this study integrates the theory of human–land coordination with the theory of sustainable development to evaluate land use function suitability. The evaluation is approached from three perspectives: production function, living function, and ecological function. An indicator system is developed, considering both natural attributes and locational conditions. Furthermore, the study identifies regional land use conflict zones and assesses the intensity of these conflicts, thereby providing valuable case references for subsequent spatial planning (Figure 1).

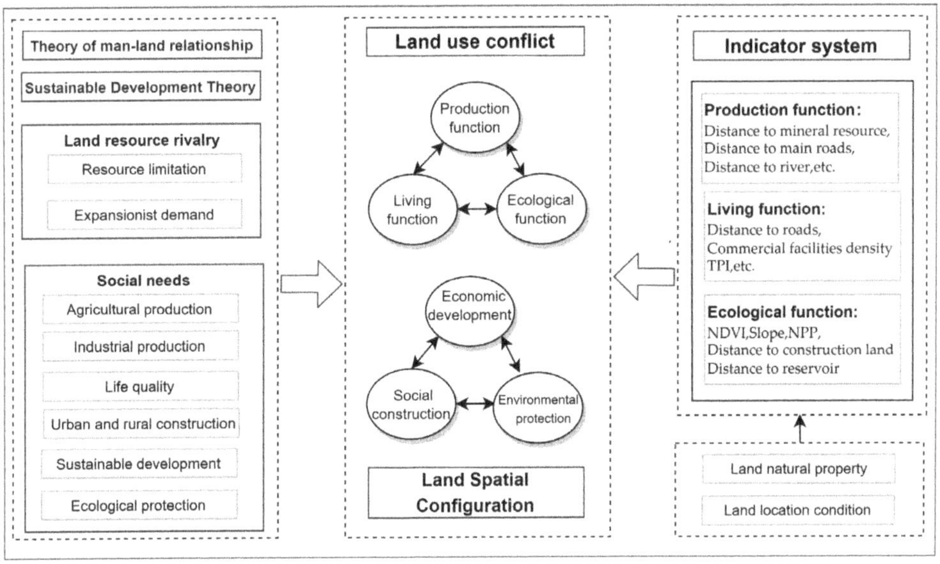

Figure 1. Theoretical analysis framework.

3. Materials and Methods

3.1. Study Area

Donghai County is part of Lianyungang City, situated in the north of Jiangsu Province. It falls under the jurisdiction of a provincial economic development zone and comprises 13 towns, 4 townships, and 2 streets. The county features a long east–west terrain and a shorter north–south expanse, with higher elevations to the west and lower elevations to the east, where the terrain is predominantly flat. Donghai County is located at the core of the national intersection of the "Belt and Road "initiative. Additionally, it serves as an overlapping area for three major strategies: the Coastal Economic Belt, the Jianghuai Ecological Zone, and the Huaihai Economic Zone (see Figure 2). Adhering to the principle of "Establishing industry to strengthen the county", Donghai County is leveraging its resource advantages to develop the new materials industry and a circular economy. The county has made significant progress in these sectors and has been recognized as one of the top 100 counties in the nation for high-quality development and for its development potential. As a significant grain functional area, Donghai County is endowed with abundant arable land resources and is recognized as a national leader in grain production. Situated in the core region of the mountainous and hilly ecological landscape in western Lianyungang City, Donghai County features a national wetland park, a provincial forest park, and several water conservation areas. In recent years, the intensity of land development and utilization in Donghai County has escalated. Industrial development and urbanization have resulted in the continuous expansion of construction land, encroaching upon other land types. Issues such as the unbalanced development of production and living spaces, as well as the deterioration of the ecological environment, have increasingly come to the forefront. The conflicts among the production, living, and ecological functions of land resources are critical factors influencing the coordinated use of land and the economic and social development of the area. Therefore, identifying potential land use conflicts is essential for rationally formulating territorial spatial planning, promoting the coordinated fulfillment of production, living, and ecological functions (PLEFs), and achieving the efficient use of land resources.

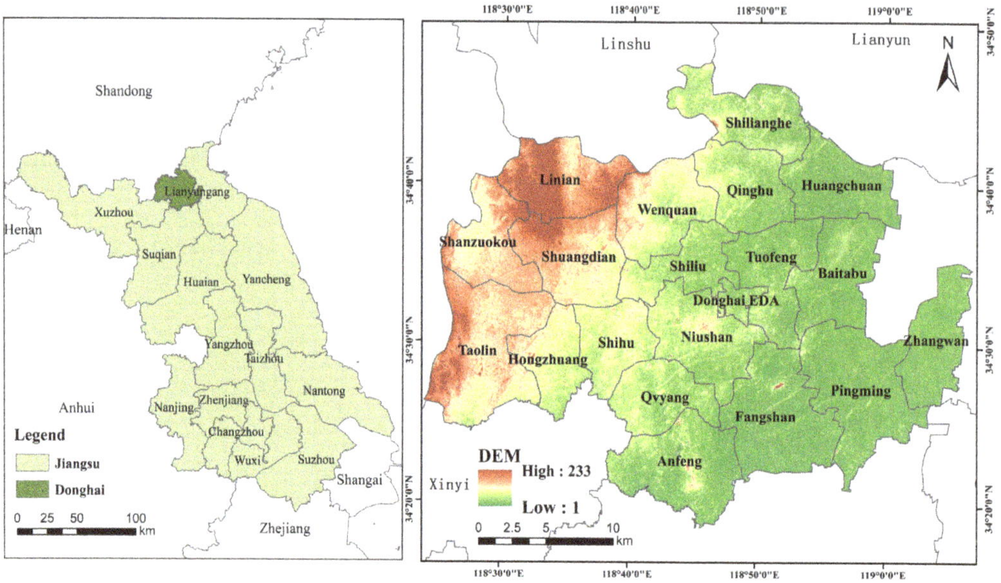

Figure 2. Map of the study area.

3.2. Research Methods

3.2.1. Topographic Position Index

The Topographic Position Index (TPI) is derived from the integration of elevation and slope, effectively representing the topographic variations within a region. A lower slope and reduced elevation result in a smaller YPI, suggesting that the challenges associated with development and construction are minimized, making these areas more suitable for habitation. The calculation formula is as follows:

$$T = \lg\left[\left(\frac{e}{E} + 1\right) \times \left(\frac{s}{S} + 1\right)\right] \quad (1)$$

In this context, T represents the Topographic Position Index (TPI) of a specific point within the area. The variable e denotes the elevation value at that point, while E signifies the average elevation value across the area. Additionally, s refers to the slope at the point, and S represents the average slope value within the area.

3.2.2. Entropy Weight Method

The entropy weight method is an objective assignment technique that draws upon the conceptual framework of information entropy. This method enables the objective calculation of the weight of each indicator, thereby mitigating the errors introduced by subjective judgment [39]. The weights derived from the entropy weight method reflect the relative change rates of the indicators within the system.

To calculate the entropy value of the jth metric e_j, the following formula was used:

$$P_{ij} = z_{ij}\sum_{i=1}^{m} z_{ij}\ i = 1,2,\ldots,m,\ j = 1,2,\ldots,n \quad (2)$$

$$e_j = -K\sum_{i=1}^{m} p_{ij}\ln\left(p_{ij}\right) \quad (3)$$

where i is the divided grid cell and j is each indicator; z_{ij} is the value of the jth indicator in the ith grid; and P_{ij} is the weight of the jth indicator in the ith grid. The constant K, which is greater than zero, is typically associated with the number of samples m. In this paper, we set the constant K to 1/ln(m).

To calculate the information utility value for the jth indicator d_j, the following formula was used:

$$d_j = 1 - e_j. \quad (4)$$

To calculate the weight of the indicator w_j, the following formula was used:

$$w_j = d_j/j\sum_{i=1}^{n} d_j \quad (5)$$

3.2.3. Calculation Results of Suitability Evaluation

The comprehensive evaluation method was used to calculate the comprehensive index of PLEF suitability, with the following formula:

$$S_i = i\sum_{j=1}^{n} w_j \times A_{ij} \quad (6)$$

where S_i is the suitability score of the ith grid and A_{ij} is the standardized score for the jth indicator of the ith raster. The higher the Si value, the higher the functional suitability of the ith grid.

3.2.4. Matrix for Identifying Land Use Conflict Types

Based on the suitability evaluation results and the existing literature [48], the suitability levels of the production function, living function, and ecological function are ranked and combined, yielding 27 distinct results. These results are categorized into 4 first-level categories and 12 second-level categories. The specific land use conflict identification matrix is presented in Table 1. The term "Production-living-ecological" indicates a potential land use conflict among the production, living, and ecological functions of land. "Production-living" denotes a potential conflict between the production and living functions, while "Production-ecological" refers to a potential conflict between the production and ecological functions. Lastly, "Living-ecological" signifies a potential conflict between the living and ecological functions of land.

Based on the perspective of PLEFs, 18 evaluation indicators were selected from two dimensions: natural and social factors that influence land use, reflecting the characteristics of different functions. The indicators were standardized to determine their weights more objectively, employing the entropy weight method. Five assignment levels were established, referencing the existing literature. A land use functional suitability evaluation index system was constructed using these indicators. Following the suitability evaluation results, land use conflicts were identified using a conflict identification matrix, enabling the diagnosis of conflict zones and their intensity. Policy suggestions will be proposed for the targeted analysis of land use conflict zones with various intensities. The overall framework of the paper is illustrated in Figure 3.

Table 1. Identification matrix of land use conflicts.

Type of Conflict		Appropriateness of PLEFs		
Conflict Type	**Secondary Conflict Type**	**Production**	**Living**	**Ecological**
PLEF Suitability Zone	Production Suitable Area	High	Medium/Low	Medium/Low
		Medium	Low	Low
	Living Suitable Area	Medium/Low	High	Medium/Low
		Low	Medium	Low
	Ecological Suitable Area	Medium/Low	Medium/Low	High
		Low	Low	Medium
Strong conflicts zone	Production–living–ecological	High	High	High
	Production–living	High	High	Medium/Low
	Production–ecological	High	Medium/Low	High
	Living–ecological	Medium/Low	High	High
Medium conflicts zone	Production–living–ecological	Medium	Medium	Medium
	Production–living	Medium	Medium	Low
	Production–ecological	Medium	Low	Medium
	Living–ecological	Low	Medium	Medium
Weak conflicts zone	Weak conflicts zone	Low	Low	Low

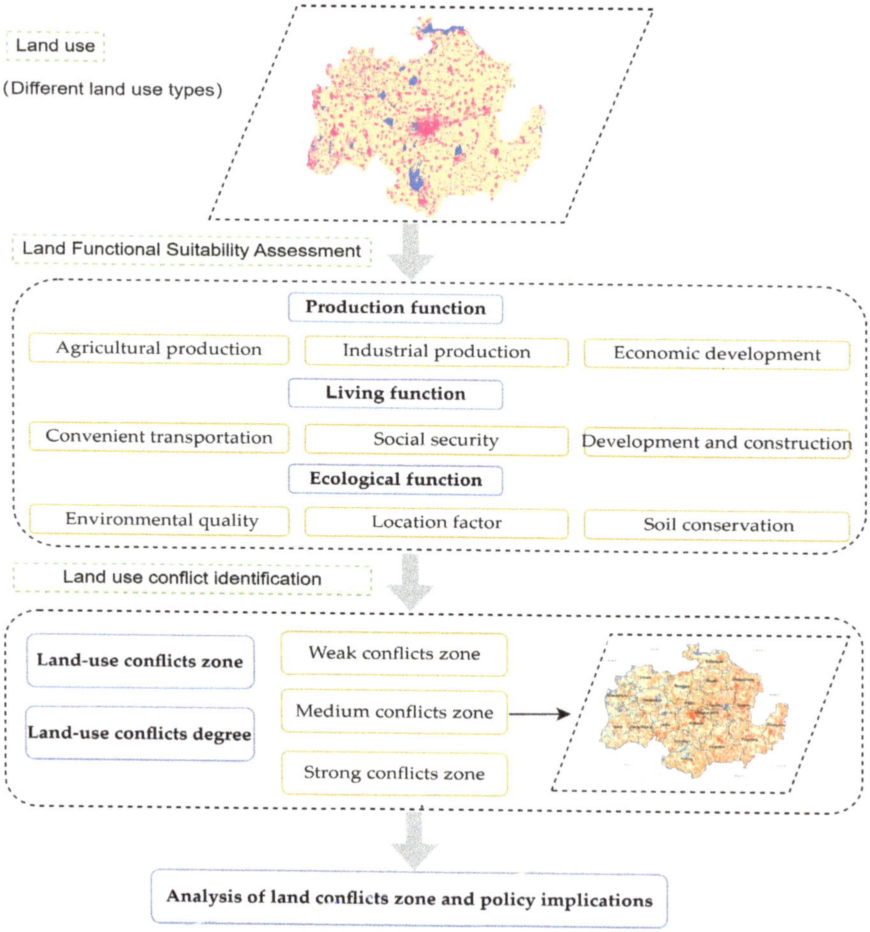

Figure 3. Flow chart of the research.

3.3. Data Sources and Processing

3.3.1. Indicator System for Evaluating the Suitability of PLEFs

The suitability of PLEFs refers to the appropriateness of land resources in fulfilling production, living, and ecological functions. Different conditions are required to achieve these distinct functions. Consequently, based on the specific circumstances of Donghai County and the requirements of each function, we selected evaluation indexes to construct an evaluation index system for assessing the suitability of Donghai County's production, living, and ecological functions, drawing on relevant studies [1,32,49] (Table 2).

Table 2. Evaluation indicator system for the suitability of PLEFs.

Target	Indicator	Indicator Classification and Score					Weight
		9	7	5	3	1	
Production Functional Suitability	Distance to mineral resource (m)	200	500	1000	1500	>1500	0.1905
	Distance to ponds and ditches (m)	100	300	500	1000	>1000	0.1410
	Soil organic matter content (g/kg)	>8	7	6	5	<4	0.1462
	Distance to river/m	100	300	500	1000	>1000	0.0979
	Slope (°)	2	6	15	25	>25	0.1423
	Industrial agglomeration degree			Natural Breaks			0.1915
	Distance to main road (m)	100	300	500	1000	>1000	0.0905
Living Functional Suitability	TPI			Natural Breaks			0.0909
	Distance to road	100	300	500	1000	>1000	0.0853
	Commercial facilities density			Natural Breaks			0.2584
	Distance to rural settlement	200	500	1000	1500	>1500	0.1094
	Educational infrastructure density			Natural Breaks			0.2304
	Medical infrastructure density			Natural Breaks			0.2255
Ecological Functional Suitability	NDVI	>0.8	0.8	0.6	0.4	<0.2	0.2658
	Slope (°)	5	8	15	25	>25	0.2320
	NPP			Natural Breaks			0.1565
	Distance to construction land (m)	1000	500	300	100	<100	0.1587
	Distance to reservoir (m)	200	500	1000	2000	>2000	0.1870

Evaluation indicators for the suitability of land use production functions were selected from both natural and social dimensions. In the natural dimension, mineral resources are crucial for industrial production in Donghai County; thus, the distance to these resources was chosen as one of the indicators. Ponds, ditches, and rivers provide essential water resources for both agricultural and industrial production, while rivers also serve important ecological functions. The content of soil organic matter directly reflects the soil's suitability for agricultural production, leading to the separate selection of distances to ponds and ditches as indicators. Additionally, slope significantly influences soil and water conservation capabilities during agricultural production, making it a key factor in this context. In the social dimension, indicators include the degree of industrial agglomeration and the distance to main roads. A higher degree of industrial agglomeration correlates with greater scale effects in production, with indicator values derived from the analysis of the density of enterprise distribution density in Donghai County. Furthermore, increased accessibility and improved traffic conditions facilitate both agricultural and industrial production, with indicator values calculated based on distances to roads and railways.

For the evaluation index for land use suitability concerning living functions, we selected the Topographic Position Index (TPI), which indicates the challenges and costs associated with residential development and construction. Proximity to the township center and the convenience of transportation correlate positively with the ease of living and overall quality of life. Additionally, the presence of commercial facilities, educational infrastructure, and medical services offers essential support for residents, thereby enhancing life convenience.

The evaluation indicators for assessing the ecological function suitability of land use include the NDVI, slope, NPP, and the proximity to construction land and reservoirs. The NDVI reflects the status of vegetation cover and provides insights into the ecological condition of the area. A steeper slope increases the risk of soil erosion, which is detrimental to ecological environmental protection. NPP represents the organic mass produced by green plants through photosynthesis, excluding the portion utilized by plant respiration. This indicator is crucial for evaluating the quality of the regional ecological environment. Proximity to construction land correlates with increased ecological impacts, while reservoirs significantly contribute to climate regulation and possess ecosystem service value; thus, closer distance to them indicates higher ecological value.

3.3.2. Data Sources

The land use data are derived from China's province-by-province yearly surface cover dataset (CLCD). The elevation data utilized in the natural environment analysis were sourced from the Geospatial Data Cloud. Slope data were processed based on the elevation data. The Topographic Position Index was calculated using both elevation and slope data. Data on rivers, ponds, ditches, and reservoirs were obtained from the third national land survey database. Soil organic matter content data and vegetation normalized index data were sourced from the National Tibetan Plateau Data Center (http://data.tpdc.ac.cn accessed on 3 July 2024). Vegetation net primary productivity data were derived from the MODIS MOD17A3HGF satellite remote monitoring dataset. In the socio-economic data category, the transportation road network data were acquired from the Open Street Map. Mineral point data were obtained from the National Mineral Points Database Version 2021 (ngac.org.cn accessed on 3 July 2024). Data on industrial points, educational facilities, medical facilities, and commercial facilities were sourced from the GOOD MAP POI Points of Interest (POIs). Location correction and cropping were performed using ArcGIS. Distribution density was calculated through kernel density analysis. The aforementioned data were converted to a 30 m × 30 m raster format following preprocessing steps, which included projection transformation, cropping, resampling, Euclidean distance analysis, and kernel density analysis.

4. Results

4.1. *Evaluation of Land Use Function Suitability*

Based on the evaluation index system of PLEFs, ArcGIS was employed for data processing and spatial analysis to derive the suitability evaluation results for production, living, and ecological functions in Donghai County.

4.1.1. Evaluation of Production Functions Suitability

From a quantitative perspective, the area classified as highly suitable for production functions is 454.45 km^2, which accounts for 22.38% of the total area of Donghai County. The area deemed moderately suitable for production functions is 999.41 km^2, representing 49.23%. Additionally, the area identified as low suitability for production functions is 576.39 km^2, making up 28.39% of the total area. Together, the highly and moderately suitable areas for production functions exceed 70% of Donghai County's total area, indicating an overall high suitability for production.

The spatial distribution of production function suitability levels in Donghai County is illustrated in Figure 4. From a spatial distribution perspective, the overall pattern exhibits a strip-like shape, with higher suitability in the central southeast and lower suitability in the northwest. The areas of high production function suitability are primarily concentrated in the central region. At the township scale, the analysis reveals that the highly suitable areas are predominantly located in the Donghai Economic Development Zone, Baitabu Township, Niushan Street, and Hongzhuang Township. Specifically, the highly suitable areas in the Donghai Economic Development Zone and Baitabu Township account for 63.81% and 47.54% of their respective total areas. Additionally, significant high-production function suitability areas are also found in Hot Spring Town, Hongzhuang Town, and Shuangdian Town. The medium suitability areas for the production function encompass the largest expanse and are mainly distributed in the southeast of Donghai County, characterized by a more fragmented distribution. The medium suitability areas primarily include Civilian Township, Huangchuan Township, Zhangwan Township, Shiliang Street, and Fangshan Township, which account for 80.79%, 59.31%, 58.23%, 57.58% and 57.69% of their total township areas, respectively. Most townships exhibit the largest proportion of production function suitable areas within their administrative boundaries.

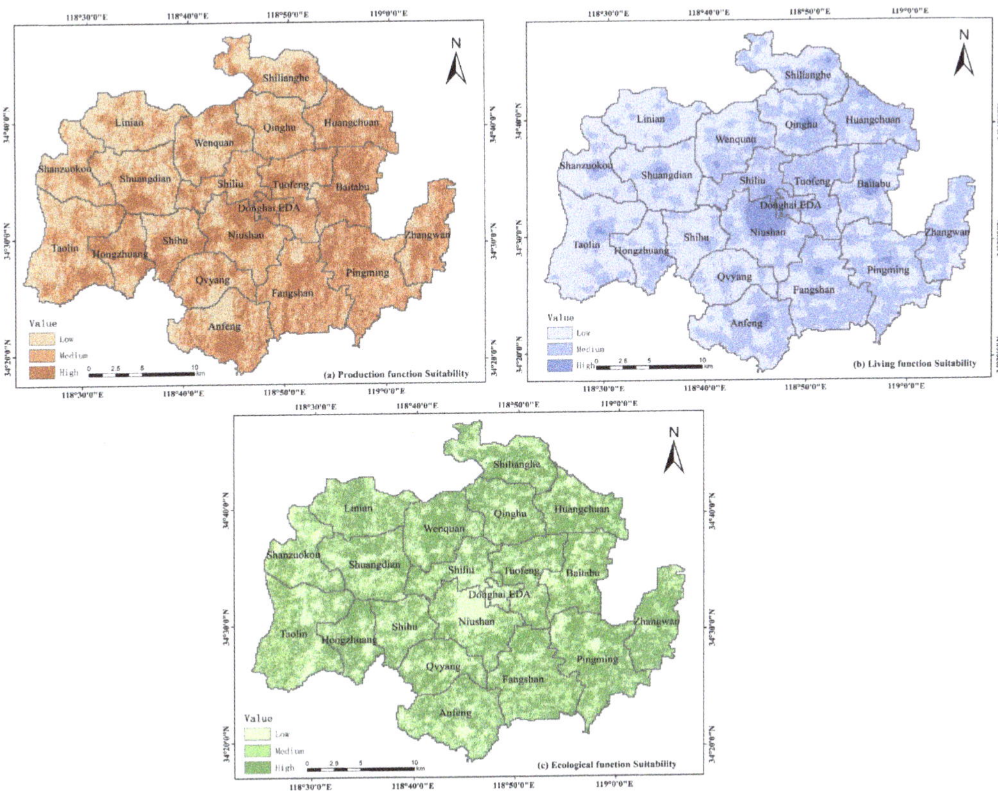

Figure 4. Spatial distribution of PLEF suitability.

4.1.2. Evaluation of Living Function Suitability

The results of the quantitative evaluation analysis regarding the suitability of living functions in Donghai County indicate that the areas designated as high, medium, and low suitability zones measure 65.24 km^2, 868.69 km^2, and 1096.32 km^2 respectively. Notably, suitability zones occupy the largest proportion, accounting for 54.00% of the entire study area, while medium suitability zones represent 42.79% of the total area. In contrast, high suitability zones constitute the smallest proportion at only 3.21%. The combined area of medium and high suitability zones in Donghai County does not exceed 50% of the total area. This suggests that the current infrastructure development and urban planning in Donghai County are insufficient to adequately support living functions.

The spatial distribution of living function suitability levels in Donghai County is illustrated in Figure 3, which reveals a cluster-like spatial arrangement primarily centered around the town center. Areas with high living function suitability are predominantly found in the Donghai Economic Development Zone and Niushan Street, collectively accounting for over 24% of the total area. These high suitability regions are situated within 200 m of the township center. Surrounding the high living function suitability areas, medium and low suitability regions are distributed in concentric circles. A statistical analysis of the living space suitability proportions across each township indicates that the medium living function suitable areas are predominantly located in Huangchuan Township, Camel Peak Township, Shidu Street, and Fangshan Township. Notably, the proportion of medium living function suitable area exceeds 50% of the total area in eight townships.

4.1.3. Evaluation of Ecological Function Suitability

The ecological function suitability results for Donghai County can be categorized into high suitability, medium suitability, and low suitability areas using the natural breakpoint method. Statistical findings indicate that the areas classified as high ecological function suitability, medium suitability, and low suitability areas of Donghai County are 816.21 km^2, 856.10 km^2, and 357.94 km^2, respectively, accounting for 40.20%, 42.17%, and 17.63% of the total area. Notably, the combined proportion of areas with high and medium ecological function suitability exceeds 80% of Donghai County's total area. Overall, the area classified as highly suitable for ecological functions is larger than that designated for high production functions and high living functions, indicating a generally high level of ecological suitability.

The spatial distribution of ecological function suitability levels in Donghai County is illustrated in Figure 4. Low suitability zones are primarily concentrated in the center and are also scattered throughout the region. In contrast, high and medium suitability zones are distributed around the center, with high suitability zones predominantly located in the eastern part, accounting for a larger proportion. Medium suitability zones are found in the west and are more widely distributed. Analyzing the spatial distribution structure of ecological function suitability at the township scale reveals that highly suitable ecological function areas are mainly concentrated in Zhangwan Township, Pingming Township, Hot Spring Township, Huangchuan Township, and Fangshan Township in the east, with proportions of 60.24%, 52.22%, 50.18%, 49.38%, and 44.08%, respectively. Medium ecological suitability areas are primarily situated in Anfeng Township, Fangshan Township, Li Communities, Shanzokou Township, Taolin Township, and Shihu Township, all of which account for more than 45% of the total township area. The low ecological suitability zones are smaller in size and are primarily located in the Donghai Economic Development Zone and Nushan Street.

4.2. *Land Use Conflict Analysis*

The distribution map of PLEF suitability (Figure 4), reveals overlapping areas of PLEF suitability, indicating the potential land use conflicts in Donghai County. Based on the suitability evaluation of production, living, and ecological functions in Donghai County, an overlay analysis was conducted using ArcGIS, resulting in 27 distinct functional combinations, which were subsequently merged into similar types. The area and percentage of each conflict type are presented in Table 3 and Figure 5.

Table 3. Statistics on the area of land use conflict zones in Donghai County.

Conflict Type	Area (km^2)	Proportion	Secondary Conflict Type	Area (km^2)	Proportion
Suitability Zone	1194.37	58.83%	Production	301.36	14.84%
			Living	65.43	3.22%
			Ecological	827.58	40.76%
Strong conflict area	215.66	10.62%	Production–living–ecological	1.65	0.08%
			Production–living	33.26	1.64%
			Production–ecological	179.43	8.84%
			Living–ecological	1.32	0.06%
Medium conflict zone	534.17	26.31%	Production–living–ecological	192.58	9.49%
			Production–living	65.79	3.24%
			Production–ecological	209.47	10.32%
			Living–ecological	66.33	3.27%
Weak conflict zone	86.05	4.24%	Production–living–ecological	86.05	4.24%

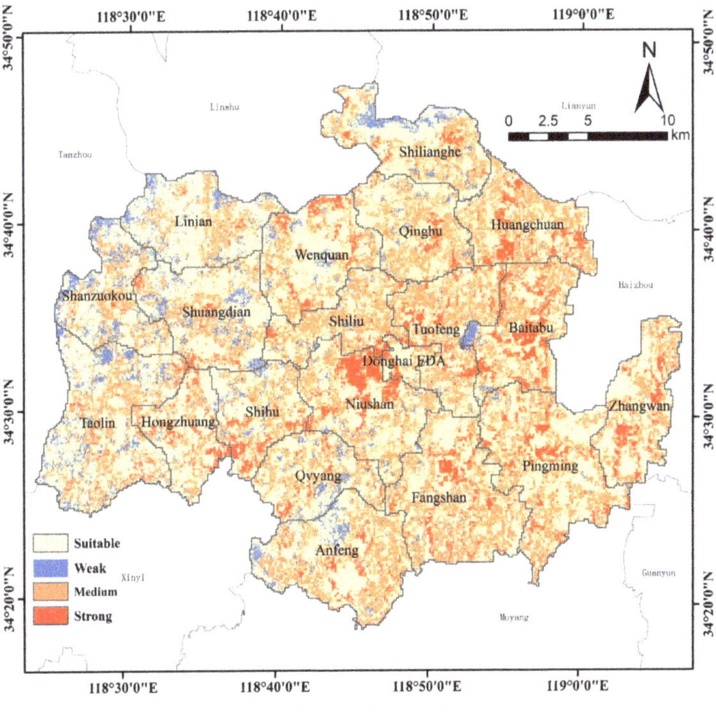

Figure 5. Spatial distribution of land use conflicts in Donghai County.

4.2.1. Strong Conflict Zone Analysis

The strong conflict zone comprises areas highly suitable for production, living, and ecological functions, or combinations of these functions, where land use conflicts are likely to arise in future land use processes. The identification results indicate that the strong conflict zone covers an area of 215.66 square kilometers, accounting for 10.62% of the total study area. Among these zones, the production–ecological function strong conflict zone occupies the largest area at 179.43 square kilometers, significantly exceeding that of other types, followed by the production–life function strong conflict zone at 33.26 square kilometers. The production–ecological function strong conflict area is predominantly located in the eastern part of Donghai County. An analysis at the township scale reveals that it is primarily situated in Baitabu Township, Huangchuan Township, Zhangwan Township, Hongzhuang Township, and Fangshan Township, which account for 20.07%,

16.36%, 15.29%, 12.85%, and 10.90% of the total area, respectively. Some strong conflict zones are distributed along both sides of the road. In Baitabu Township and Huangchuan Township, the production–ecological function of strong conflict zones is found in areas with high soil organic matter content, which is conducive to agricultural production. When conflicts arise between production and ecological functions, the ecological space may be easily encroached upon by production activities, thereby threatening the quality of the ecological environment. The production–life function conflict zones are mainly concentrated in the center of Donghai County and are sporadically distributed across various townships. From a township perspective, the production–life conflict zones are primarily located in Niushan Street, Baitabu Township, and Jiangsu Donghai Economic Development Zone, with areas of 16.24 square kilometers, 3.58 square kilometers, and 2.31 square kilometers, respectively.

4.2.2. Medium Conflict Zone Analysis

The medium conflict zone in Donghai County encompasses an area of 534.17 square kilometers, representing 26.31% of the total land area, thereby constituting the largest proportion. The predominant land use conflicts in Donghai County are classified as medium conflicts. This medium conflict zone serves as an intermediate buffer between the strong conflict zone and the weak conflict zone. Although potential conflicts exist within this zone, they remain within a controllable range. The medium conflict zones include the production–ecology medium conflict zone and the production–life–ecology function conflict zone, which cover areas of 209.47 square kilometers and 192.58 square kilometers respectively, accounting for 10.32% and 9.49% of the total area. The production–ecological function medium conflict zone is primarily located in Shihu Township, Taolin Township, Quyang Township, Shuangdian Township, and Shankouzuo Township, predominantly concentrated in the western and southwestern regions of Donghai County. The total area of the production–life–ecology function conflict zone is 192.58 km^2, representing 9.49% of the County's area. They are spatially distributed and scattered, with a higher density observed in the northern and southwestern parts of the county. Township-scale analyses reveal that eight townships contain production–life–ecology function conflict zones that exceed 10% of their total area, with Shiduajie, Huangchuan Township, Fangshan Township, and Mofeng Township accounting for the majority.

4.2.3. Weak Conflict Zone Analysis

The weak conflict zone is characterized by overlapping areas with low PLEF suitability. It is primarily located on the edges of townships in the western part of Donghai County, as illustrated in Table 3. This weak conflict zone encompasses an area of 86.05 square kilometers, which represents 4.24 percent of the total land area. It is the smallest zone in terms of area and is distributed sporadically. The weak conflict zones, situated at higher elevations and featuring a wide distribution of mountains and hills, are located far from township centers, which complicates development and construction efforts.

5. Discussion

5.1. Evaluation and Analysis of Land Use Function Suitability

Geographical location is a crucial factor influencing land use suitability, consistent with the findings of previous studies [34,47]. Each land use type is affected by distinct factors. Areas with high production function suitability are predominantly situated in the central region. The central and south-central slopes are characterized by relatively low topography, with less undulating terrain that favors agricultural production [43]. Additionally, the degree of industrial agglomeration is significant, particularly in the Economic Development Zone and Niushan Street. The high level of accessibility further supports the transportation needs of both agriculture and industry, thereby enhancing production suitability. Furthermore, regions with a denser distribution of mineral resources are likely to develop production advantages, a phenomenon attributed to the unique resource characteristics of the study area. The production function suitable area constitutes the largest segment, indicating that the current natural environment and basic public facilities in Donghai County are conducive to production. The largest area suitable for living function is located in the center of the study area, characterized by high transportation convenience, superior commercial facilities, and a greater density of medical and basic educational services [50]. In contrast, areas with low suitability for living functions are predominantly situated in mountainous and hilly terrains, which exhibit significant topographic relief. These regions are distanced from the town center, resulting in challenges for the provision of various public infrastructure services and facilities. Conversely, areas with high ecological function suitability boast extensive forest coverage, substantial ecosystem service value, dense distribution of reservoirs, and numerous water conservation areas, such as Anfeng Mountain and Fangshan Mountain. Conversely, regions with low ecological function suitability are heavily developed, with dense land use, leading to frequent disruptions of ecological functions due to human activity.

5.2. Analysis of Major Land Use Conflict Zones

We developed an analytical framework for land use conflict from the perspective of PLEF suitability, and the empirical study was conducted at the county level. Cartography was employed to illustrate the extent of the conflict, as noted in other literature sources [32]. Our focus is on the county scale, where we analyze the relevant factors in detail. The primary conflict zones are identified as strong production–ecological function conflict zones and strong production–life function conflict zones. High production–ecological function conflicts are predominantly concentrated in the eastern part of Donghai County, Primarily located near transportation arteries and industrial and mining areas. The presence of convenient transportation infrastructure serves as an advantageous space for future production. Additionally, numerous rivers traverse the eastern part of Donghai County, which features high vegetation coverage and several water conservation areas, such as Fangshan. These dynamics result in a conflict between production and ecological considerations. A significant area of this conflict is situated in the center of Donghai County, which serves as the primary development zone. The dense road network facilitates industrial agglomeration, while the flat terrain supports production and construction [40]. Additionally, a variety of medical, scientific, educational, and cultural facilities contribute to a high-density concentration of living resources, enhancing convenience and quality of life in the area. The infrastructure surrounding the center is well developed, offering convenient transportation options that further elevate the suitability for living. Concurrently, the degree of industrial agglomeration is high, leading to an intensified conflict between production and residential needs.

The main conflict zones include production–ecology conflicts and production–life–ecology conflicts. The production–ecology conflict zones are located near major traffic arteries, as well as industrial and mining sites. In the western part of Donghai County, lakes, water surfaces, and reservoirs are densely populated with water sources, making them susceptible to production–ecology conflicts due to the anticipated continuous development of industrial production. Conflict zones in PLEFs are mainly situated around townships and central towns. These areas are characterized by their proximity to urban centers, convenient transportation, high industrial density, and advantages in production and living conditions. Consequently, potential conflicts within the production–living–ecological function have emerged, particularly in key development and construction areas for the future.

5.3. Land Use Conflicts and Policy Implications

The analysis of land use conflicts based on the evaluation of land suitability through PLEFs is instrumental in mitigating these conflicts in a more targeted manner. It provides valuable insights for land use planning and the high-quality development of regions. The ecological quality of the environment in Donghai County has significantly improved due to ongoing protection and restoration efforts. To achieve the coordinated and sustainable development of urban and rural areas, prioritizing ecological environmental protection is essential. Unlike other literature sources that analyze policy direction from the perspective of urban–rural relationships [32], this study examines the spatial pattern of PLEFs through the lens of national spatial planning. In instances of conflict between land ecological functions and other land uses, greater emphasis should be placed on ecological functions. The realization of living functions is contingent upon the local economy and the level of social development; therefore, relevant policies should be formulated with consideration for the specific circumstances of each location.

Strong conflict zones primarily encompass production–ecology conflicts and production–life conflicts. With the advancement of urbanization, industrial production inevitably impacts both ecology and people's lives [51], which is a critical issue to consider in land use planning and policy making. In agricultural production, it is essential to prioritize the construction of high-standard farmland and enhance agricultural comprehensive production capacity. In industrial production, a rational layout and planning of industrial production zones are necessary. For instance, promoting the concentration of various production factors within industrial parks can leverage agglomeration effects to improve resource utilization efficiency and production quality. Additionally, it is vital to elevate the level of fine management of construction land, ensuring that its development and utilization are both economical and intensive. It is crucial to control the encroachment of ecological and residential land caused by the expansion of construction land. Furthermore, delineating and strictly protecting the regional ecological environment while increasing efforts in ecological restoration is imperative. Initiatives centered on the construction of forest parks, such as Fangshan Forest and Anfeng Mountain Forest, should be implemented to enhance water conservation and soil conservation capacities, thereby safeguarding the ecological and living environment.

The medium conflict zone is primarily characterized by the production–ecology conflict and the PLEF conflict, both of which fall within a controllable range. However, this conflict zone represents the largest proportion of various land use conflicts. It is crucial to implement necessary measures to prevent its escalation into a strong conflict zone, a key concern during the land use planning process. The production–ecology medium conflict zone is predominantly located in the western part of Donghai County, in contrast to the production–ecology strong conflict zone. The townships in this western region are abundant in mineral resources, necessitating a prioritization of ecological environmental protection during the production process. For instance, green belts should be established along major traffic routes in conjunction with industrial development. Additionally, the development of an "ecology-culture-tourism" initiative could leverage local resources such as forests, geothermal hot springs, and historical and cultural assets. In PLEF conflict zones, efforts should focus on enhancing ecological functions, increasing the per capita area of green space in urban areas, and improving the quality of industrial development.

A weak conflict zone is characterized by a smaller size and greater stability, resulting in a reduced risk of land use conflicts. Relevant authorities should maintain a certain level of oversight to enhance the intensive and economical use of land. Each land use suitability zone has a dominant function that should be recognized. It is recommended that each township concentrate on and strengthen its dominant function within the land use process.

6. Conclusions

In this paper, we constructed a suitability evaluation index system for production–life–ecological functional suitability from the perspective of ecological sustainability. We identified and analyzed land use conflicts. The conclusions are as follows: (1) the structural composition and spatial distribution of production, life, and ecological suitability in Donghai County exhibit distinct characteristics. The distribution of production suitability is linear, with high suitability observed in the central and southeastern regions of the county, while the northwestern area shows lower suitability levels; the high suitability area constitutes 22.38% of the total. Living suitability is clustered around the town center, with high suitability areas accounting for 3.21%. In contrast, the distribution of ecological suitability is characterized by low values in the middle and high values on all sides, with high suitability land covering 40.20%. (2) In Donghai County, the areas of land use are categorized as follows: suitable area (58.83%), strong conflict area (10.62%), medium conflict area (26.31%), and weak conflict area (4.24%). The strong conflict zones are primarily characterized by production–ecological and production–life function conflicts, predominantly located in regions with dense road networks, industrial agglomeration, and flat terrain with high soil organic matter content. In these areas, the suitability for each function is relatively high, which increases the likelihood of conflicts. Conversely, the medium conflict zones are mainly identified as production–ecology and production–life–ecology conflict zones, primarily situated near traffic arteries and industrial or mining lands. In these locations, the development and construction processes may adversely affect both production and ecological functions.

Through the identification of land use conflicts, this study presents the distribution structure of potential land use conflicts across various regions, providing valuable references for governmental departments aiming to optimize the spatial use of national territory. This approach is intended to enhance the economic, social, and ecological benefits of land, thereby facilitating high-quality sustainable development. Additionally, we have conducted an exploration into the identification and intensity diagnosis of land use conflicts at the county scale. This expansion of the research scale offers greater accuracy and practical guidance, which is of significant importance to China's ongoing urbanization efforts and integrated development plans. Furthermore, it will aid in the land use planning of small- and medium-sized cities and enhance the coordination of land use structure.

Despite these innovations, there remains significant potential for further enhancement of the land use functional suitability evaluation index system. Future efforts should incorporate policy and social factors to refine this evaluation process. Land use conflicts can be identified from various perspectives, utilizing relevant and appropriate methodologies, such as the modulation model, in combination to address future sustainable development demands. This approach aims to optimize the spatial structure of the country and promote efficient resource utilization.

Author Contributions: Conceptualization, J.X. and L.C.; methodology, J.D.; software, J.D. and J.X.; validation, J.X. and Y.S.; formal analysis, J.D., J.X. and L.C.; investigation, J.X. and J.D.; resources, L.C. and J.X.; data curation, J.D.; writing—original draft preparation, J.D. and J.X.; writing—review and editing, L.C. and Y.S.; visualization, J.D. and J.X.; funding acquisition, J.X. All authors have read and agreed to the published version of the manuscript.

Funding: This research was funded by the National Social Science Foundation of China, grant No. 20BJY119, and the Fundamental Research Funds of the Central Universities, grant No. 2024JCXKSK05.

Data Availability Statement: The original contributions presented in the study are included in the article; further inquiries can be directed to the corresponding author.

Conflicts of Interest: The authors declare no conflicts of interest.

References

1. Chen, X.; Wu, S.; Wu, J. Characteristics and formation mechanism of Land use conflicts in northern Anhui: A Case study of Funan county. *Heliyo* **2024**, *10*, e22923. [CrossRef]
2. Zhang, Y.; Long, H.; Tu, S.; Tu, S.; Ge, D.; Ma, L.; Wang, L. Spatial identification of land use functions and their tradeoffs/synergies in China: Implications for sustainable land management. *Ecol. Indic.* **2019**, *107*, 105550. [CrossRef]
3. Cui, X.; Li, F.; de Vries, W.T. Smart land use planning: New theories, new tools and new practice. *Land* **2023**, *12*, 1315. [CrossRef]
4. Jiang, S.; Meng, J.; Zhu, L. Spatial and temporal analyses of potential land use conflict under the constraints of water resources in the middle reaches of the Heihe River. *Land Use Policy* **2020**, *97*, 104773. [CrossRef]
5. Peerzado, M.B.; Magsi, H.; Sheikh, M.J. Land use conflicts and urban sprawl: Conversion of agriculture lands into urbanization in Hyderabad, Pakistan. *J. Saudi Soc. Agric. Sci.* **2019**, *18*, 423–428. [CrossRef]
6. Ianoş, I.; Sorensen, A.; Merciu, C. Incoherence of urban planning policy in Bucharest: Its potential for land use conflict. *Land Use Policy* **2017**, *60*, 101–112. [CrossRef]
7. Halleux, J.M.; Marcinczak, S.; van der Krabben, E. The adaptive efficiency of land use planning measured by the control of urban sprawl. the cases of the Netherlands, Belgium and Poland. *Land Use Policy* **2012**, *29*, 887–898. [CrossRef]

8. de Jong, L.; De Bruin, S.; Knoop, J.; van Vliet, J. Understanding land-use change conflict: A systematic review of case studies. *J. Land Use Sci.* **2021**, *16*, 223–239. [CrossRef]
9. Delgado-Matas, C.; Mola-Yudego, B.; Gritten, D.; Kiala-Kalusinga, D.; Pukkala, T. Land use evolution and management under recurrent conflict conditions: Umbundu agroforestry system in the Angolan Highlands. *Land Use Policy* **2015**, *42*, 460–470. [CrossRef]
10. Tan, S.; Tong, B.; Zhang, J. How did the land disputes evolve? Evidence from the Yangtze River economic belt, China. *Land* **2023**, *12*, 1334. [CrossRef]
11. Mann, C.; Jeanneaux, P. Two approaches for understanding land-use conflict to improve rural planning and management. *J. Rural Community Dev.* **2009**, *4*, 119–139.
12. Petrescu-Mag, R.M.; Petrescu, D.C.; Azadi, H.; Petrescu-Mag, I.V. Agricultural land use conflict management-Vulnerabilities, law restrictions and negotiation frames. a wake-up call. *Land Use Policy* **2018**, *76*, 600–610. [CrossRef]
13. Moein, M.; Asgarian, A.; Sakieh, Y.; Soffianianian, A. Scenario-based analysis of land-use competition in central Iran: Finding the trade-off between urban growth patterns and agricultural productivity. *Sustain. Cities Soc.* **2018**, *39*, 557–567. [CrossRef]
14. Meimei, W.; Zizhen, J.; Tengbiao, L.; Yongchun, Y.; Zhuo, J. Analysis on absolute conflict and relative conflict of land use in mining metropolitan area under different scenarios in 2030 by PLUS and PFCI. *Cities* **2023**, *137*, 104314. [CrossRef]
15. Karimi, A.; Hockings, M. A socio-ecological approach to land-use conflict to inform regional and conservation planning and management. *Landsc. Ecol.* **2018**, *33*, 691–710. [CrossRef]
16. Cieślak, I. Identification of areas exposed to land use conflict with the use of multiple-criteria decision-making methods. *Land Use Policy* **2019**, *89*, 104225. [CrossRef]
17. Qu, Y.; Wang, S.; Tian, Y.; Jiang, G.; Zhou, T.; Meng, L. Territorial spatial planning for regional high-quality development-An analytical framework for the identification, mediation and transmission of potential land utilisation conflicts in the Yellow River Delta. *Land Use Policy* **2023**, *125*, 106462. [CrossRef]
18. Lin, Q.; Tan, S.; Zhang, L.; Wang, S.; Wei, C.; Li, Y. Conflicts of land expropriation in China during 2006-2016: An overview and its spatio-temporal characteristics. *Land Use Policy* **2018**, *76*, 246–251. [CrossRef]
19. Adam, Y.O.; Pretzsch, J.; Darr, D. Land use conflicts in central Sudan: Perception and local coping mechanisms. *Land Use Policy* **2015**, *42*, 1–6. [CrossRef]
20. Dong, G.; Liu, Z.; Niu, Y.; Jiang, W. Identification of Land Use Conflicts in Shandong Province from an Ecological Security Perspective. *Land* **2022**, *11*, 2196. [CrossRef]
21. Brown, G.; Raymond, C.M. Methods for identifying land use conflict potential using participatory mapping. *Landsc. Urban Plan.* **2014**, *122*, 196–208. [CrossRef]
22. Karimi, A.; Brown, G. Assessing multiple approaches for modelling land-use conflict potential from participatory mapping data. *Land Use Policy* **2017**, *67*, 253–267. [CrossRef]
23. Aghmashhadi, A.H.; Zahedi, S.; Kazemi, A.; Fürst, C.; Cirella, G.T. Conflict analysis of physical industrial land development policy using game theory and graph model for conflict resolution in Markazi Province. *Land* **2022**, *11*, 501. [CrossRef]
24. Reuveny, R.; Maxwell, J.W.; Davis, J. On conflict over natural resources. *Ecol. Econ.* **2011**, *70*, 698–712. [CrossRef]
25. Duraiappah, A.K.; Ikiara, G.; Manundu, M.; Nyangena, W.; Sinange, R. *Land Tenure, Land Use, Environmental Degradation and Conflict Resolution: A PASIR Analysis for the Narok District, Kenya*; International Institute for Environment and Development: London, UK; Institute for Environmental Studies: Amsterdam, The Netherlands, 2000; pp. 1–33. Available online: https://www.iied.org/sites/default/files/pdfs/migrate/8141IIED.pdf? (accessed on 9 October 2024).
26. Cheng, H.; Zhu, L.; Meng, J. Fuzzy evaluation of the ecological security of land resources in mainland China based on the Pressure-State-Response framework. *Sci. Total Environ.* **2022**, *804*, 150053. [CrossRef]
27. Bao, W.; Yang, Y.; Zou, L. How to reconcile land use conflicts in mega urban agglomeration? A scenario-based study in the Beijing-Tianjin-Hebei region, China. *J. Environ. Manag.* **2021**, *296*, 113168. [CrossRef]
28. Moomen, A.W. Strategies for managing large-scale mining sector land use conflicts in the global south. *Resour. Policy* **2017**, *51*, 85–93. [CrossRef]
29. Zhou, D.; Lin, Z.; Lim, S.H. Spatial characteristics and risk factor identification for land use spatial conflicts in a rapid urbanisation region in China. *Environ. Monit. Assess.* **2019**, *191*, 677. [CrossRef] [PubMed]
30. Iojă, C.I.; Niţă, M.R.; Vânău, G.O.; Onose, D.A.; Gavrilidis, A.A. Using multi-criteria analysis for the identification of spatial land-use conflicts in the Bucharest Metropolitan Area. *Ecol. Indic.* **2014**, *42*, 112–121. [CrossRef]
31. Carr, M.H.; Zwick, P. Using GIS suitability analysis to identify potential future land use conflicts in North Central Florida. *J. Conserv. Plan.* **2005**, *1*, 89–105.
32. Zou, L.; Liu, Y.; Wang, J.; Yang, Y. An analysis of land use conflict potentials based on ecological-production-living function in the southeast coastal area of China. *Ecol. Indic.* **2021**, *122*, 107297. [CrossRef]
33. Miller, W.; Collins, M.G.; Steiner, F.R.; Cook, E. An approach for greenway suitability analysis. *Landsc. Urban Plan.* **1998**, *42*, 91–105. [CrossRef]
34. Rahaman, S.A.; Aruchamy, S. Land suitability evaluation of tea (*Camellia sinensis* L.) plantation in Kallar watershed of Nilgiri Bioreserve, India. *Geographies* **2022**, *2*, 701–723. [CrossRef]
35. Aymen, A.T.; Al-husban, Y.; Farhan, I. Land suitability evaluation for agricultural use using GIS and remote sensing techniques: The case study of Ma'an Governorate, Jordan. *Egypt. J. Remote Sens. Space Sci.* **2021**, *24*, 109–117.

36. Morales, F., Jr.; de Vries, W.T. Establishment of land use suitability mapping criteria using analytic hierarchy process (AHP) with practitioners and beneficiaries. *Land* **2021**, *10*, 235. [CrossRef]
37. Dong, G.; Ge, Y.; Jia, H.; Sun, C.; Pan, S. Land use multi-suitability, land resource scarcity and diversity of human needs: A new framework for land use conflict identification. *Land* **2021**, *10*, 1003. [CrossRef]
38. Azizi, A.; Malekmohammadi, B.; Jafari, H.R.; Nasiri, H.; Parsa, V.A. Land suitability assessment for wind power plant site selection using ANP-DEMATEL in a GIS environment: Case study of Ardabil province, Iran. *Environ. Monit. Assess.* **2014**, *186*, 6695–6709. [CrossRef]
39. Cheng, Z.; Zhang, Y.; Wang, L.; Wei, L.; Wu, X. An analysis of land-use conflict potential based on the perspective of production–living–ecological function. *Sustainability* **2022**, *14*, 5936. [CrossRef]
40. Hua, L.; Squires, V.R. Managing China's pastoral lands: Current problems and future prospects. *Land Use Policy* **2015**, *43*, 129–137. [CrossRef]
41. Huang, A.D.; Zhao, M.S.; Gao, M. Spatial-temporal evolution characteristics of land use in Anhui province from 1980 to 2020. *Sci. Technol. Eng.* **2022**, *22*, 4627–4635.
42. Zhou, D.; Xu, J.; Lin, Z. Conflict or coordination? Assessing land use multi-functionalization using production-living-ecology analysis. *Sci. Total Environ.* **2017**, *577*, 136–147. [CrossRef] [PubMed]
43. Cui, J.; Kong, X.; Chen, J.; Sun, J.; Zhu, Y. Spatially explicit evaluation and driving factor identification of land use conflict in yangtze river economic belt. *Land* **2021**, *10*, 43. [CrossRef]
44. Wang, C.; Wang, H.; Wu, J.; He, X.; Luo, K.; Yi, S. Identifying and warning against spatial conflicts of land use from an ecological environment perspective: A case study of the Ili River Valley, China. *J. Environ. Manag.* **2024**, *351*, 119757. [CrossRef]
45. Owusu, G.; Oteng-Ababio, M.; Afutu-Kotey, R.L. Conflicts and governance of landfills in a developing country city, Accra. *Landsc. Urban Plan.* **2012**, *104*, 105–113. [CrossRef]
46. Bernués, A.; Rodríguez-Ortega, T.; Alfnes, F.; Morten Clemetsen, M.; Eik, L.O. Quantifying the multifunctionality of fjord and mountain agriculture by means of sociocultural and economic valuation of ecosystem services. *Land Use Policy* **2015**, *48*, 170–178. [CrossRef]
47. Long, H. Theorizing land use transitions: A human geography perspective. *Habitat Int.* **2022**, *128*, 102669. [CrossRef]
48. Zhao, J.; Ji, G.; Tian, Y.; Chen, Y.; Wang, Z. Environmental vulnerability assessment for mainland China based on entropy method. *Ecol. Indic.* **2018**, *91*, 410–422. [CrossRef]
49. Jing, W.; Yu, K.; Wu, L.; Luo, P. Potential land use conflict identification based on improved multi-objective suitability evaluation. *Remote Sens.* **2021**, *13*, 2416. [CrossRef]
50. Chen, M.; Liu, W.; Lu, D. Challenges and the way forward in China's new-type urbanization. *Land Use Policy* **2016**, *55*, 334–339. [CrossRef]
51. Koellner, T.; Schröter, M.; Schulp, C.J.; Verburg, P.H. Global flows of ecosystem services. *Ecosyst. Serv.* **2018**, *31 Pt B*, 229–230. [CrossRef]

land

MDPI

Article

Achieving Sustainable Land Use Allocation in High-Altitude Area by 2030: Insights from Circle Structure and Scenario Predictions for Production–Living–Ecological Land in Xining Marginal Area, China

Zizhen Jiang [1,2,†], Yuxuan Luo [2,†], Qi Wen [3], Mingjie Shi [4], Ramamoorthy Ayyamperumal [2] and Meimei Wang [1,2,*]

1 Key Laboratory of Earth Surface System and Human-Earth Relations, Ministry of Natural Resources of China, Shenzhen 518055, China
2 College of Earth and Environmental Sciences, Lanzhou University, Lanzhou 730000, China
3 School of Architecture, Ningxia University, Yinchuan 750021, China; wenq98@163.com
4 College of Resources and Environment, Xinjiang Agricultural University, Urumqi 830052, China
* Correspondence: wangmm@lzu.edu.cn
† These authors contributed equally to this work.

Abstract: The paper focused on the Xining marginal area, providing the concept of land use competitive advantage, employing the PLUS and PFCI model to simulate production–living–ecological (PLE) land in 2030, and revealing the relationship between regional land interactions and sustainable land allocation. The results indicate that the following: (1) By 2030, the land use of the Xining marginal area is primarily production and ecological land, with approximately 0.1% of living land; however, living and production land will increase while ecological land will decrease in general, and the growth momentum of urban and other living space in government-seated regions is stronger. (2) The PLE land does not exhibit a piecemeal expansion pattern, as it is influenced by mountains and rivers. Agricultural production land and grassland ecological land have advantages for development, whereas urban living land has just development potential. (3) Developing the corresponding lands in the dominant regions can result in sustainable land allocation, and five nexus approaches are proposed for the sustainable allocation of PLE land in the Xining marginal region. The study addresses the interaction of different land use types across regions rather than examining them separately, and we provide significant insight into whether the Qinghai Tibet Plateau should be urbanized.

Keywords: patch-generating land use simulation (PLUS) model; Pythagorean fuzzy conflict information (PFCI); production–living–ecological (PLE) land; sustainable development; land use allocation; Xining marginal area

1. Introduction

Sustainable development has become a common concern in metropolitan and remote areas due to the continual growth of urbanization and the stronger use of space resources [1]. It has become an important global issue as all nations strive to coordinate their economies, societies, and environments [2]. The sustainable development goals (SDGs) were released by the UN in 2015 with the purpose of helping nations develop sustainable development plans [3]. In fact, land use plays a crucial role in achieving the SDGs, as the two are in good agreement and the evolution of land use has a deep impact on the realization of SDGs [4]. As a carbon source or sink, land use serves as one of the leading factors responsible for global climate change (SDG 13). Besides, it also acts as a pollution source and sink, and thereby has a great impact on ecological protection (SDG 14–15). Land is the main carrier of food and plays a role that cannot be underestimated in ensuring food security (SDG 2) [5].

The Qinghai–Tibet Plateau is one of the largest carbon sequestration regions in China and an important control area for global atmospheric circulation and water cycle. The protection and allocation of its land resources will contribute to the mitigation of global climate issues and the achievement of SDGs [6]. Xining, as the locomotive driving the development of the Qinghai–Tibet Plateau, has witnessed a gradual acceleration in its urbanization process, leading to an increased demand for the utilization of spatial resources. In response, research proposes the concept of urban marginal areas, which refers to the transition area between the city and the countryside as well as the sub-center or new urban area [7], and it is also the area most affected by urban elements as the main space for population and land use growth [8]. Marginal area faces multiple challenges such as the daunting task of coordinating land use with construction and ecological protection, where different spatial structures and multiple land use types intertwine in a subtle manner, with a mixture of rural, suburban, industrial parks. wholesale markets and undeveloped interstitial spaces, presenting a situation of uncontrolled living space, disorderly production space, and imbalanced ecological space. Therefore, the Chinese government has proposed production–living–ecological (PLE) land as the basic framework for land use planning and management to achieve sustainable social, economic, and ecological development [9]. The conflicts between PLE land essentially manifest as competition and conflict between spatial resources for different purposes and functions within the same area, resulting from the interaction of human–environment relationships. Inconsistency between PLE land exacerbates the contradictions between economic development, ecological protection, and cultivated land conservation, seriously threatening the sustainable use of land. Will the Xining marginal area embark on the way of rapid urbanization? How to optimize the allocation of land use patterns to achieve social, economic, and ecological balance and stability? These have become key issues to be solved urgently for sustainable development in the Xining marginal area.

Scholars have conducted in-depth studies on the impacts of land use change or allocation on sustainable development, with the efforts mainly on risk identification and spatial optimization. From the perspective of risk identification, some scholars have engaged in ecological risk analysis based on future land use projections [10], and the assessment of land use policies on sustainable development [11]. From the perspective of spatial optimization, land use optimization is based on the theory of comparative advantage, which adjusts and optimizes the proportion of land types under certain constraints to improve the overall benefits of land use [12]. Scholars have studied the optimal allocation of land use at different scales such as global [13], continental [14], national [15], regional [16], watershed [17], and farm [18], as well as single ecosystems such as agriculture [19] and urban [20]. Mathematical models are the frontiers of land use optimization, which can be divided into two categories: quantity structure optimization models and spatial distribution optimization models. Firstly, due to the complex non-linear relationship between PLE land and driving factors, it is hard to construct an optimization model with traditional quantity structure methods. Secondly, spatial distribution optimization models mainly solve the spatial configuration problem of land use to achieve optimal suitability, such as the CLUE-S model [21], Cellular automata (CA) [22], multi-agent systems (MAS) [23], and PLUS (patch-generating land use simulation) model [24]. With an adaptive inertia mechanism and a roulette wheel selection mechanism, the PLUS model improves simulation accuracy based on the FLUS model [25]. In addition, it also offers the development probability of each category with the Random Forest Algorithm, making up for the deficiency of the CA-Markov model in exploring the change law of land use, and overcoming the difficulty in dynamically modeling the patch-level changes in natural land types in time and land. However, all the land types interact with each other and affect all the SDGs, but each is often studied in isolation. On the one hand, developing a land or achieving a sub-goal may result in the impossibility to balance other lands or achieve other sub-goals [26]. On the other hand, different entities or regions have different development objectives and different spatial trends, and stakeholders may contradict each other in achieving policy objectives in different regions [27,28]. Therefore, it is imperative to take the interests and endowments of different regions into account for integrated allocation, and to consider interactions among more sectors, across scales, and between adjacent and distant places, and linkages with SDGs [26].

In this article, we emphasize nexus approaches [29] to understand the connections and synergies on spatial distribution optimization, and propose the concept of land use competitive advantages, i.e., the dominance of a region in developing a certain land use type. Correspondingly, we propose Pythagorean fuzzy conflict information (PFCI), a combination of the Pythagorean theorem and fuzzy theory, to calculate competitive advantages. As the mismatch between one-size-fits-all policies and complicated regional relationships violates the principle of nexus land use allocation, this article aims to address the above problem from the perspective of interactions between different land types and different regions, rather than considering them in isolation. Given the particularity of the Xining marginal area and sustainable development goals, we set up four scenarios, namely, natural development, urban development (SDG8, economic development), cultivated land conservation (SDG2, Zero Hunger), and ecological protection (SDG13, Climate action) to simulate the 2030 spatial pattern of PLE land in the Xining marginal area by the PLUS model. Then, we further propose the concept of land use competitive advantages and improve PFCI by introducing concepts such as the maximum doughty coalition, the trisections of four types of sets, and viable strategies. Based on this, we identify sets of advantageous land use types and regions. Ultimately, we put forward policy references to achieve nexus approaches to land use sustainable allocation based on regional endowments. In summary, our research aims to answer the following questions:

(1) Will the Xining marginal area embark on the path of rapid urbanization?

(2) How do the advantages of certain land use types and regional endowments reveal the sustainable allocation of land use?

(3) What nexus approaches should we take to achieve sustainable land use allocation in the Xining marginal area?

2. Materials and Methods

2.1. Study Area

The Xining marginal area is located in Qinghai Province, including three administrative regions: Haibei Tibetan Autonomous Prefecture (Menyuan Hui Autonomous County, Qilian County, Haiyan County, and Gangcha County), Hainan Tibetan Autonomous Prefecture (Gonghe County, Tongde County, Guide County, Xinghai County, and Guinan County), and Huangnan Tibetan Autonomous Prefecture (Tongren City, Jianzha County, Zeku County, and Henan Mongol Autonomous County). Among them, Haiyan County, Gonghe County, and Tongren City are government-seated regions (Figure 1).

2.2. Data Sources

The data in this paper consist of a land dataset, a socio-economic dataset, and a natural dataset.

The input datasets for the PLUS model are the 30 m land use for the years 2015 and 2020, and they have been reclassified as production–living–ecological (PLE) land, which is based on their dominant and secondary functions [30–33]. Ecological land refers to the land that maintains human survival, including climate regulation, water regulation, the mitigation of emergencies, and soil conservation. Therefore, its evolution is an effective way to reveal the implementation of SDG 13 (Climate action). Production land refers to the land where products and services can be directly obtained as labor objects or produced. It functions as a carrier for social production, including food supply, raw material production, energy, and mineral production [31]. Therefore, its evolution is an effective way to reveal the implementation of SDG 2 (Zero Hunger). Living land is primarily aimed at human production, leisure, and living. While providing resources and giving impetus to economic growth, it also creates a comfortable living and entertainment environment [21]. Therefore, its evolution is an effective way to reveal the implementation of SDG 8 (economic growth). It should be noted that there are no paddy fields and beaches in the study area (Table 1).

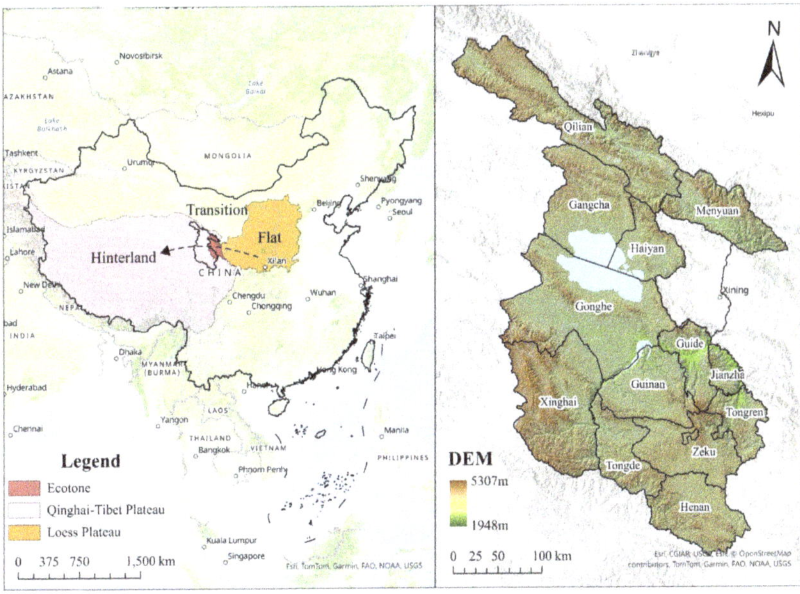

Figure 1. Position of Xining marginal area.

Table 1. Classification of lands and their sub-lands.

Land	Sub-Land	Corresponding Land Use Type
Living land	Urban living land Rural living land Other living land	Urban built-up land Rural residential land Other built-up land
Production land	Agricultural production land	Dry land; Canal
Ecological land	Forest ecological land Grassland ecological land Water ecological land Other living land	Forest; shrub land; wood land; other forest High grassland; mid grassland; low grassland Lake; reservoir-pond; snow; shallow Sand; gobi; saline; swamp; barren land Rock; others

The process of land cover change in the Xining marginal area has a fundamentally different driving mechanism compared to inland regions. Firstly, there is an active adaptation of human activities and urban construction to the fragile ecological environment of the high-altitude, cold, and oxygen-deficient plateau. In this active adaptation process, elevation serves as the primary constraint on the expansion of human activities, with most of the development of the plateau towns and agricultural activities limited to the river valleys. As an important evaluation criterion for urban development and cultivated land conservation, the slope profoundly affects the evolution of built-up land and cultivated land. Therefore, among the natural driving factors, we have selected these two indicators. Secondly, there are progressive processes, as well as external driving processes. The progressive process mainly considers the inertia of development based on the existing towns, as the scale of towns in the Xining marginal area is generally small. New construction land is primarily concentrated around the existing towns where the population density and industrial activities are relatively vibrant. Moreover, the driving force for urbanization in the Qinghai–Tibet Plateau is mainly top-down government-led initiatives. Thus, we have chosen population density, GDP, and government locations as the indicators for such processes. The external driving process primarily includes the impact of tourism and targeted assistance, both of which depend on the transportation infrastructure conditions. Additionally, when operating the PLUS model, it is essential to consider the volume of data. After multiple attempts and assessments of the factors' driving force, we ultimately selected "distance to primary roads" as the indicator for transportation as an external driving force (Table 2).

Table 2. Data sources.

Data	Sub-Data	Year(s)	Resolution	Sources
Land use dataset	PLE land classification in Table 1	2015, 2020	30 m	https://www.resdc.cn/ accessed on 4 August 2022
Socio-economic dataset	Population	2020	1 km	https://www.resdc.cn/ accessed on 4 August 2022
	GDP	2020	1 km	https://www.resdc.cn/ accessed on 4 August 2022
	Primary roads	2020	Vector data	Open Street Map
	Seat of county government	2015	Vector data	http://www.dsac.cn/ accessed on 4 August 2022
Natural dataset	Elevation	2015	1 km	https://www.resdc.cn/ accessed on 4 August 2022
	Slope	2015	1 km	https://www.resdc.cn/ accessed on 4 August 2022

2.3. *Patch-Generating Land Use Simulation (PLUS) Model*

The PLUS (patch-generating land use simulation) model was proposed by Liang et al. [24] for simulating land use change. We set up a natural growth scenario (scenario A) and urban development scenario (scenario B) in this paper for simulating the land use pattern of the Xining marginal area according to the existing trend and rapid urban development. In addition, based on the previous studies [34], we set up the sustainable development goals and policies currently implemented in China for the "red line of ecological protection, the red line of cultivated land, and the urban development boundary", the cultivated land conservation scenario (scenario C), and the ecological protection scenario (scenario D), and the transition matrix of each scenario is shown in Table 3.

Table 3. Transition matrix in PLUS.

	Scenario A						Scenario B						Scenario C						Scenario D					
	a	b	c	d	e	f	a	b	c	d	e	f	a	b	c	d	e	f	a	b	c	d	e	f
a	1	0	1	0	1	1	1	0	0	0	1	1	1	0	0	0	0	0	1	1	1	1	1	1
b	0	1	1	0	0	1	1	1	0	0	1	1	1	1	1	0	0	1	0	1	0	0	0	0
c	1	1	1	1	1	1	1	0	1	0	1	0	1	1	1	1	1	1	0	1	1	1	0	0
d	1	0	1	1	0	1	0	1	1	1	1	0	1	0	1	1	0	1	0	0	0	1	0	0
e	0	0	0	0	1	0	0	0	0	0	1	0	0	0	0	0	1	0	0	0	0	0	1	0
f	1	1	1	1	1	1	1	1	1	1	1	1	1	1	1	1	1	1	1	1	1	1	1	1

Notes: (1) According to the classification system of land use monitoring using remote sensing of the Chinese Academy of Sciences, we combined the land use types into cultivated land, forest, grass, water, built-up land, and bare land in the transition matrix. In the table, a represents cultivated land (including dry land), b represents forest (including forest, shrub land, wood land, and other forest), c represents grass (including high grassland, mid grassland, and low grassland), d represents water (including canal, lake, reservoir-pond, snow, and shallow), e represents built-up land (including urban built-up land, rural built-up land, and other built-up land), and f represents bare land (including sand, gobi, saline, swamp, barren land, rock, and others). (2) The 0 means conversion to another land type is prohibited, and 1 means conversion to another land type is allowed.

The neighborhood weight parameter indicates the strength of the expansion capacity of the land type. According to Wang et al. [35], the change in TA (Total Area) of each land type at the same time scale can better reflect its expansion intensity. The dimensionless value of TA change conforms to the parameter requirements of the model neighborhood weight in terms of data meaning and data structure. Therefore, the calculation of the PLUS model neighborhood weight in this study is as follows:

$$W = \frac{\Delta TA_i - \Delta TA_{min}}{\Delta TA_{max} - \Delta TA_{min}}$$

where ΔTA_i represents the amount of TA change in the land type; ΔTA_{min} represents the land type with the smallest amount of change; ΔTA_{max} represents the land type with the largest amount of change. Based on the TA changes in all the land use types in the Xining marginal area from 2015 to 2020, we calculated the neighborhood weight of each type as shown in Table 4.

Table 4. Weight of each land use type according to ΔTA.

Variety	Dry Land	Forest	Shrub Land	Wood Land	Other Forest	High Grassland
ΔTA Weight	−193.77 0.61	35.46 0.62	−1743.39 0.59	−150.84 0.62	9.27 0.62	22,166.10 1
Variety	Mid grassland	Low grassland	Canal	Lake	Reservoir-pond	Snow
ΔTA Weight	−9190.26 0.46	−35,812.26 0	−1224 0.60	11,317.32 0.81	6207.93 0.72	1429.20 0.64
Variety	Shallow	Urban built-up land	Rural residential land	Other built-up land	Sand	Gobi
ΔTA Weight	−74.34 0.62	827.55 0.63	519.75 0.63	15,283.26 0.88	−8804.97 0.47	18.09 0.62
Variety	Saline	Swamp	Barren land	Rock	Others	
ΔTA Weight	101.97 0.62	−1710.72 0.59	51.30 0.62	943.38 0.63	−6.03 0.62	

2.4. Construction of Pythagorean Fuzzy Conflict Information (PFCI) Model

Pythagorean fuzzy conflict information (PFCI) is an effective tool for modeling real-world decision-making problems involving information uncertainty. It originated from the fuzzy sets (FSs) proposed by Zadeh [36]; Yager [37] proposed PFSs in 2014. Afterwards, he introduced q-ROFSs to retain more imprecise information [37]. Looser restrictions allow for more flexible application of q-ROFSs, which can be used to solve land use conflicts. By this, the calculation of competitive advantages is possible by defining the conflict distance and conflict function to describe the uncertainty in the conflict, and then calculating and ranking the scoring function based on the feasible strategies and rough set theory. It is possible to find the internal causes of the conflict and find a feasible solution. The calculation process is as follows:

Definition: A q-ROFS B in terms of a finite universal set X is defined as follows:

$$\mathcal{B} = \{\langle x, \mu_b(x), v_b(x)\rangle | x \in X\} \quad (1)$$

where $\mu_b(x)$ and $v_b(x)$ represent the degree of the membership and non-membership of element x with respect to the set $\mathcal{B}$; simultaneously, $\mu_b(x)$ and $v_b(x)$ satisfy the restriction $q \geq 1$, $0 \leq (\mu_b(x))^q \leq 1$, $0 \leq (v_b(x))^q \leq 1$ and $0 \leq (\mu_b(x))^q + (v_b(x))^q \leq 1$, and $\langle \mu_b(x), v_b(x)\rangle$ is a q-ROF number (q-ROFN) denoted by $\rho = \langle \mu_b, v_b\rangle$. In addition, $r_b(x) = \sqrt[q]{(\mu_b(x))^q + (v_b(x))^q}$ represents the degree of confidence, reflecting the strength of commitment, and its influence is related to the angle $\theta_b(x)$ between $r_b(x)$ and $\mu_b(x)$; $d_b(x) = 1 - \frac{2\theta_b(x)}{\pi} \in [0,1]$ indicates the direction of the confidence. Assume $\vec{\chi_{ij}} = (\chi_{ij1}, \chi_{ij2}, \chi_{ij3}, \chi_{ij4}) = (\mu_{ij}{}^q, v_{ij}{}^q, r_{ij}{}^q, d_{ij})$ represents the q-ROF attitudes of agent u_i toward competitive advantage a_j. Then, one can see that the four parameters of $\vec{\chi_{ij}}$ can completely describe q-ROFSs. In addition, the q-ROF number can effectively model support, opposition, and neutral components in real-world conflict problems by dividing them into three levels of conflict relationships. Therefore, in a known conflict information system, the specific steps to determine the degree of conflict in each competitive advantage and find feasible strategies are as follows:

Input: Agent set $L = \{l_1, l_2, \cdots, l_n\}$, competitive advantage set $B = \{b_1, b_2, \cdots, b_m\}$, parameter q, and thresholds $(\varsigma_\star, \varsigma^\star)$ and $(\varsigma_\diamond, \varsigma^\diamond)$.

Initialization: Aggregation function $A(x,y) = O_{mM}(x,y) = \min(x,y)\max(x^2,y^2)$.

Step 1: By combining the Hamming distance with the four parameters of $\vec{\chi_{ij}}$, the absolute conflict distance between the two agents l_i and l_k regarding the competitive advantage b_j is calculated by the following formula:

$$\zeta^1_{b_j}(l_i, l_k) = \frac{1}{4}\sum_{h=1}^{4}\left|\chi_{ijh} - \chi_{kjh}\right| = \frac{1}{4}\left(\left|\mu^q_{ij} - \mu^q_{kj}\right| + \left|v^q_{ij} - v^q_{kj}\right| + \left|r^q_{ij} - r^q_{kj}\right| + \left|d_{ij} - d_{kj}\right|\right) \quad (2)$$

Step 2: Based on the dual operation of similarity measure and distance, the relative conflict distance between the two agents l_i and l_k regarding the competitive advantage b_j is calculated by the following formula:

$$\zeta^2_{b_j}(l_i, l_k) = 1 - \frac{\vec{\chi_{ij}} \cdot \vec{\chi_{kj}}}{max\left(\left|\vec{\chi_{ij}}\right|^2, \left|\vec{\chi_{kj}}\right|^2\right)} \quad (3)$$

where $\cdot$ represents the scalar product of vectors and $|\cdot|$ represents the norm of vectors.

Step 3: The comprehensive conflict distance between the two agents l_i and l_k regarding the competitive advantage b_j is calculated by the following formula:

$$\zeta_{b_j}(l_i, l_k) = A\left(\zeta^1_{b_j}(l_i, l_k), \zeta^2_{b_j}(l_i, l_k)\right) \quad (4)$$

Step 4: ω_j is used to calculate the weight of the competitive advantage b_j, and the formula $CF_B(l_i, l_k)$ is used to calculate the conflict function of the two agents l_i and l_k regarding the competitive advantage set B:

$$\omega_j = \frac{\sum_{i=1}^{n}\sum_{k=1}^{n}\zeta_{b_j}(l_i, l_k)}{\sum_{j=1}^{m}\sum_{i=1}^{n}\sum_{k=1}^{n}\zeta_{b_j}(l_i, l_k)} \quad (5)$$

$$CF_B(l_i, l_k) = \sum_{j=1}^{m}\omega_j\zeta_{b_j}(l_i, l_k) \quad (6)$$

Step 5: The conflict degree of each agent regarding the competitive advantage b is calculated by the following formula:

$$CB(b) = \frac{\sum_{x,y \in L, x \neq y}\zeta_b(x,y)}{|L|(|L|-1)} \quad (7)$$

Step 6: In order to show the complexity of the conflicts among the agents and facilitate the search for alliances, the strong-, weak-, and non-conflict sets of the agent x with regard to the single competitive advantage b are defined as follows:

$$\mathbb{DR}^{\varsigma_\star, \varsigma^\star}_b(x) = \left\{y \in L \middle| \zeta_{b_j}(x,y) \geq \varsigma^\star\right\} \quad (8)$$

$$\mathbb{WR}^{\varsigma_\star, \varsigma^\star}_b(x) = \left\{y \in L \middle| \varsigma_\star < \zeta_{b_j}(x,y) < \varsigma^\star\right\} \quad (9)$$

$$\mathbb{NR}^{\varsigma_\star, \varsigma^\star}_b(x) = \left\{y \in L \middle| \zeta_{b_j}(x,y) \leq \varsigma_\star\right\} \quad (10)$$

Step 7: In order to better understand and analyze competitive advantages, the strong-, weak-, and non-competitive advantage sets of the agent x with regard to the single competitive advantage b are defined as follows:

$$\mathbb{DB}_B^{\varsigma_\diamond,\varsigma^\diamond}(x) = \{b \in B | CB(b) \geq \varsigma^\diamond\} \tag{11}$$

$$\mathbb{WB}_B^{\varsigma_\diamond,\varsigma^\diamond}(x) = \{b \in B | \varsigma_\diamond < CB(b) < \varsigma^\diamond\} \tag{12}$$

$$\mathbb{NB}_B^{\varsigma_\diamond,\varsigma^\diamond}(x) = \{b \in B | CB(b) \leq \varsigma_\diamond\} \tag{13}$$

Output: The comprehensive conflict distance ζ_{b_j}, the conflict function CF_B, a three-level conflict set, and a three-level competitive advantage set.

The agents' attitudes towards all the competitive advantages constitute a conflict situation. Conflict strategy is an important factor affecting the evolution of conflict situations, assuming that the best strategy can maximize the relative advantage of a single competitive advantage, and the agents can achieve the most valuable resource plunder based on this advantage. On this basis, the division of three levels of competitive advantage sets showcases the conflict advantage levels of each competitive advantage, achieving intuitive conflict analysis. The division of the three levels of conflict sets showcases the degree of conflict among each agent regarding a single competitive advantage and determines feasible strategies for selecting the most suitable agent to handle the corresponding competitive advantage. The overall framework of the paper is shown in Figure 2.

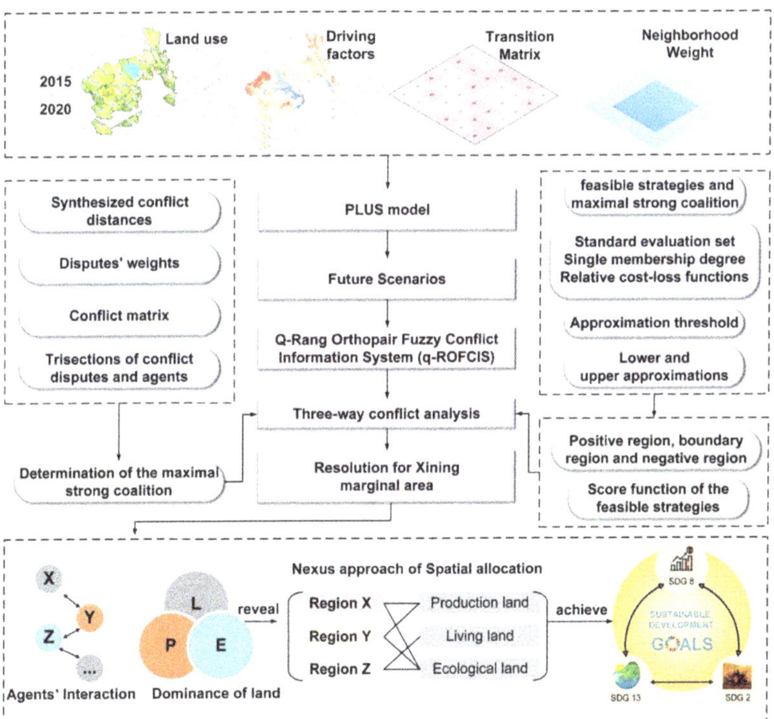

Figure 2. Flowchart of the research.

3. Scenario Prediction on PLE Land in Xining Marginal Area

3.1. Scenario Prediction on PLE Land

PLE land in 2030 is generally similar to that in 2020 under the scenarios of natural growth (A) and cultivated land conservation (C). Production land is the smallest in the urban development scenario (B). Under the ecological protection scenario (D), the production and ecological land are the largest, while the living land is the smallest compared with other scenarios. Under the natural growth scenario, the production and living land increase slightly, while the ecological land decreases significantly, indicating human land use expansion (Table 5).

Table 5. Comparison of PLE land each year/scenario (Unit: km^2).

	2015	2020	2030A	2030B	2030C	2030D
Production	3493.3536	3479.1759	3491.1495	3461.9418	3491.154	3501.3375
Ecological	92,416.092	92,263.964	92,125.742	92,156.133	92,125.723	92,233.14
Living	190.5417	356.8473	482.0958	481.9131	483.111	365.5098

Under the urban development scenario, the production and ecological land decrease slightly, whereas the living land increases significantly. Under the cultivated land conservation scenario, the production land expands slightly, ecological land decreases slightly, and living land grows similarly to the first two scenarios, achieving the goal of protecting farmland. Under the ecological protection scenario, the production land increases slightly, while the living land increases less than in the previous three scenarios, and the ecological land loses slightly, but less than in 2020 (Table 5).

3.2. Scenario Prediction on PLE Sub-Land

In 2015, the Xining marginal area had 3493.35 km^2 of agricultural production land, 11,520.62 km^2 of forest ecological land, 59,935.19 km^2 of grassland ecological land, 6009.86 km^2 of water ecological land, 34.22 km^2 of urban living land, 128.65 km^2 of rural living land, 27.68 km^2 of other living land, and 14,950.42 km^2 of other ecological land. In contrast, the Xining marginal area in 2020 had 3479.18 km^2 (99.59% in 2015) of agricultural production land, 11,502.13 km^2 (99.84% in 2015) of forest ecological land, 59,706.83 km^2 (99.62% in 2015) of grassland ecological land, (103.14% in 2015) of water ecological land, 42.49 km^2 (124.19% in 2015) of urban living land, 133.85 km^2 (104.04% in 2015) of rural living land area, 180.51 km^2 (652.17% in 2015) of other living land, and 14,856.35 km^2 (99.37% in 2015) of other ecological land, as shown in Figure 3.

We simulated the evolution of the ecological–production–living land in 2030 by PLUS, and the ecological–production–living land in natural growth scenario (A), urban development scenario (B), cultivated land conservation scenario (C), and ecological protection scenario (D) is shown in Figure 4.

The natural growth scenario (scenario A) indicates a trend toward decreasing ecological land as compared to 2020, with the exception of an increase in the water ecological land in the counties of Haiyan, Gangcha, and Gonghe. While the other living land in Gonghe County has increased significantly, the living land and production land have only slightly increased. In the Xining marginal area, scenario A corresponds with an increase in land use without external interference.

The urban development scenario (scenario B) suggests that the production land decreases, with the highest decrease in Henan County, while the living land will expand overall, with Gonghe County experiencing the fastest growth in the other living land. Regarding the ecological land, the grassland ecological land in Haiyan, Gonghe, and Guinan Counties will be significantly decreased, as will the other ecological land also reduce. The water ecological land changes little; Haiyan County and Gonghe County have increased, whereas the forest ecological land will increase in general. The widespread expansion of the living land demonstrates the prioritization of urban development, which to some extent occupies the grassland ecological land.

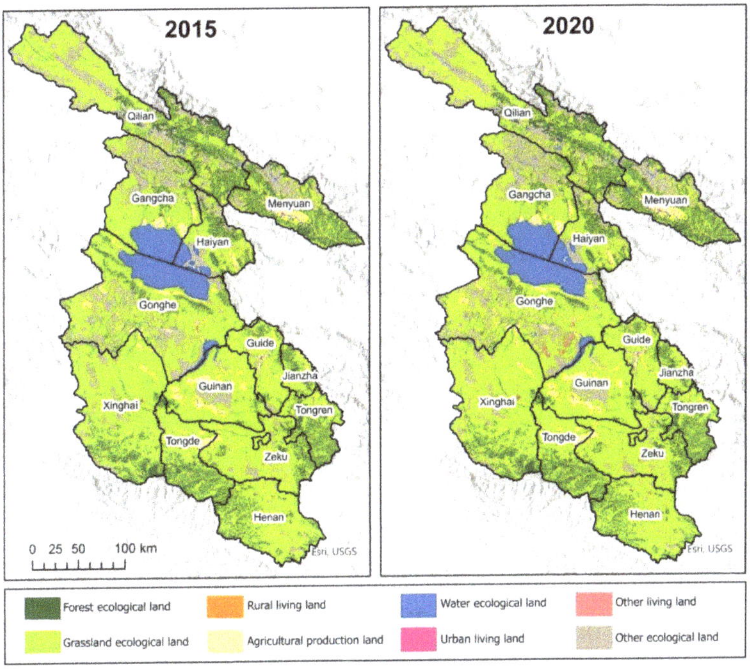

Figure 3. Distribution of different lands in 2015 and 2020.

Under the cultivated land conservation scenario (scenario C), the production land will grow minimally, whereas the agricultural production land in Gangcha and Gonghe Counties will grow more rapidly. The living land will also increase, with most of it occurring in Gonghe County. Except for the water ecological land, which will tend to increase, the remaining ecological land will decrease significantly. In this scenario, the goal of the cultivated land conservation is still not entirely achieved. In the ecological protection scenario (scenario D), the production land will grow minimally while the living land will remain largely unchanged. The grassland ecological land and other ecological land will decrease, while the forest ecological land and water ecological land will increase from 2020. Minor changes in the production and living lands, as well as the transition from the grassland ecological land and other ecological land to forest ecological land and water ecological land, demonstrate that conservation will on the top of the agenda in this scenario (Table 6).

To summarize, the agricultural production land is predicted to increase by 2030, except for the urban growth scenarios. Except for the modest growth in the ecological protection scenario, the rural living land will increase at nearly the same rate (7.44%) in all the other scenarios. Except for the ecological protection scenario, all the scenarios predict an increase in the urban living land, with the urban development scenario showing the most significant growth rates. The other living land will also see significant expansion, with the ecological protection scenario indicating a slight increase and all the other scenarios exhibiting increases of about 60%. The forest ecological land will decrease slightly in the natural growth and cultivated land conservation scenarios while increasing in the urban development and ecological protection scenarios. The water ecological land is predicted to grow in all the scenarios, particularly in the natural growth and cultivated land conservation scenarios. All four scenarios will result in a decrease in the grassland ecological land, with urban development having the greatest effect. In all the scenarios, there will be decreases in the other ecological land, with the largest decrease in the ecological protection scenario.

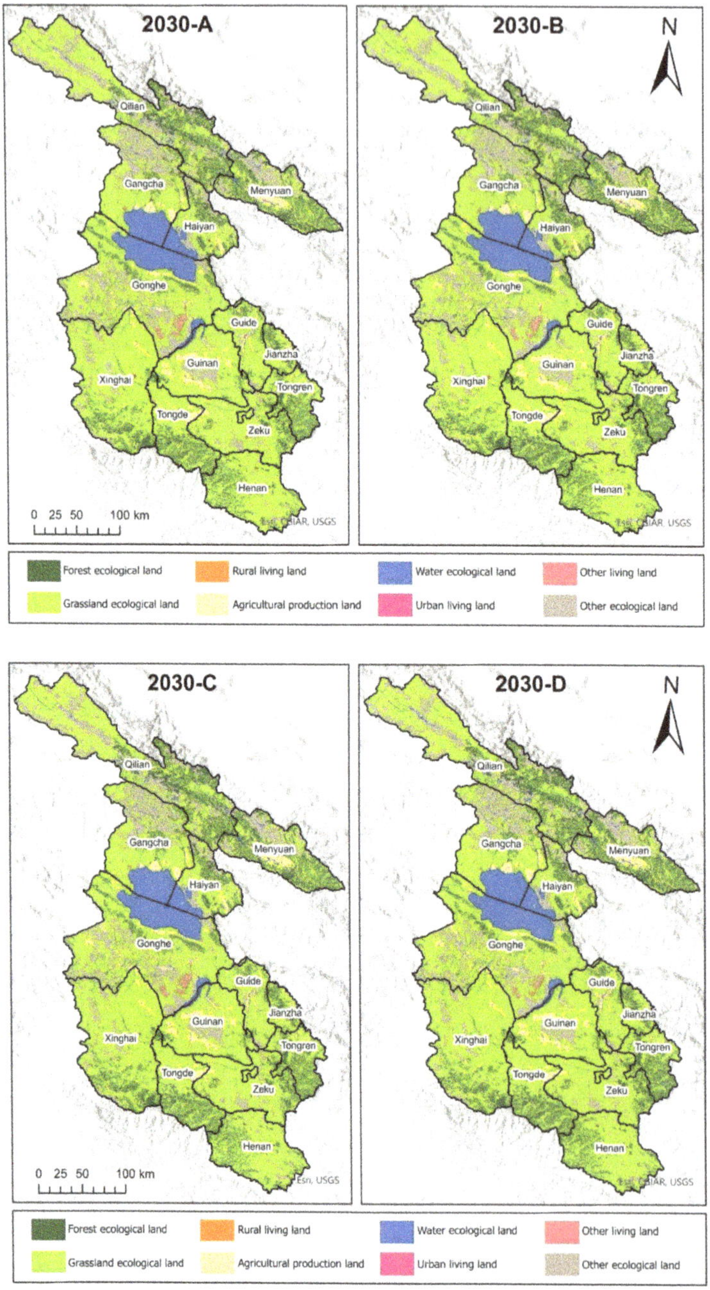

Figure 4. Distribution of different lands under 2030 scenarios.

Table 6. Area and growth rate of each land use type compared to 2020.

	2015	2020	2030-A		2030-B		2030-C		2030-D	
	Area	Area	Area	Rate	Area	Rate	Area	Rate	Area	Rate
b1	3493.35	3479.18	3491.15	0.34	3461.94	−0.50	3491.15	0.34	3501.34	0.64
b2	128.65	133.85	143.80	7.44	143.80	7.44	143.80	7.44	134.19	0.26
b3	34.32	42.49	49.22	15.84	51.48	21.15	49.28	15.97	41.97	−1.23
b4	11,520.62	11,502.13	11,471.60	−0.27	11,935.99	3.77	11,471.60	−0.27	11,957.16	3.96
b5	6009.86	6198.66	6331.42	2.14	6265.49	1.08	6330.24	2.12	6303.99	1.70
b6	59,935.19	59,706.83	59,504.39	−0.34	59,313.26	−0.66	59,504.63	−0.34	59,458.24	−0.42
b7	14,950.42	14,856.35	14,819.34	−0.25	14,641.39	−1.45	14,819.26	−0.25	14,513.76	−2.31
b8	27.68	180.51	289.08	60.14	286.64	58.79	290.04	60.67	189.35	4.90

Note: b1 refers to the agricultural production land, b2 refers to the rural living land, b3 refers to the urban living land, b4 refers to the forest ecological land, b5 refers to the water ecological land, b6 refers to the grassland ecological land, b7 refers to the other ecological land, and b8 refers to the other living land. The same applies below.

3.3. Overall Pattern of Scenario Prediction on PLE Land in Each County

In the 2015, 2020, and 2030 scenarios, the land use in Xining's marginal area is primarily production and ecological land, with a relatively low proportion of living land (about 0.1%). The grassland ecological land has the largest proportion, covering an area around 7–8 times that of production land, followed by the forest ecological land, which covers an area of 1.5–2 times that of production land. Specifically: (1) Forest ecological land is mainly distributed in Menyuan County, Tongren City, Jianzha County, and Tongde County, accounting for 20–22% of the total land use area of the counties. (2) Water ecological land is distributed relatively high in Haiyan County, Gangcha County, and Gonghe County. (3) Grassland ecological land accounts for a large proportion in all 14 counties, with the smallest proportion in Menyuan County, ranging from 38.14% to 39.37%. Except for Menyuan and Haiyan Counties, the proportion in all the other counties exceeds 50% of the total land use area, with Henan County having the highest at 79.01–79.77%. (4) Menyuan County, Qilian County, Haiyan County, Gangcha County, and Gonghe County are the counties with the highest proportions of other ecological land. (5) Menyuan, Tongde, Guide, and Guinan Counties contribute 8% to 12% of agricultural production land. (6) The proportions of rural living land, urban living land, and other living land in the 14 counties are quite low, with a total of approximately 1%, with only other living land in Gonghe County accounting for more than 0.1%. Figure 5 depicts the pattern of scenario predicted for each county's PLE land in 2015, 2020, and 2030.

3.4. Core–Edge Mode on PLE Land in Xining Marginal Area

The Xining marginal area shows an evident core–edge pattern. We investigate the government seat as the core area (Haiyan County, Gonghe County, and Tongren City), and the remaining counties as external areas. The living land will increase overall, particularly the urban living land and other living land in the core areas, while the rural living land will see a similar growth rate. The agricultural production land will not grow significantly and may perhaps decrease in both the core and edge areas under scenario B. The grass ecological land and other ecological land may diminish more in the core areas under all the situations. Under scenarios A and C, the forest ecological land will decrease more in the core areas while under scenarios B and D, the core areas will witness more increase. The water ecological land will slightly increase in both the core and edge areas (Table 7).

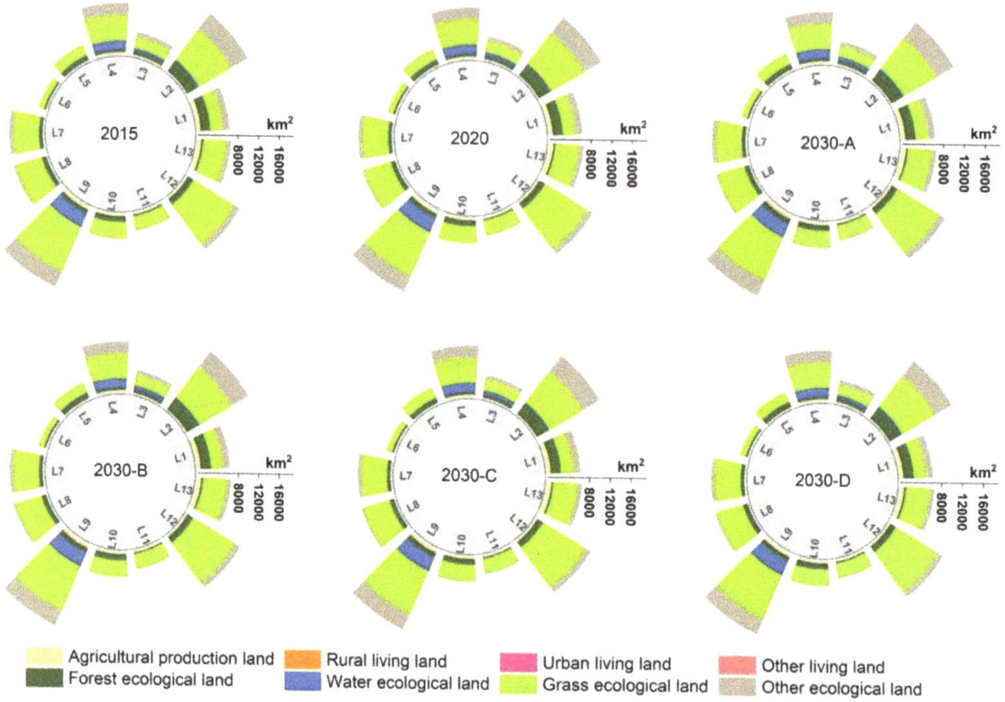

Figure 5. Patterns of scenario prediction on PLE land of each county. Note: scenario A means natural growth, scenario B means urban development, scenario C means cultivated land conservation, and scenario D means ecological protection.

Table 7. Growth rate and core–edge mode of sub-land (%).

	2030-A		2030-B		2030-C		2030-D		Conclusion
	Core	Edge	Core	Edge	Core	Edge	Core	Edge	
b1	0.455	0.316	−0.122	−0.591	0.454	0.316	0.439	0.688	all slightly increase except scenario B
b2	6.880	7.668	6.318	7.901	6.917	7.652	0.216	0.274	increase overall except scenario D
b3	36.658	8.292	43.500	13.050	36.705	8.457	0.016	−1.682	more growth in the core area
b4	−0.310	−0.254	5.504	3.316	−0.307	−0.255	5.535	3.541	varying according to the scenario
b5	2.745	1.352	1.808	0.123	2.715	1.348	1.272	2.258	all slightly increase
b6	−1.097	−0.120	−1.098	−0.532	−1.091	−0.121	−0.524	−0.385	all slightly decrease
b7	−1.109	0.082	−3.485	−0.663	−1.107	0.080	−2.878	−2.086	more decrease in the core area
b8	63.892	44.063	60.806	50.150	63.937	46.679	5.809	0.992	more growth in the core area

4. Spatial Optimization Strategies of PLE Land in Xining Marginal Area

4.1. Competitive Advantage Sets of PLE Land Under Each Scenario

The PFCI results trivialize the collection of competitive advantages in the sub-land in each scenario. Furthermore, the division of strong, weak, and non-competitive advantage sets influences the future direction of land use in the Xining marginal area (Table 8).

The significant competitive advantage set consists mainly of the grassland ecological land (b6) and other ecological land (b7). This finding corresponds with the natural characteristics and endowments of the Xining marginal area; however, the other ecological land (b7) has a relatively low development value. As a result, focus should be on developing the grassland ecological land (b6).

Table 8. Competitive advantage sets of PLE land in different scenarios.

	Scenario A	Scenario B	Scenario C	Scenario D
Strong competitive set	{b6, b7}	{b6}	{b6, b7}	{b6, b7}
Weak competitive set	{b1, b5, b8}	{b1, b5, b7, b8}	{b1, b5, b8}	{b1, b2, b3, b5, b8}
Noncompetitive set	{b2, b3, b4}	{b2, b3, b4}	{b2, b3, b4}	{b4}

Note: Ranges of strong-, weak-, and non-conflict are [0.09, 1], [0.08, 0.09), and [0, 0.08), respectively. Scenario A means natural growth, scenario B means urban development, scenario C means cultivated land conservation, and scenario D means ecological protection.

The weak competitive advantage set mainly includes the agricultural production land (b1), water ecological land (b5), and other living land (b8). The agricultural production land is critical for guaranteeing regional food security while the water ecological land is basically invariable. The other living land (b8) is construction land, which includes road land, land for other units, reserved natural villages or unbuildable land, and so on. They have no development advantages in high-altitude, sparsely populated, and underdeveloped areas.

The non-advantage set mainly includes the rural living land (b2), urban living land (b3), and forest ecological land (b4), as the proportion of these types of land is relatively small, and they have no development advantages over the other land use types in the Xining marginal area. Among them, the urban living land (b3) shows a relatively clear spatial differentiation and the most dramatic changes, which make it worth being concerned about.

In Xining's marginal area, the strong competitive advantage set is the grassland ecological land (b6), the weak competitive advantage set is the agricultural production land (b1), and the potential competitive advantage set is the urban living land (b3).

4.2. Competitive Advantage Sets between Counties

Conflicts between the different regions of the Xining marginal area range from 0 to 0.531 in every scenario. Based on the relative conflict values, we defined a strong conflict relation as [0.4, 0.531], a weak conflict relation as [0.2, 0.4), and a non-relative conflict relation as [0, 0.2).

(1) Strong competitive advantage set: grassland ecological land (b6)

Xinghai County and Gonghe County are the primary sources of the grassland ecological land (b6) conflict, and they have the most influence in developing the grassland ecological land in all the scenarios. To be more specific, Xinghai County has a strong conflict relationship with Menyuan, Haiyan, Jianzha, Tongde, and Guide Counties, and Tongren City, but a weak conflict relationship with Gangcha, Zeku, Henan, and Guinan Counties. Gonghe County has weakdispute relationships with Menyuan, Haiyan, Jianzha, Tongde, and Guide Counties, and Tongren City (Figure 6). As a result, Xinghai and Gonghe Counties dominate in the development of the grassland ecological land.

(2) Weak competitive advantage set: agricultural production land (b1)

Guinan County is the main area of the agricultural production land (b1) conflict. To be more specific, Guinan County has a major conflict with Qilian, Haiyan, Jianzha, Zeku, Henan, Guide, and Xinghai Counties, and Tongren City, as well as a weak conflict relationship with Gangcha and Tongde Counties (Figure 7). As a result, Guinan County dominates the development of the agricultural production land.

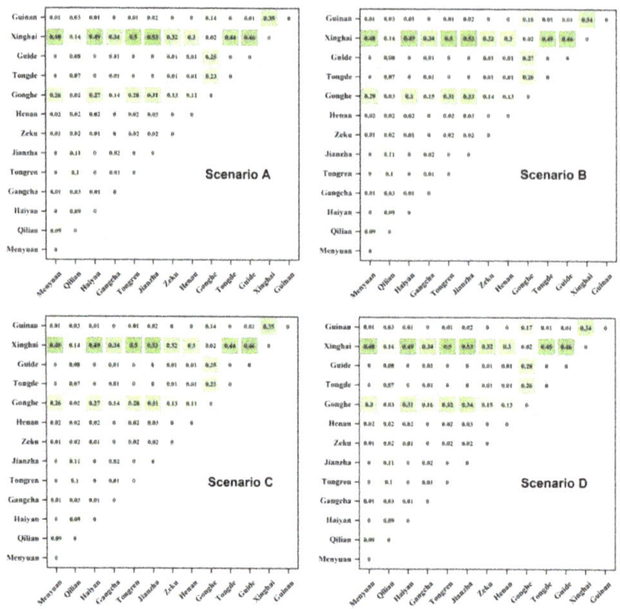

Figure 6. Relative conflict of grassland ecological land between different regions. Note: Scenario A means natural growth, scenario B means urban development, scenario C means cultivated land conservation, and scenario D means ecological protection.

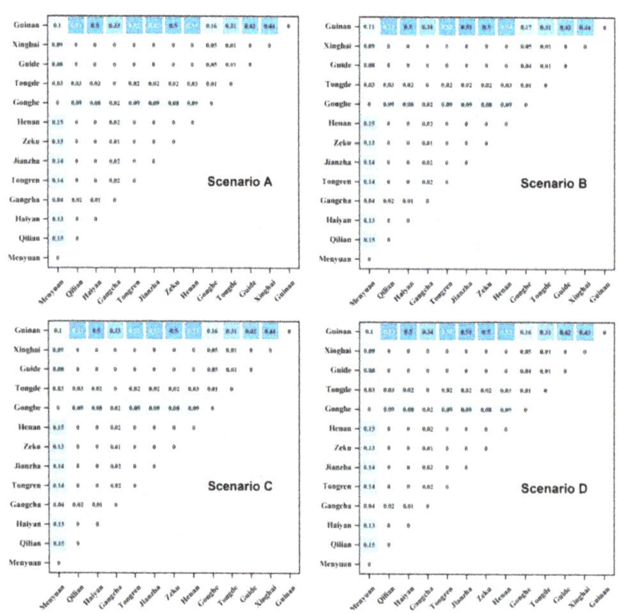

Figure 7. Relative conflict of agricultural production land between different regions. Note: Scenario A means natural growth, scenario B means urban development, scenario C means cultivated land conservation, and scenario D means ecological protection.

(3) Potential competitive advantage set: urban living land (b3)

The urban living land (b3) may not have enough area to support the various forms of development, but as urbanization develops rapidly, its advantages will become more obvious. Additionally, it will become more crucial for regional development, appear to be a significant spatial conflict, and have a great deal of potential for economic development in the Xining marginal area. Haiyan County leads the development of the urban living land in the natural growth scenario (A), urban development scenario (B), and cultivated land conservation scenario (C), while Xinghai County dominates the development of the urban living land in the ecological protection scenario. Haiyan County has a significant conflict relationship with Qilian, Gangcha, Jianzha, Gonghe, Tongde, Guide, and Guinan Counties, and Tongren City, but a weak conflict relationship with Menyuan, Zeku, and Henan Counties. However, in the context of the ecological protection scenario, Xinghai County is the primary source of the urban living land conflict. Xinghai has strong conflict relationships with Qilian, Jianzha, Zeku, Tongde, Guide, and Guinan Counties, and Tongren City, but a weak conflict relationship with Gangcha and Gonghe Counties (Figure 8).

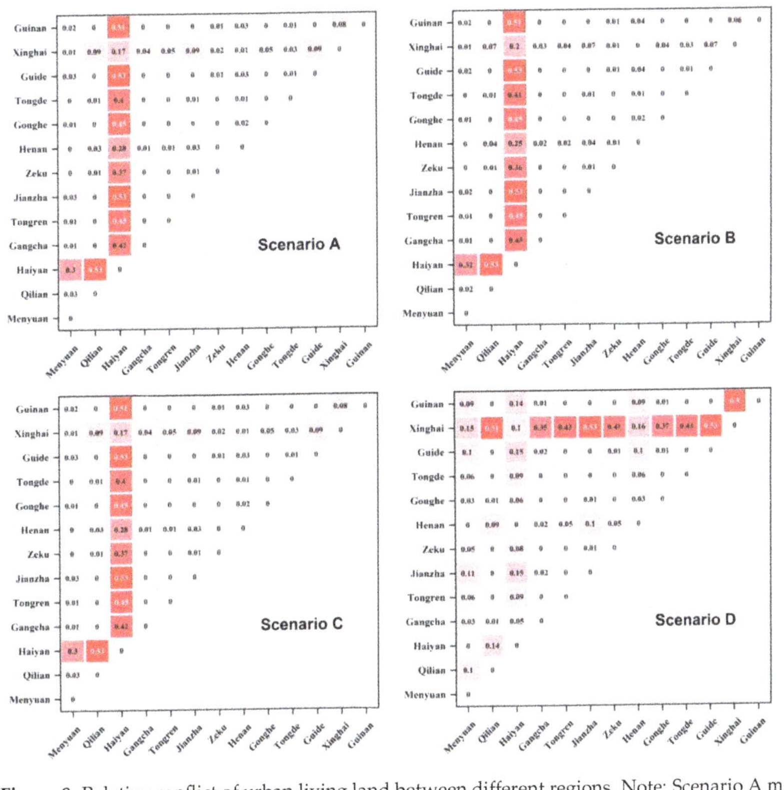

Figure 8. Relative conflict of urban living land between different regions. Note: Scenario A means natural growth, scenario B means urban development, scenario C means cultivated land conservation, and scenario D means ecological protection.

4.3. Spatial Distribution Optimization Strategies on PLE Land

Qinghai Province, as a plateau province, has complex terrain and an arid climate, but it also has abundant natural resources such as grassland and plateau lakes, providing conditions for the development of agriculture and animal husbandry. It is vital to consider factors such as land availability, resource endowment, and environmental protection to maximize the utilization of land resources and achieve optimal allocation. According to the PFCI, the Xining marginal area should develop the agricultural production land and grassland ecological land to effectively alleviate the overall conflicts. Therefore, as the periphery of the provincial capital, for the Xining marginal area, constrained by factors such as transportation and market, it is a better choice to develop agriculture and animal husbandry in the context of balancing economic development and environmental protection.

According to Table 9, in terms of the grassland ecological land, Gonghe County (L9) and Xinghai County (L12) are high-conflict areas, while Qilian County (L2) is a low-conflict area. Zeku County (L7) and Gonghe County (L9) differ from the other three scenarios only in terms of cultivated land protection, but the overall situation is very identical. As a result, developing the grassland ecological land is a top priority in Gonghe and Xinghai Counties, followed by Qilian County. Menyuan County (L1) and Guinan County (L13) are both high-conflict areas in terms of the agricultural production land, indicating that Menyuan and Guinan Counties should be prioritized for the agricultural production land development (Table 10).

Table 9. Sets of grassland ecological land.

X	A	B	C	D
L1	{L9, L12}	{L9, L12}	{L9, L12}	{L9, L12}
L2	{L12}	{L12}	{L12}	{L12}
L3	{L9, L12}	{L9, L12}	{L9, L12}	{L9, L12}
L4	{L9, L12}	{L9, L12}	{L9, L12}	{L9, L12}
L5	{L9, L12}	{L9, L12}	{L9, L12}	{L9, L12}
L6	{L9, L12}	{L9, L12}	{L9, L12}	{L9, L12}
L7	{ L12}	{L9, L12}	{L12}	{L9, L12}
L8	{L12}	{L12}	{L12}	{L12}
L9	{L1, L3, L4, L5, L6, L10, L11, L13}	{L1, L3, L4, L5, L6, L7, L10, L11, L13}	{L1, L3, L4, L5, L6, L10, L11, L13}	{L1, L3, L4, L5, L6, L7, L10, L11, L13}
L10	{L9, L12}	{L9, L12}	{L9, L12}	{L9, L12}
L11	{L9, L12}	{L9, L12}	{L9, L12}	{L9, L12}
L12	{L1, L2, L3, L4, L5, L6, L7, L8, L10, L11, L13}	{L1, L2, L3, L4, L5, L6, L7, L8, L10, L11, L13}	{L1, L2, L3, L4, L5, L6, L7, L8, L10, L11, L13}	{L1, L2, L3, L4, L5, L6, L7, L8, L10, L11, L13}
L13	{L9, L12}	{L9, L12}	{L9, L12}	{L9, L12}

Notes: (1) X means county/city, while L1 refers to Menyuan Hui Autonomous County, L2 refers to Qilian County, L3 refers to Haiyan County, L4 refers to Gangcha County, L5 refers to Tongren City, L6 refers to Jianzha County, L7 refers to Zeku County, L8 refers to Henan Mongol Autonomous County, L9 refers to Gonghe County, L10 refers to Tongde County, L11 refers to Guide County, L12 refers to Xinghai County, and L13 refers to Guinan County. The same applies below.

Table 10. Sets of agricultural production land.

X	A	B	C	D
L1	{L2, L5, L6, L8}	{L2, L5, L6, L8}	{L2, L5, L6, L8}	{L2, L5, L6, L8}
L2	{L1, L13}	{L1, L13}	{L1, L13}	{L1, L13}
L3	{L13}	{L13}	{L13}	{L13}
L4	{L13}	{L13}	{L13}	{L13}
L5	{L1, L13}	{L1, L13}	{L1, L13}	{L1, L13}
L6	{L1, L13}	{L1, L13}	{L1, L13}	{L1, L13}
L7	{L13}	{L13}	{L13}	{L13}
L8	{L1, L13}	{L1, L13}	{L1, L13}	{L1, L13}
L9	{L13}	{L13}	{L13}	{L13}
L10	{L13}	{L13}	{L13}	{L13}
L11	{L13}	{L13}	{L13}	{L13}
L12	{L13}	{L13}	{L13}	{L13}
L13	{L2, L3, L4, L5, L6, L7, L8, L9, L10, L11, L12}	{L2, L3, L4, L5, L6, L7, L8, L9, L10, L11, L12}	{L2, L3, L4, L5, L6, L7, L8, L9, L10, L11, L12}	{L2, L3, L4, L5, L6, L7, L8, L9, L10, L11, L12}

The strong competitive advantage sets of the urban living land (b3) in the natural growth scenario (A), urban development scenario (B), and cultivated land conservation scenario (C) show that Haiyan (L3) is a strong conflict area, while Xinghai County (L12) is a strong conflict area in the ecological protection scenario (D). Therefore, Haiyan County has advantages in developing the urban living land, which is followed by Xinghai County (Table 11).

Table 11. Sets of urban living land.

X	A	B	C	D
L1	{L3}	{L3}	{L3}	{L12}
L2	{L3}	{L3}	{L3}	{L3, L12}
L3	{L1, L2, L4, L5, L6, L7, L8, L9, L10, L11, L12, L13}	{L1, L2, L4, L5, L6, L7, L8, L9, L10, L11, L12, L13}	{L1, L2, L4, L5, L6, L7, L8, L9, L10, L11, L12, L13}	{L2, L6, L11, L13}
L4	{L3}	{L3}	{L3}	{L12}
L5	{L3}	{L3}	{L3}	{L12}
L6	{L3}	{L3}	{L3}	{L3, L12}
L7	{L3}	{L3}	{L3}	{L12}
L8	{L3}	{L3}	{L3}	{L12}
L9	{L3}	{L3}	{L3}	{L12}
L10	{L3}	{L3}	{L3}	{L12}
L11	{L3}	{L3}	{L3}	{L3, L12}
L12	{L3}	{L3}	{L3}	{L1, L2, L4, L5, L6, L7, L8, L9, L10, L11, L12, L13}
L13	{L3}	{L3}	{L3}	{L3, L12}

5. Discussion and Conclusions

Qinghai, a plateau region with diverse topography and an arid climate, provides abundant grassland resources, plateau lakes, and other natural resources essential to agricultural and animal husbandry. The small population, poor natural environment, and restricted carrying capacity of the Xining marginal area have reduced the importance of expanding the urban and rural living land.

5.1. Discussion

By 2030, the Qinghai Tibet Plateau will not embark on the path of urbanization. The Xining marginal area is mostly used for production and ecological purposes, with a very small proportion of land designated for living and a unique core–edge pattern. The urban living land in the county where the government is located will increase more significantly, and the overall trend in PLE land is an increase in the living and production land, combined with a decrease in the ecological land, indicating that the Xining marginal area is still in the economic aggregation stage, but the relatively developed core areas have a limited impact on the other areas.

The development strategy of land use in the Xining Marginal area is significantly influenced by the natural environment. Overall, the counties and cities suitable for agricultural production land, grassland ecological land, and urban living land present a trend around the Qinghai Lake Basin at an altitude between 3000 and 3300 m. According to the PFCI, priority should be given to the development of the grassland ecological land in the counties of Gonghe and Xinghai, followed by Qilian County. Menyuan County and Guinan County should prioritize the development of the agricultural production land, followed by Gonghe County. Haiyan County has advantages in developing the urban living land in scenarios A, B, and C, followed by Xinghai County in scenario D. These counties are located around Qinghai Lake and can leverage their geographical advantages and resources to drive the development of agriculture, animal husbandry, and subsequently, the integrated development of secondary and tertiary industries.

From a county perspective, Haiyan and Gonghe Counties in the Qinghai Lake Basin have an absolute development advantage. Gonghe County surrounds Qinghai Lake on its east, south, and west sides, which is adjacent to the Three-river Headwaters National Nature Reserve and serves as the administrative center of Hainan Tibetan Autonomous Prefecture. It has comprehensive geographical development advantages and is suitable for the development of agricultural production land, grassland ecological land, and urban living land. The development of the grassland ecological land in this area has an absolute advantage in promoting ecological tourism around Qinghai Lake. Haiyan County, located northeast of Qinghai Lake, has abundant tourism resources. In recent years, with the impetus of urbanization and industrialization, there have been advantages in developing the urban living land. The majority of Gangcha is above 3300–3800 meters in altitude, and its development advantage is relatively weak.

The PLE land in the Xining Marginal area can be summarized by the following model. The PLE land does not exhibit a piecemeal expansion pattern, as it is influenced by mountains and rivers. The agricultural production land and grassland ecological land have advantages for development, whereas the urban living land has just development potential. Furthermore, the development pattern of the Qinghai Lake Basin exhibits a layered structure. The urban living land, agricultural production land, and grassland ecological land all expand outward in that sequence to the north of the Qinghai Lake Basin, with the Qilian Mountains forming the outermost circle. The urban living land, agricultural production land, and grassland ecological land all expand eastward to the south, constrained by the headwaters of three rivers. To the west of the Qinghai Lake Basin, only the development of the grassland ecological land is feasible at elevations above 3300 m. The Qilian Mountains, which are to the east, are suitable for the development of ecological lands with a grassland and forest ecological land (Figure 9).

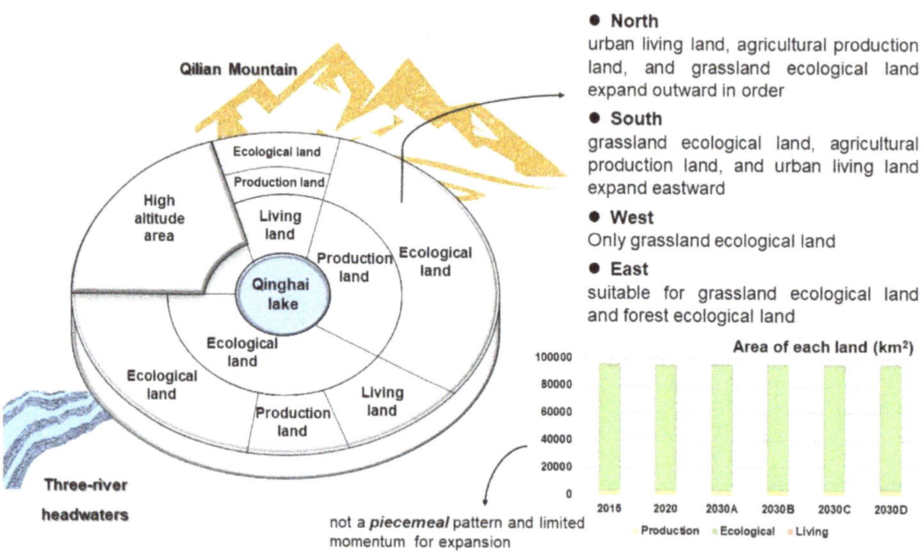

Figure 9. Layered structure of PLE land in Xining marginal area.

Generally, the human–environment relationship system is essentially a complex system, and many global challenges are interconnected, and addressing only one aspect may exacerbate another. The intricate human–environment relationship system in the Xining Marginal area requires planners to reconcile the conflicts between traditional agricultural livelihoods and modern development, as well as the pressures of ecological environment protection and social development goals. This process involves multi-stakeholder models and cross-regional management, with the ultimate goal of enhancing human well-being. Similar issues exist in many regions of the Global South, particularly in areas that are impoverished, ecologically vulnerable, and in urgent need of social well-being enhancement, such as the Ganges River basin in India, the Amazon rainforest, and Southeast Africa.

This study provides ideas for sustainable development path design for these regions from three perspectives: land use prediction, land use interaction, and administrative unit interaction. It is essential to consider the future interactions between different spatial relationships and the competitive dynamics among different regional units. By adopting a zoning approach, strategically allocating advantageous land use types in favorable regions can promote regional sustainable development.

5.2. Nexus Approach for the Sustainable Development in Xining Marginal Area

Qinghai Province's relatively slow economic development and uneven population distribution may result in development differences and competitiveness among different regions. Simultaneously, it is distinguished by a remote geographical location with inconvenient transportation, which may limit economic development. Whereas the National Territory Development Planning System requires the delineation of urban development boundaries to assure the spatial capacity of future urban development and construction, the arrangement of urban development time series results in a significant loss of rural development rights. In practice, local governments continue to implement indiscriminate policy supply for the region, and they are unable to timely change the system and policies that constrain the use of PLE land. The mismatch between policy supply and rural development stage will undoubtedly result in a decrease or even stagnation of the existing economies of scale, with the result that marginal areas will violate development laws. As a result, more specific and realistic actions must be implemented in the Xining marginal area to achieve a balance between economic development and environmental preservation.

The urbanization of the Qinghai–Tibet Plateau should be based on water tower protection and green development. Given the unique characteristics of plateau urbanization and the maximum permissible urban population size, the Qinghai–Tibet Plateau's urbanization rate can be increased to 57.25%. Compared to the current urbanization rate of 47.58% in 2020, the Qinghai–Tibet Plateau may only expect a 9.67% increase in future urbanization [38]. Furthermore, the severe and fragile ecology of the Qinghai–Tibet Plateau necessitates a relatively small urban population. The plateau's low carrying threshold prevents large-scale urbanization and development. The urbanization of the plateau does not conform to the law of stage development, and there is no need to significantly improve the urbanization of the plateau. Therefore, for counties with room to grow, urbanization can only take place at a low speed and with high quality.

The land use methods are designed to optimize the type and spatial combination of land use to protect land resources from damage; to gain the best integrated economic, social, and ecological benefits; and to maintain the long-term stability of such benefits [39]. We calculated the dominance of the different regions in developing different lands and designed a feasible nexus method for the sustainable development of the Xining marginal area (Figure 10). In this section, we propose five ways to reconcile the conflicts in the Xining marginal area, as shown below:

(1) Low-speed development and high-quality urbanization

This approach applies to Haiyan County, Gonghe County, and Tongren City. According to Section 4.2, Haiyan County has dominance in developing urban living land, while Tongren City is a regional central city on the Qinghai–Tibet Plateau, acting as a point for driving social development and consolidating borders of the plateau [38]. In addition, according to Section 4.2, the living land in state government locations has a stronger growth momentum. Therefore, it is reasonable for Haiyan County and Tongren City to develop urban living lands. However, Gonghe County shows a diversity of appropriate characteristics. It has dominance in developing rural living land, other living land, and developing grassland ecological land. Therefore, it is worth exploring what type of lands Gonghe County should focus on constructing. We calculated the average conflict values of Gonghe County to develop the above lands under different scenarios. According to PFCI, in all the scenarios, Gonghe County has greater dominance in developing the rural living land and other living land, both of which are largely outside of the same competitive advantage set (see Section 4.2). Therefore, it is feasible for Gonghe County to focus on the rural living land and other living land based on its original foundation, with the grass ecological land surrounding it to protect the Qinghai Lake.

To conclude, Haiyan County, Gonghe County, and Tongren City are suitable for taking the approach of low-speed and high-quality urbanization. Fang also proposed to promote the conversion of Gonghe and Haiyan from counties into cities [38], which further verifies the views of this paper.

(2) Mountain forest conservation

This approach applies to Qilian County. According to Section 4.3, Qilian County has dominance in developing forest ecological land and other ecological land. And given that the Qilian Mountain Nature Reserve lies in Qilian County, the path of mountain forest protection is advisable.

(3) Plateau characteristic agriculture

This approach applies to Guinan and Menyuan, which have dominance in developing land for agricultural production. They should advocate the inheritance of traditional agricultural civilization and the moderate development of modern agriculture with highland characteristics on the premise of guaranteeing food security [40].

(4) Water tower protection

This approach applies to Xinghai, Tongde, and Zeku Counties. Xinghai County has dominance in developing grassland ecological land. In addition, the unique alpine vegetation system in three-river headwaters plays a pivotal role in global climate change. The vegetation in the region determines the local ecology and animal husbandry production, and has a significant impact on the ecological security of China and Asia as a whole [41]. However, in the process of urbanization and industrialization in the three-river headwaters from 2012 to 2016, the environmental stress presented a point-like effect, the agricultural and animal husbandry production presented a planar stress, and the tourism and transportation presented a linear stress [42]. Therefore, the counties located in the three-river headwaters should take the approach of water town protection.

(5) Live in town, pasture/farming in country

This approach applies to the counties characterized by no obvious dominance in development, including Gangcha, Guide, Jianzha, and Henan Counties. All these counties have a smaller center, surrounded by grassland ecological land or agricultural production land. According to Section 4.2, the development of grassland ecological land and agricultural production land may alleviate the overall conflict. It is appropriate for residents to live in the town, and pasture or farm in the country.

5.3. Limitations

Firstly, due to limitations in data acquisition and data volume, the influential factors we added to the PLUS model are not convincing enough. In addition, we only simulated the evolution pattern of the PLE lands in the different scenarios without specifying specific locations. For example, we did not add constrained areas or urban master plans to calculate the dominance of the development of different regions in the face of policy disruptions [43]. Secondly, due to the volume of data, we only considered the conflicts between the different lands from a holistic perspective, which was much larger in scale. We did not measure the conflicts between the different lands on a more microscopic scale, such as the spatial conflicts based on landscape ecology [44,45] or map multi-factor overlay [46,47].

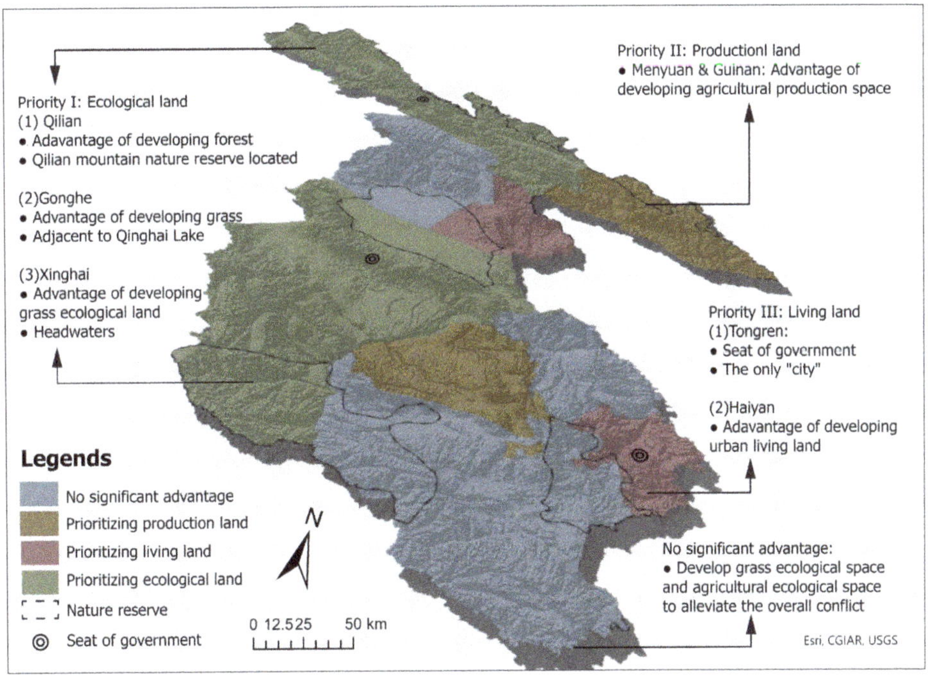

Figure 10. Nexus approaches for sustainable development of Xining marginal area.

5.4. Conclusions

In this study, we established the scenarios of natural growth, urban development, cultivated land conservation, and ecological protection based on the PLUS model, and calculated the competitive advantages between the different lands and regions in the Xining marginal area using PFCI. In conclusion, the Qinghai Tibet Plateau will not embark on the path of rapid urbanization by 2030. In any situation, the PLE land in the Xining marginal area is mainly focused on production and ecological land, with a very low proportion of living land (around 0.1%). The Xining marginal area has formed and will continue to maintain the core–edge pattern, namely, the seats of government will see stronger growth in the urban living land and other living land. All these indicate that the Xining marginal area is still in the early stages of economic development while the influence of the relatively developed areas exerted on other regions is still quite limited. The strong competitive advantage set includes the grassland ecological land and other ecological land, which indicates that in the Xining marginal area, the grassland ecological land and other ecological land have a dominant role or are in a dominant position. Regardless of the future development scenarios, the urban living land always plays a secondary role [47]. Gonghe County has dominance in developing rural living land and other living land, Xinghai County has dominance in developing grassland ecological land, Guinan County has dominance in developing agricultural production land, and Haiyan County has dominance in developing urban living land. On that basis, we analyzed the sustainable allocation strategies of the different lands and found that developing the corresponding lands in the dominant regions is the best option. Finally, we established nexus approaches to harmonizing conflicts for the Xining marginal area, aiming to provide a reference for the ecological highland. We hope that the Xining marginal area will thrive and become more lingering.

Author Contributions: Z.J.: conceptualization, original draft, and review and editing. Y.L.: formal analysis, visualization, and review and editing. Q.W.: formal analysis and methodology. M.S.: software and methodology. R.A.: formal analysis and methodology. M.W.: resources and conceptualization. All authors have read and agreed to the published version of the manuscript.

Funding: This research was funded by the Key Laboratory of Earth Surface System and Human-Earth Relations, Ministry of Natural Resources of China (LBXT2023YB09), the National Natural Science Foundation of China (42201198), the Youth Science and Technology Fund of Gansu Province (22JR5RA518), the Hui-Chun Chin and Tsung-Dao Lee Chinese Undergraduate Research Endowment (CURE LZU-JZH2738), and the 2024 Gansu Province Youth Doctoral Support Project (2024QB-001). The APC was funded by the Key Laboratory of Earth Surface System and Human-Earth Relations, the Ministry of Natural Resources of China (LBXT2023YB09).

Data Availability Statement: The data presented in this study are available on request from the corresponding author. The data are not publicly available due to privacy and commercial restrictions.

Conflicts of Interest: The authors declare that they have no known competing financial interests or personal relationships that could have appeared to influence the work reported in this paper.

References

1. Hurtt, G.C.; Chini, L.P.; Frolking, S.; Betts, R.A.; Feddema, J.; Fischer, G.; Fisk, J.P.; Hibbard, K.; Houghton, R.A.; Janetos, A.; et al. Harmonization of land-use scenarios for the period 1500–2100: 600 years of global gridded annual land-use transitions, wood harvest, and resulting secondary lands. *Clim. Chang.* **2011**, *109*, 117–161. [CrossRef]
2. Haliscelik, E.; Soytas, M.A. Sustainable development from millennium 2015 to Sustainable Development Goals 2030. *Sustain. Dev.* **2019**, *27*, 545–572. [CrossRef]
3. Caiado, R.G.G.; Filho, W.L.; Quelhas, O.L.G.; Nascimento, D.L.d.M.; Ávila, L.V. A literature-based review on potentials and constraints in the implementation of the sustainable development goals. *J. Clean. Prod.* **2018**, *198*, 1276–1288. [CrossRef]
4. Lu, X.H.; Zhang, Y.W.; Lin, C.R.; Wu, F. Analysis and comprehensive evaluation of sustainable land use in China: Based on sustainable development goals framework. *J. Clean. Prod.* **2021**, *310*, 127205. [CrossRef]
5. Acheampong, M.; Yu, Q.Y.; Enomah, L.D.; Anchang, J.; Eduful, M. Land use/cover change in Ghana's oil city: Assessing the impact of neoliberal economic policies and implications for sustainable development goal number one—A remote sensing and GIS approach. *Land Use Policy* **2018**, *73*, 373–384. [CrossRef]
6. Sun, H.L.; Zheng, D.; Yao, T.D.; Zhang, Y.L. Protection and Construction of the National Ecological Security Shelter Zone on Tibetan Plateau. *Acta Geogr. Sin.* **2012**, *67*, 3–12.
7. Han, W.; Cai, J.M.; Zhao, Y.F. Structure, mechanism, and paths of spatial governance in metropolitan fringe with the participation of multi-subjects. *Prog. Geogr.* **2021**, *40*, 1730–1745. [CrossRef]
8. Liao, G.; He, P.; Gao, X.; Lin, Z.; Huang, C.; Zhou, W.; Deng, O.; Xu, C.; Deng, L. Land use optimization of rural production–living–ecological land at different scales based on the BP–ANN and CLUE–S models. *Ecol. Indic.* **2022**, *137*, 108710. [CrossRef]
9. Cui, J.X.; Gu, J.; Sun, J.W.; Luo, J. The spatial pattern and evolution characteristics of the production, living and ecological space in Hubei provence. *Chin. Land Sci.* **2018**, *32*, 67–73.
10. Gao, L.A.; Tao, F.; Liu, R.; Wang, Z.L.; Leng, H.J.; Zhou, T. Multi-scenario simulation and ecological risk analysis of land use based on the PLUS model: A case study of Nanjing. *Sustain. Cities Soc.* **2022**, *85*, 104055. [CrossRef]
11. Reidsma, P.; Konig, H.J.; Feng, S.Y.; Bezlepkina, I.; Nesheim, I.; Bonin, M.; Sghaier, M.; Purushothaman, S.; Sieber, S.; Van, I.; et al. Methods and tools for integrated assessment of land use policies on sustainable development in developing countries. *Land Use Policy* **2011**, *28*, 604–617. [CrossRef]
12. Drechsler, M.; Surun, C. Land-use and species tipping points in a coupled ecological-economic model. *Ecol. Complex.* **2018**, *36*, 86–91. [CrossRef]
13. Hasegawa, T.; Fujimori, S.; Ito, A.; Takahashi, K.; Masui, T. Global land-use allocation model linked to an integrated assessment model. *Sci. Total Environ.* **2017**, *580*, 787–796. [CrossRef] [PubMed]
14. Verburg, P.H.; Tabeau, A.; Hatna, E. Assessing spatial uncertainties of land allocation using a scenario approach and sensitivity analysis: A study for land use in Europe. *J. Environ. Manag.* **2013**, *127*, 132–144. [CrossRef] [PubMed]
15. Chakir, R.; Le Gallo, J. Predicting land use allocation in France: A spatial panel data analysis. *Ecol. Econ.* **2013**, *92*, 114–125. [CrossRef]
16. Peltonen-Sainio, P.; Jauhiainen, L.; Laurila, H.; Sorvali, J.; Honkavaara, E.; Wittke, S.; Karjalainen, M.; Puttonen, E. Land use optimization tool for sustainable intensification of high-latitude agricultural systems. *Land Use Policy* **2019**, *88*, 104104. [CrossRef]
17. Sadeghi, S.H.R.; Jalili, K.; Nikkami, D. Land use optimization in watershed scale. *Land Use Policy* **2009**, *26*, 186–193. [CrossRef]
18. Arciniegas, G.; Janssen, R.; Omtzigt, N. Map-based multicriteria analysis to support interactive land use allocation. *Int. J. Geogr. Inf. Sci.* **2011**, *25*, 1931–1947. [CrossRef]

19. Kaim, A.; Cord, A.F.; Volk, M. A review of multi-criteria optimization techniques for agricultural land use allocation. *Environ. Model. Softw.* **2018**, *105*, 79–93. [CrossRef]
20. Rahman, M.M.; Szabo, G.A. Geospatial Approach to Measure Social Benefits in Urban Land Use Optimization Problem. *Land* **2021**, *10*, 1398. [CrossRef]
21. Zhao, X.; Tang, F.; Zhang, P.T.; Hu, B.Y.; Xu, L. Dynamic simulation and characteristic analysis of County production-living-ecological spatial conflicts based on CLUE-S model. *Acta Ecol. Sin.* **2019**, *39*, 5897–5908.
22. Gharaibeh, A.A.; Ali, M.H.; Abo-Hammour, Z.S.; Al Saaideh, M. Improving Genetic Algorithms for Optimal Land-Use Allocation. *J. Urban Plan. Dev.* **2021**, *147*, 04021049. [CrossRef]
23. Berger, T.; Schreinemachers, P. Creating agents and landscapes for multiagent systems from random samples. *Ecol. Soc.* **2006**, *11*, 19. [CrossRef]
24. Liang, X.; Guan, Q.F.; Clarke, K.C.; Liu, S.; Wang, B.Y.; Yao, Y. Understanding the drivers of sustainable land expansion using a patch-generating land use simulation (PLUS) model: A case study in Wuhan, China. *Comput. Environ. Urban Syst.* **2021**, *85*, 101569. [CrossRef]
25. Liu, X.P.; Liang, X.; Li, X.; Xu, X.C.; Ou, J.P.; Chen, Y.M.; Li, S.; Wang, S.; Pei, F. A future land use simulation model (FLUS) for simulating multiple land use scenarios by coupling human and natural effects. *Landsc. Urban Plan.* **2017**, *168*, 94–116. [CrossRef]
26. Liu, J.G.; Hull, V.; Godfray, H.C.J.; Tilman, D.; Gleick, P.; Hoff, H.; Pahl-Wostl, C.; Xu, Z.; Chung, M.G.; Sun, J.; et al. Nexus approaches to global sustainable development. *Nat. Sustain.* **2018**, *1*, 466–476. [CrossRef]
27. Flammini, A.; Puri, M.; Pluschke, L.; Dubois, O. Walking the nexus talk: Assessing the water-energy-food nexus in the context of the sustainable energy for all initiative. *J. Phys. Chem. A* **2014**, *115*, 7869–7870.
28. Rasul, G. Food, water, and energy security in South Asia: A nexus perspective from the Hindu Kush Himalayan region. *Environ. Sci. Policy* **2014**, *39*, 35–48. [CrossRef]
29. Shi, M.J.; Wu, H.Q.; Jiang, P.A.; Zheng, K.; Liu, Z.; Dong, T.; He, P.; Fan, X. Food-water-land-ecosystem nexus in typical Chinese dryland under different future scenarios. *Sci. Total Environ.* **2023**, *880*, 163183. [CrossRef]
30. He, X.H. Analysis on Land Use Structure of Ecological-production-living land in Shaanxi Province. *Remote Sens. Inf.* **2021**, *36*, 120–124.
31. Li, G.D.; Fang, C.L. Quantitative function identification and analysis of urban ecological-production-living lands. *Acta Geogr. Sin.* **2016**, *71*, 49–65.
32. Liu, J.L.; Liu, Y.S.; Li, Y.R. Classification evaluation and spatial-temporal analysis of "production-living-ecological" lands in China. *Acta Geogr. Sin.* **2017**, *72*, 1290–1304.
33. Zhang, H.Q.; Xu, E.Q.; Zhu, H.Y. An ecological-living-industrial land classification system and its spatial distribution in China. *Resour. Sci.* **2015**, *37*, 1332–1338.
34. Wang, J.N.; Zhang, Z. Land Use Change and Simulation Analysis in the Northern Margin of the Qaidam Basin Based on Markov-PLUS Model. *J. Northwest For. Univ.* **2022**, *37*, 139–148.
35. Wang, B.S.; Liao, J.F.; Zhu, W.; Qiu, Q.Y.; Wang, L.; Tang, L. The weight of neighborhood setting of the FLUS model based on a historical scenario: A case study of land use simulation of urban agglomeration of the Golden Triangle of Southern Fujian in 2030. *Acta Ecol. Sin.* **2019**, *39*, 4284–4298.
36. Zadeh, L.A. Fuzzy sets. *Inf. Control.* **1965**, *8*, 338–353. [CrossRef]
37. Yager, R.R. Pythagorean Membership Grades in Multicriteria Decision Making. *IEEE Trans. Fuzzy Syst.* **2014**, *22*, 958–965. [CrossRef]
38. Fang, C.L. Special thinking and green development path of urbanization in Qinghai-Tibet Plateau. *Acta Geogr. Sin.* **2022**, *77*, 1907–1919.
39. Giuliani, G.; Mazzetti, P.; Santoro, M.; Nativi, S.; Van Bemmelen, J.; Colangeli, G.; Lehmann, A. Knowledge generation using satellite earth observations to support sustainable development goals (SDG): A use case on Land degradation. *Int. J. Appl. Earth Obs. Geoinf.* **2020**, *88*, 102068. [CrossRef]
40. Liu, Y.S.; Zhang, Z.W.; Wang, J.Y. Regional differentiation and comprehensive regionalization scheme of modern agriculture in China. *Acta Geogr. Sin.* **2018**, *73*, 203–218.
41. Li, H.X.; Liu, G.H.; Fo, B.J. Response of vegetation to climate change and human activity based on NDVI in the Three-River Headwaters region. *Acta Ecol. Sin.* **2011**, *31*, 5495–5504.
42. Zhou, K.; Liu, H.C.; Fan, J.; Yu, H. Environmental stress intensity of human activities and its spatial effects in the Qinghai-Tibet Plateau national park cluster: A case study in Sanjiangyuan region. *Acta Ecol. Sin.* **2021**, *41*, 268–279.
43. Zhou, L.; Dang, X.W.; Sun, Q.K.; Wang, S.H. Multi-scenario simulation of urban land change in Shanghai by random forest and CA-Markov model. *Sustain. Cities Soc.* **2020**, *55*, 102045. [CrossRef]
44. Bao, W.K.; Yang, Y.Y.; Zou, L.L. How to reconcile land use conflicts in mega urban agglomeration? A scenario-based study in the Beijing-Tianjin-Hebei region, China. *J. Environ. Manag.* **2021**, *296*, 113168. [CrossRef] [PubMed]
45. Jiang, S.; Meng, J.J.; Zhu, L.K.; Cheng, H.R. Spatial-temporal pattern of land use conflict in China and its multilevel driving mechanisms. *Sci. Total Environ.* **2021**, *801*, 149697. [CrossRef] [PubMed]

46. Zou, L.L.; Liu, Y.S.; Wang, J.Y.; Yang, Y.Y. An analysis of land use conflict potentials based on ecological-production-living function in the southeast coastal area of China. *Ecol. Indic.* **2021**, *122*, 107297. [CrossRef]
47. Jiang, S.; Meng, J.J.; Zhu, L.K. Spatial and temporal analyses of potential land use conflict under the constraints of water resources in the middle reaches of the Heihe River. *Land Use Policy* **2020**, *97*, 104773. [CrossRef]

land

MDPI

Article

Dynamics of Built-Up Areas and Challenges of Planning and Development of Urban Zone of Greater Lomé in Togo, West Africa

Têtou-Houyo Blakime [1], Kossi Adjonou [2,*], Kossi Komi [3,4], Atsu K. Dogbeda Hlovor [2], Kodjovi Senanou Gbafa [1], Jean-Bosco Benewinde Zoungrana [5], Botolisam Polorigni [2] and Kouami Kokou [2]

1 Polytechnic School of Lome, University of Lome, Lome 01 BP 1515, Togo; matcheloblak@yahoo.fr (T.-H.B.); sgbafa@univ-lome.tg (K.S.G.)
2 Forestry Research Laboratory (LRF), Faculty of Science, University of Lome, Lome 01 BP 1515, Togo; patrickhlovor@gmail.com (A.K.D.H.); botolisam@yahoo.fr (B.P.); kkokou@univ-lome.tg (K.K.)
3 Regional Center of Excellence on Sustainable Cities in Africa (CERViDA-DOUNEDON), University of Lome, Lome 01 BP 1515, Togo; kossi.komi@cervida-togo.org
4 Research Laboratory on Spaces, Exchanges and Human Security (LaREESH), University of Lome, Lome 01 BP 1515, Togo
5 WASCAL Master Research Programme in Informatics for Climate Change, University Joseph KI-ZERBO, Ouagadougou 03 BP 7021, Burkina Faso; zoungrana.b@wascal.org
* Correspondence: kadjonou@univ-lome.tg; Tel.: +228-90244301

Abstract: The expansion of African cities leads to the occupation of peripheral urban areas without respecting planning rules. The Greater Lomé (Togo) is no exception to this phenomenon of high-speed horizontal spreading, which causes recurrent flooding. The objective of this research is to understand the spatio-temporal changes in the dynamics of built-up areas in Greater Lomé. The methodology used is based on the analysis of Landsat images from the years 2007, 2012, 2016, and 2020 coupled with direct field observations and a literature review. The results showed an increase in residential areas to the detriment of the other land use/cover types. Estimated at 15,481 ha in 2007, the built-up area reached 35,521 ha in 2020, an increase of 33% to the detriment of vegetation and cultivation areas. This increase was marked by constructions in the floodplain of the Zio River. The field surveys revealed an increase in the density of most of the agglomerations. From 1863 ha in 2007, they increased to 14,485 ha in 2020, an increase of 12,622 ha or approximately 33%. These results indicate that attention needs to be paid to both the planning and control of the development of spaces in the outlying areas of Greater Lomé.

Keywords: urban sprawl; residential areas; urban planning; Greater Lomé; Togo

1. Introduction

Urban sprawl, broadly defined as dispersed, excessive, and wasteful urban growth, characterized by the excessive use of land for the building of single-family houses in the suburbs [1], is increasingly observed in recent years in African cities. This urban explosion has placed the problem of the surge of populations on the urban outskirts at the center of debates on the city. While the developed countries are collapsing in the face of the crisis in the suburbs marked by violence and bad living, the population of the countries of the South exerts strong pressure on the outskirts of cities [2], since the actual infrastructure built without any official urban plan do not provide for any urban service let alone the preservation of the surrounding ecology [3]. The main cause of these pressures is the rapid increase in local populations [4] combined with the absence of city extension policies or the implementation of certain master plans [5].

Africa's high population growth rate makes the demographic explosion one of the most important causes of land use changes in African cities. The latter is characterized by a peri-urban ring, which is a transition between the rural environment and the urban environment [6–8]. Under the weight of vertiginous demographic growth, the peri-urban zone is receding, giving way to an urbanized area and, in turn, transforming its periphery, which was once rural. The urbanization of peripheral zones appears to be the essential form of growth for West African cities. It is manifested everywhere by a sprawl of residential areas, which sometimes pushes the limits of the city to considerable distances from the urban center [9]. Despite multifaceted consequences, such as the housing and transport crisis, precarious employment, and a lack of sanitation [10], climatic factors in general and heat islands in particular [11] are considered today as consequences of a reduction in vegetation cover in residential areas.

Many studies on the dynamics of land use and land cover in urban areas have been performed during the last decades in various parts of the world. For instance, in Latin America, ref. [12] analyzed the land cover dynamics along the urban–rural gradient of the Port-au-Prince agglomeration (Republic of Haiti) from 1986 to 2021 and found that the landscape has undergone significant changes because of the "high demand for housing" while in Asia, ref. [13] estimated, determined the patterns, and identified the potential drivers of land-use changes during 1995–2015 in an urbanizing tropical watershed in Indonesia. They found, among other results, a major change from agricultural to urban areas in the study area. In Europe, the relationships between the spatial and temporal dynamics of land use and land cover (LULC), the hydro-geomorphological processes, and their impacts were evaluated by [14]. They showed a highlighted increase in artificial areas for the period 1958–2018. In West Africa, land use and land cover dynamics were analyzed in Calabar Metropolis (Nigeria) by [15] using a combined approach of remote sensing and a geographic information system. Their studies showed an increased trend in built-up areas from 2002 to 2016. Moreover, by analyzing the global satellite data of 120 cities, ref. [16] found that cities "fragmented" a large area of landscapes. With the urban extension that can be observed everywhere in the world, the monitoring of territorial dynamics has taken an important place in the context of urban planning. It then appeared necessary to have reliable, precise, and continuously updated data on the evolution of the territory [17,18].

However, little has been done to understand land use and land cover dynamics in the urban areas of Greater Lomé. Thus, the main objective of this work is to understand the spatial and temporal dynamics of built-up areas in Greater Lomé with a view to providing guidelines for sustainable urban planning in the study area. Specifically, it aims to analyze (i) land use and land cover changes in relation to the evolution of buildings as well as (ii) trends in the annual rate and change matrix of LULC.

The city of Lomé is today the largest city in Togo whose development exceeds all forecasts. The observation is the strong spatial growth due to demographic pressure and the need for city dwellers to find housing. The majority of housing is built through the informal sector in the city center or on its outskirts, which gives rise to spontaneous outlying districts. The urban policies of Togo are those where the public power of urbanization is out of phase with the occupation of spaces by the population. In Lomé, the development of outlying districts was the work of customary landowners outside of any control by the state and local authorities. It was favored by the housing problems, which continue to worsen. These owners did not comply with the subdivision procedures provided for by law, in particular, obtaining the agreement of the minister in charge of town planning before any fragmentation [19,20]. This should help to monitor compliance with planning and has resulted in a dramatic increase in the area of the city. Thus, since the 1970s, the extension of Greater Lomé towards its peripheral margins has started attracting the attention of researchers, who have not hesitated to develop research themes within the framework of numerous scientific works. The literature indicates that an increase in the population results in accelerated demand for natural resources, resulting in ecosystem and landscape degradation [21].

Finally, it is important to note that the ecosystem and associated landscapes provide important services, such as oxygen production, carbon sequestration, flood control, food, and cultural services. These landscapes provide urban dwellers with opportunities for tourism and recreation [22]. This growth is characterized by the extension of the outskirts described as an "unfinished landscape" where facilities are lacking [23,24].

2. Materials and Methods

2.1. Study Area

This study was performed in the agglomeration and the peripheral areas of Greater Lomé, the capital town of Togo. The study area is located between 6°06′–6°25′ North latitudes and 1°15′–1°45′ East longitudes (Figure 1). It is composed of 152 neighborhoods [25] and has a population of 2,188,376 [26]. The district of Greater Lomé has an area of 61,315 ha and is composed of two prefectures (six communes for the Agoènyivé prefecture and seven communes for the Gulf prefecture) as well as a few localities of the Zio and Avé prefectures (Djagblé, Aveta, and Akepe).

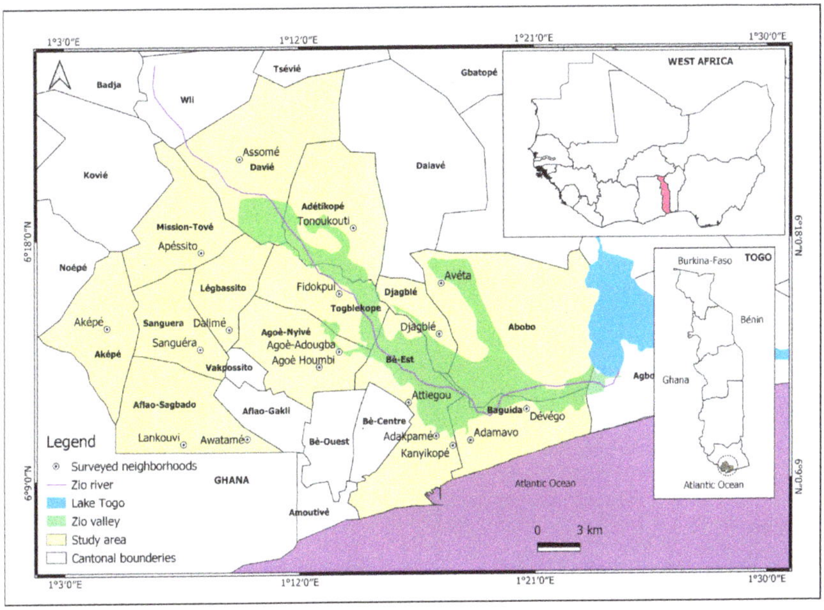

Figure 1. Location of Greater Lomé.

2.2. Data Acquisition

The data used in this study were derived from Landsat images (30 m × 30 m resolution) at the end of the dry season (January–February) in order to obtain satellite images with very low cloud cover (less than 10%). These satellite images were obtained from the Center for Earth Resources Observation and Science (https://earthexplorer.usgs.gov/, accessed on 1 January 2020). The images without cloud cover and characteristics not affected by seasonality were considered in this study. The use of data from the same season has the advantage of providing homogeneous spectral and radiometric characteristics. It reduces the seasonal variation in the spectral reflectance of the different land cover types [27,28]. The level-2 products (surface reflectance) of the Landsat 7 and Landsat 8 satellites from 2012, 2016, and 2020 were selected in order to obtain geometric, radiometric, and atmospheric corrected data (Table 1). Moreover, the Landsat images were used for data analysis for several reasons: (i) the availability of several images up to 2002, (ii) open-source data, (iii) good spectral resolution, and (iv) the spatial resolution (30 m) is sufficient to distinguish forest/non-forest classes and, thus, limits the amount of data to be processed. In addition, Landsat images have good radiometric and geometric qualities to carry out land use dynamics analyses [29].

Table 1. Characteristics of the images collected.

Year	Acquisition Date	Sensor
2020	16 February 2020	Landsat 8/OLI
2016	25 January 2017	Landsat 8/OLI
2012	04 January 2012	Landsat 7/ETM + (SLC-off)
2007	22 January 2007	Landsat 7/ETM + (SLC-off)

2.3. *Determination of Building Occupancy Classes*

The land use and land cover (LULC) classes were defined in two stages. Initially, two LULC classes were defined: built-up areas and other land occupations, such as water bodies and vegetated areas. Secondly, it was useful to differentiate in each class the "housing zone": three (03 subclasses (Table 2).

Table 2. Land use and land cover classification applied in the study area.

No.	Class Name	Description
1	Dense zone	Areas covered by more than 90% of built-up area
2	Moderate-density zone	Areas covered by built-up area between 75% and 90%
3	Low-density zone	Areas that covered less than 75% of built-up area
4	Other land use/land cover	Natural vegetation, watering holes, cultivated areas

2.4. *Processing of Satellite Data and Classification*

The classification was performed using the Random Forest (RF) algorithm based on 312 reference data (occupation classes). This algorithm was selected for its good predictive abilities of land cover [30,31] and for temporal analysis [32]. Moreover, the RF provides means to estimate missing values and perform multiple types of data analysis, including regression, classification, and unsupervised learning [33].

In order to maximize the band information and improve the discrimination of the land use and land cover (LULC) classes, a principal component analysis (PCA) was calculated on the indices derived from the primary channels of the satellite images. The PCA was then used to perform a classification based on the training plots collected during the field survey. These indices included the normalized building difference index (NDBI), the soil adjustment vegetation index (SAVI), and the normalized humidity difference index (MNDWI) [34].

The NDBI values vary between −1.0 and +1.0. The highest value represents built-up areas, the lowest value indicates vegetation, and the negative value signifies water bodies. In NDBI methods, it is assumed that all positive NDVI and NDBI values represent vegetation and built-up areas. This approach is prone to many errors. Consequently, the Built-Up Index (BU) minimizes this error by subtracting the NDVI from the NDBI.

The Soil Adjustment Vegetation Index (SAVI) is a transformation technique used to minimize the influence of soil brightness from spectral vegetation indices involving red and near-infrared (NIR) wavelengths. The SAVI is calculated using the following Equation (1):

$$\text{SAVI} = ((\text{NIR} - \text{Red})/(\text{NIR} + \text{Red} + \text{L})) \times (1 + \text{L}) \tag{1}$$

where L denotes a correction factor for soil brightness. To take into account the ground brightness for most land use and land cover types, L types are defined as 0.5.

The normalized humidity difference index (MNDWI) is a remote-sensing-based indicator that is sensitive to changes in leaf water content [35]. The MNDWI mitigates the errors of the NDWI by extracting the water content from remote sensing data. The Modified Normalized Difference Index for water uses the green and Short-Wave Infrared Red (SWIR) bands to highlight the characteristics of open water dominated by built-up areas. It removes noise from built-up areas, vegetation, and soil. The NDWI value ranges from −1.0 to +1.0. In general, positive values above 0.5 indicate bodies of water, while lower values of 0–0.2 indicate built-up areas and negative values indicate vegetation.

Using the three (03) different indices (NDBI, SAVI, and MNDWI), a PCA was carried out for the 4 years considered in this study, namely 2007, 2012, 2016, and 2020. The results of this analysis were used to assess the quality of the classification (Figure 2).

(a) 2007

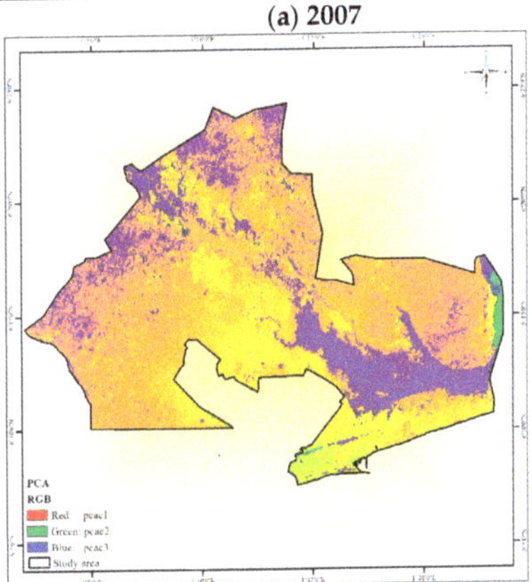

Bands	PCA1	PCA2	PCA3
MNDWI	0.486	0.637	0.5955
NDBI	0.7704	0.6357	−0.0469
SAVI	0.4084	−0.4358	0.801

(b) 2012

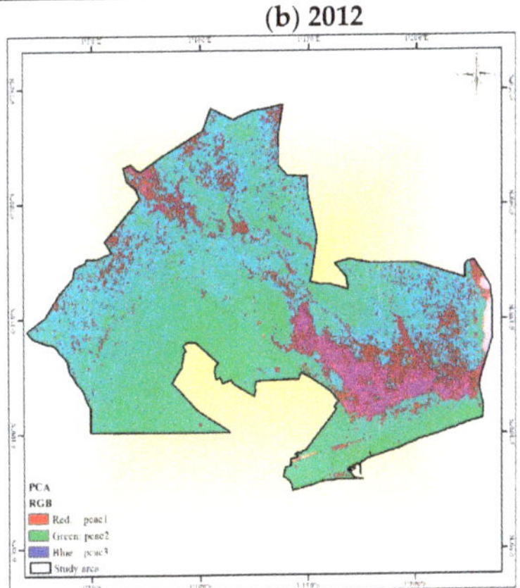

Bands	PCA1	PCA2	PCA3
MNDWI	0.523	0.3087	0.7943
NDBI	−0.785	0.5366	0.3085
SAVI	−0.331	−0.785	0.5232

(c) 2016

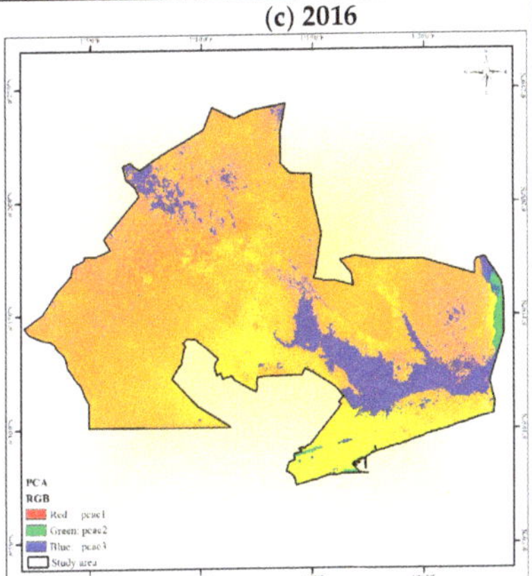

Bands	PCA1	PCA2	PCA3
MNDWI	−0.5065	0.8416	0.1871
NDBI	0.6176	0.2027	0.7598
SAVI	−0.6016	−0.5005	0.6225

(d) 2020

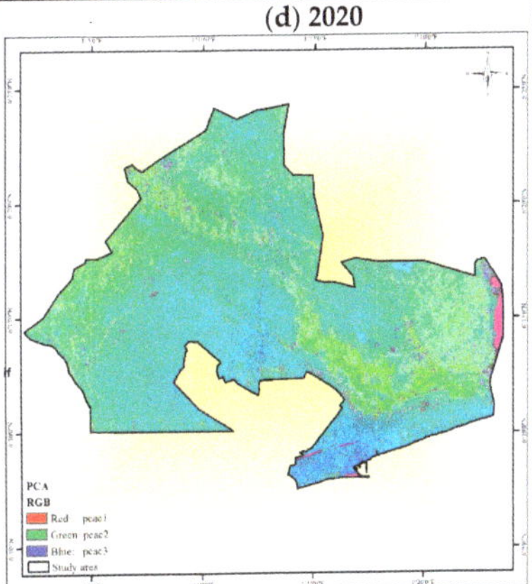

Bands	PCA1	PCA2	PCA3
MNDWI	0.0125	−0.6202	−0.7842
NDBI	0.9993	−0.0181	0.0303
SAVI	0.0330	0.7841	−0.6196

Figure 2. Principal Component Analysis results on index bands.

2.5. Accuracy Assessment

An evaluation of the quality of the classifications is performed by computing the confusion matrix [36] and the Kappa (K) coefficient expressed as the probability of correct classification [37,38]. The Kappa (K) coefficient, developed by Cohen [39], is a powerful and widely used statistical measure to assess the inter-raster agreement between variables [40].

The Kappa coefficient is extensively used because each element in the classification error matrix contributes to its calculation [41]. It lies between 0 and 1. The latter indicates total agreement and is often multiplied by 100 to give a percentage measure of the classification accuracy. Moreover, the Kappa values are subdivided into 3 groups: strong agreement (Kappa > 80%), moderate agreement (40% ≤ Kappa ≤ 80%), and poor agreement (Kappa < 40%) [41].

A stratified random sampling method was used in the accuracy assessment based on the observed data and a visual interpretation (expert knowledge). In addition, 300 points were generated for each of the classified images. Each point had specific color and pixel values, which were considered reference values. All the points that were randomly generated were then identified by the user and assigned to different LULC classes.

As suggested by [42], the overall accuracy of the two maps must be multiplied in order to evaluate the overall accuracy of the overlaid classification and verify the mis- classified errors. Furthermore, [42] emphasises the need not to proceed with the analysis when the computed accuracy is not acceptable. To determine whether the comparison is still useful, the accuracy value should always be compared to a previously defined threshold of acceptance. In this study, we adopted the accuracy threshold of 75% used by [42] for the final product of the two overlaid classifications.

2.6. Change Detection

The classified maps were compared using a change matrix in Orfeo ToolBox (version 8.1.2) and QGIS software (version 3.30) in order to obtain the changes in the different classes for the different periods considered in this study [43]. Moreover, the results of the LULC area distribution were used to compute the LULC trends, net change, percent change, and rate of LULC between the years 2007 and 2012, 2012 and 2016, and 2020 as well as for the periods 2007 and 2020. In order to calculate the percentage change (%), the initial and final LULC area coverages were compared using the following Equation (2):

$$\text{Rate of change}(\%) = (\text{Present LULC area} - \text{Previous LULC area})/(\text{Previous LULC area}) \times 100 \quad (2)$$

To obtain the annual rate of change for each LULC type, the rate of change of the final year was subtracted from the one of the initial year and then divided by the total number of years using the following Equation (3):

$$\text{Annual Rate of Change} = (\text{Final Year} - \text{Initial Year})/(\text{total number of Years}) \quad (3)$$

A post-classification change matrix was further used to analyze these changes. This post-classification change detection technique provides important information about the spatial distribution of LULC [44]. A land use change matrix displaying the LULC was generated from the classified images of 2007, 2012, 2016, and 2020. Finally, a change matrix from 2007 to 2020 was generated to assess the overall changes in the LULC classes between 2007 and 2020 for Greater Lomé.

3. Results

3.1. Land Use/Cover Dynamics in Relation to the Evolution of Buildings

The outskirts of the city of Lomé have undergone a remarkable change in LULC in relation to the evolution of buildings. The results of the pixel-level accuracy assessment are presented in Table 3.

Table 3. Accuracy of the Random Forest classification.

Year	Kappa Coefficient	Overall Accuracy
2020	0.93	95.62%
2016	0.92	94.34
2012	0.91	93.45%
2007	0.89	91.19%

Since the overall accuracy values are between 91% and 95%, we can confirm that the classification is quite good for an analysis of the dynamics of land use in relation to the evolution of buildings (Table 3).

In general, the proportion of the urbanized zone has increased from 2007 to 2020 to the detriment of the vegetation and agricultural zones of land use (Table 4). This progression is pronounced as one advances in time (Figure 3).

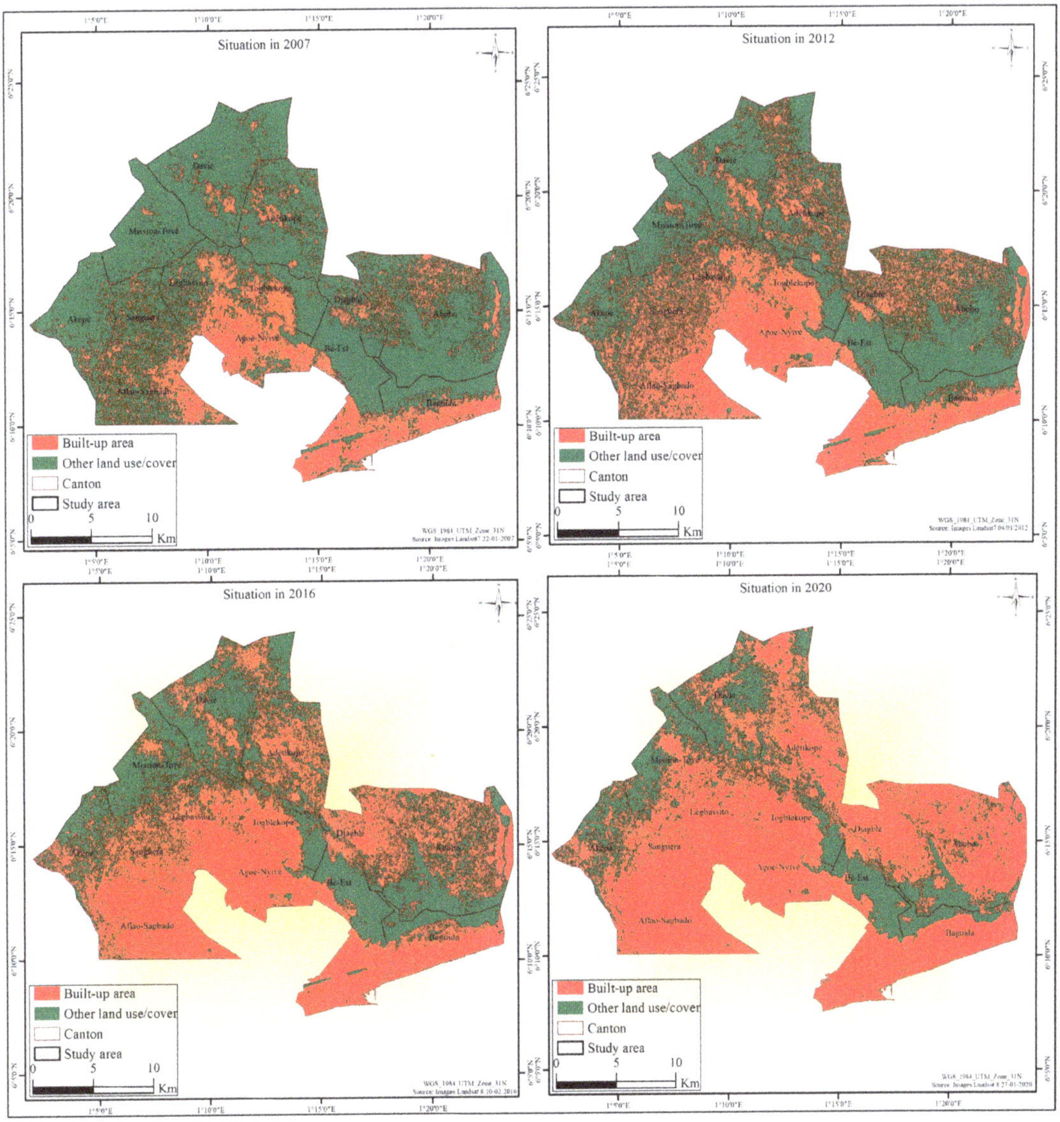

Figure 3. Land use/cover in relation to the evolution of buildings in Greater Lomé.

The analysis of the spatial distribution of built-up areas showed that the moderately built-up areas (areas with moderate density) have increased at the expense of the sparsely built-up areas (low-density areas) in some localities, such as Aflao Gakli, Togblékopé, Agoè-Nyivé, Adétikopé, and Djablé, which were peripheral areas at this period. At the same time, new sparsely built areas (low-density areas) appeared in these localities in addition to those that appeared in the outermost areas of greater Lomé, such as the Mission de Tové, Davié, Aképé, and Abobo during the same period. This situation compensates for the lightly built-up areas that have evolved into moderately built-up areas. The medium and low-density classes also increased slightly (Figure 4).

Table 4. Evolution of buildings on the outskirts of district of Greater Lomé.

Occupation	2007		2012		2016		2020	
Housing area	Area (ha)	Rate (%)	Area (ha)	Rate (%)	Area (ha)	Rate (%)	Area (ha)	Rate (%)
	15,481	25.25	20,924	34.12	25,624	41.79	35,521	57.93
Other classes	45,826	74.75	40,377	65.88	35,678	50.21	25,783	42.17

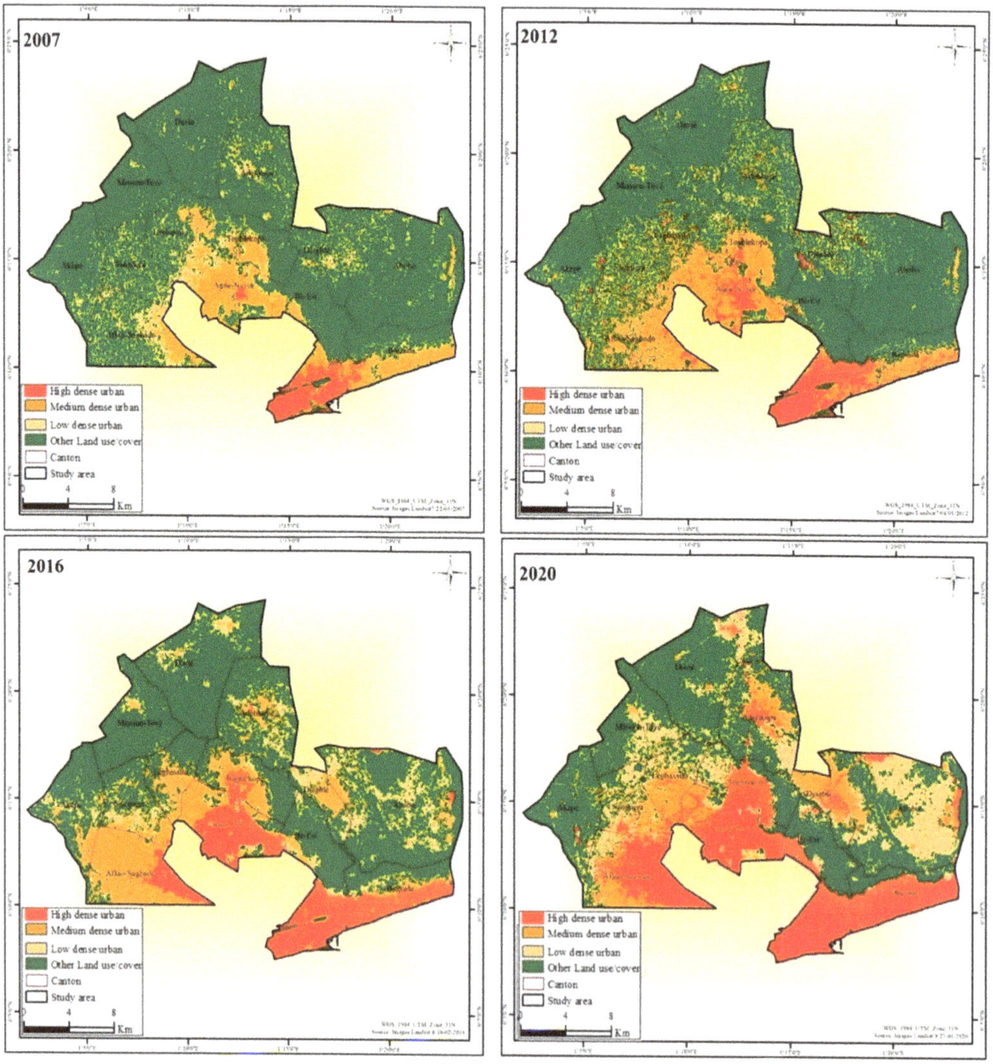

Figure 4. Dynamics of land use/cover centered on building subclasses.

In addition, the analysis of the LUCLC dynamics showed that the proportions of highly dense built-up areas increased significantly between 2007 and 2020, increasing from 1863 hectares to 14,485 hectares. From 2012 to 2016, the proportion of highly built-up, moderately built-up, and weakly built-up areas has increased, respectively, from 5956 to 6299 ha, 8207 to 10,674 ha, and 6761 to 8649 ha (Table 5).

Table 5. Land use/cover areas centered on the subclasses of buildings.

Type of Occupation	Area in Hectares and Percentage							
	2007 (%)		2012 (%)		2016 (%)		2020 (%)	
Dense zone	1863	3.04	5956	9.72	6299	10.28	14,485	23.63
Moderate-density zone	6137	10.01	8207	13.39	10,675	17.41	6058	9.88
Low-density zone	7479	12.20	6761	11.03	8649	14.11	14,978	24.43
Other land use/cover	45,826	74.75	40,377	65.87	35,679	58.20	25,783	42.06
Total area	61,305	100	61,301	100	61,302	100	61,302	100

It appears that from 2007 to 2020, there were significant conversions from other land cover classes (water, bare soil, vegetated areas, crops, and pasture) to those of built-up areas as a whole. On the other hand, there were no significant conversions of built-up areas into other classes of land use during the entire study period, and these have evolved over time from downtown Lomé to the peripheral areas of Greater Lomé. A gradual regression of green spaces was observed throughout the study area. This situation is largely due to the progression of urbanized spaces to the detriment of woodlots, fields, etc. These maps show that land use trends progress in the direction of urban sprawl. At the same time, we are witnessing the densification of buildings from the city center to the outskirts of Greater Lomé. The minimal conversions that were observed from built-up areas to other LULC classes can be justified for places where houses have been washed away by floods, rendering these places uninhabitable and subsequently occupied by vegetation.

Some direct observations during the fieldwork have enabled the identification of dwellings in some flood-prone areas of the Zio River (Figure 5). These houses were abandoned by the owners and remain dilapidated during the rainy season (Figure 6).

Figure 5. Constructions in the flood zone of the lower Zio Valley in Djagblé. Source: Fieldwork (2022).

Figure 6. Houses located in the lowlands and flooded in Baguida. Source: Fieldwork (2022).

Other areas that appeared to be lowlands but built by the population have also been listed. The inhabitants of these houses are forced to leave their homes during the rainy season and return to them during the dry season when the waters recede.

3.2. Trends in Annual Rate and Change Matrix of Land/Cover in Greater Lomé

The study revealed that there was an increase in the "High-density area" during the years 2007–2020, which shows an increase in the rate of change (22.79 ha/year) (Figure 7). The same trend of increase was observed in the "Low-density area" for the same period (2007–2020), with an estimated rate of change of 7.72 ha/year, except for the years 2007–2012 when a regression was observed (−1.12 ha/year).

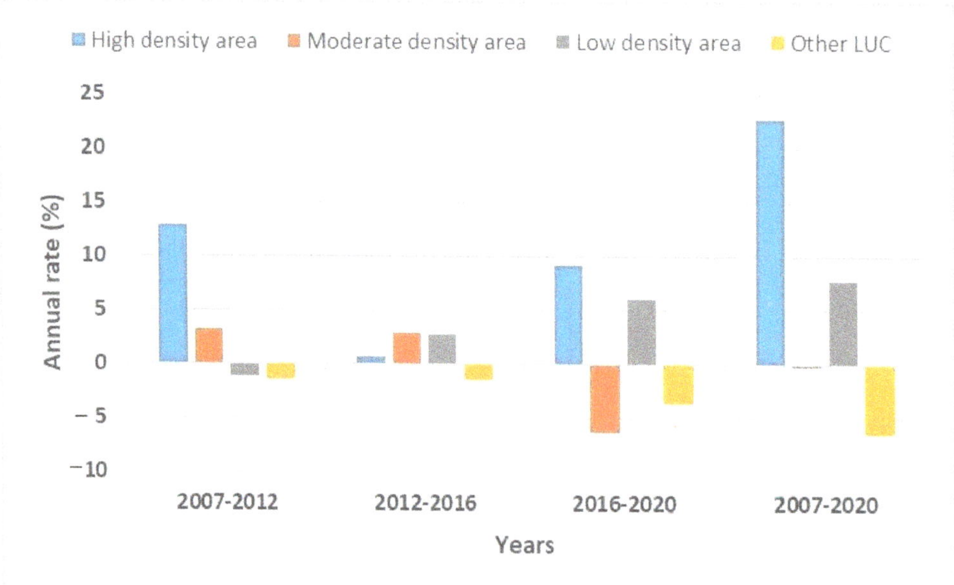

Figure 7. Trends in land use/cover.

The maximum change rate (−6.39 ha/year) was recorded for "Other classes" during the period 2007–2020. The "Moderate-density areas" were characterized by trends, with an increase in the built-up areas from 2007 to 2016 and an annual rate of change estimated at 3.23 ha/year (2007–2012) and 2.92 ha/year (2012–2016), respectively. However, the 2016–2020 period was marked by a regression in this land use, with an estimated annual rate of −6.29 ha/year. In summary, for the areas characterized by "Moderate-density area", the overall regression was estimated at −0.14 ha/year during the period 2007–2020 (Figure 7).

The LULC change matrix of Greater Lomé during the period 2007–2020 showed the conversion of one LULC class to another type of LULC. The increase in the magnitude of the "High-density area" from 2007 to 2020 is mainly due to the conversion of the "Moderate-density area" and "Other classes" (Table 6). The "Low-density areas" are the main land use/cover classes that converted to "High-density areas" from 2007 to 2012, resulting in a net increase of 1.62 ha/year of "High-density area". The decrease in "Other classes" of LULC is attributed to the conversion of this land use/cover type to other major classes (i.e., "High-density area", "Moderate-density area", and "Low-density area" (Table 6).

Table 6. Land use and land cover change matrix in Greater Lomé 2007–2020.

Years	LULC	Dense Zone (Ha)	Moderate-Density Zone (Ha)	LowDensity Zone (Ha)	Other Land Use/Cover (Ha)	Total Areas (Ha)
2007–2012	**Dense zone (Ha) (ha)**	1196.64	172.53	1.62	5508.99	6879.78
	Moderate-density zone (ha)	2716.29	3732.03	537.3	3155.49	10,141.11
	Low-density zone (ha)	419.58	1413.09	1286.82	1250.46	4369.95
	Other land use/cover (ha)	4158.54	831.96	54.36	34,880.31	39,925.17
	Total areas (ha)	**8491.05**	**6149.61**	**1880.1**	**44,795.25**	**61,316.01**
2012–2016	**Dense zone (Ha) (ha)**	1410.84	542.79	221.85	6490.62	8666.1
	Moderate-density zone (ha)	1159.65	7468.11	2106.09	3851.19	14,585.04
	Low-density zone (ha)	12.06	775.44	1311.75	109.26	2208.51
	Other land use/cover (ha)	4297.23	1354.77	730.26	29,474.1	35,856.36
	Total areas (Ha)	**6879.78**	**10,141.11**	**4369.95**	**39,925.17**	**61,316.01**
2016–2020	**Dense zone (Ha) (ha)**	3503.61	1435.41	0.18	10,121.4	15,060.6
	Moderate-density zone (ha)	818.37	3120.21	3.06	2017.8	5959.44
	Low-density zone (ha)	1513.53	8873.37	2205.18	2398.95	14,991.03
	Other land use/cover (ha)	2830.59	1156.05	0.09	21,318.21	25,304.94
	Total areas (ha)	**8666.1**	**14,585.04**	**2208.51**	**35,856.36**	**61,316.01**
2007–2020	**Dense zone (Ha) (ha)**	2279.07	357.75	6.3	12,417.48	15,060.6
	Moderate-density zone (ha)	1645.29	411.39	8.73	3894.03	5959.44
	Low-density zone (ha)	2648.88	5006.34	1857.15	5478.66	14,991.03
	Other land use/cover (ha)	1917.81	374.13	7.92	23,005.08	25,304.94
	Total areas (ha)	**8491.05**	**6149.61**	**1880.1**	**44,795.25**	**61,316.01**

4. Discussion

This study revealed that the changes in LULC concern three (03) categories: dense zone (heavily built-up area), moderate-density zone (moderately built-up area), and low-density zone (weakly built-up area). Indeed, as one progresses from the countryside towards the center of a city, one can observe a tightening of the plot design, a convergence and densification of the communication networks, and a change in the assignments of which the most spectacular is undoubtedly the densification of buildings [45]. Similar results were obtained during the study of landscape dynamics in the upper Ouémé basin (Benin Republic) using Landsat imagery [46]. Land use and land cover change is one of the major driving forces of global environmental change and is of major concern because of its impacts on various sectors of the economy [47]. These changes take place temporally and spatially such as the extent of area and the intensity of LULC. It appears that human activities have caused an increase in land utilization, change, and alteration [7]. Some large areas of Greater Lomé and mainly natural vegetation have been turned into "high-density" and "low-density" areas due to an increase in pressure from building activities. This pressure leads to a considerable loss of biodiversity due to the destruction of many natural habitats.

This confirms the idea that the Earth's surface is affected by the presence of anthropogenic activities in specific areas [48,49], exacerbated by the anthropocentric perspective of several societies [50]. The increase in the built-up area observed during the study period is a result of the construction of some buildings, roads, and infrastructure development as well as the high demand for land for settlements by the growing population in Greater Lomé. The population increase is mainly due to a high influx of people from other parts of the country for jobs and income generation opportunities [7].

The rapid urbanization of Greater Lomé has also led to a reduction in the proportion of land and degraded vegetation in peripheral areas in favor of buildings. It is recognized that the extension of urban areas can be influenced, among other things, by the configuration of space, in particular their accessibility and availability [51]. In addition, ref. [52], through his study on the spatio-temporal analysis of the dynamics of landscape conversion along the urban–rural gradient in Lubumbashi, revealed that the increase in the proportion of buildings to the detriment of vegetation in the landscape of peri-urban areas makes building space more limited in these areas. This could lead to land saturation, probably followed by land conflicts.

In the outlying areas as well as in the city center of Greater Lomé, the extent of the phenomenon of urbanization is considerable. The needs regarding housing and equipment accumulate from year to year. Like the results of this study, those of [53] on the Mediterranean coast of north-eastern Morocco showed the importance of socio-economic and political factors in the artificialization of the peri-urban spaces. Indeed, urban expansion is causing a decline and relocation of agricultural activities, in particular market gardening and tree farms, a large part of whose production is intended for the market of Greater Lomé. The strong urban expansion is likely to aggravate the problems of mobility, particularly those of the populations of the outlying districts whose individual means of transport are limited. In fact, there is an imbalance in the spatial distribution of infrastructure, equipment, and services between the north, west, and east of downtown Greater Lomé. These environments constitute peripheral zones where the urban extension continues.

The observation of the geomorphological landscape of Togblé, Adétikopé clearly shows that most of the agglomerations established in the alluvial plain regularly suffer from seasonal flooding. The modification of the natural conditions of runoff caused by each human development has consequences on the dynamics of the watercourse. Clearly, the hydrology of the Zio is deeply affected by its decimetric variations, with the consequence of increasing hydrological risks, in particular the frequency of floods. Despite the frequency of these risks, the lack of recent and continuous discharge data is a serious handicap for the detailed analysis of the impacts of LULC changes on flood risk in this area. Also, the current discontinuous series of flows has enabled flood frequency analysis of the Zio River. Gracius [54], in his study on the analysis of vulnerability to flood risk and land-use planning in the Commune of Cap-Haïtien, showed that this phenomenon of peri-urbanization could lead to an upsurge in flooding.

To these harmful practices, which have repercussions on the ecology of the river, in this case, the morpho-dynamics of the bed, several other human activities are added, which constitute, in reality, factors of aggravation of the floods, particularly the construction of houses and buildings in the bed of the Zio River. The same is true for the surrounding lowlands, which change the initial geomorphological characteristics of the plain. This aspect has been mentioned by other authors, particularly the development of human activities likely to alter the environment, which remains much more evident at a distance closer to urban centers [51]. This urbanization profoundly modifies the natural conditions of the water flow, which can cause flooding in these environments.

Scientific information on the spatial dynamics of built-up areas integrating the temporal dimension in Greater Lomé is, therefore, of great importance for decision-makers evaluating urban land use and planning decisions and for the scientific community discovering the causes and effects of land use changes on the management of urban spaces in Togo. However, in this study, spatial resolution is a key factor that can affect image quality and mapping accuracy. The mapping of built-up areas from medium spatial resolution Landsat images can, therefore, limit class discrimination and affect classification accuracy. These medium spatial resolution images may not be able to provide accurate information on the density or distribution of buildings in this area, even though the PCA were calculated on images composed of indices derived from the primary image channels, maximized band information, and eliminated noise. Indeed, the quality of the classifications was assessed by calculating the confusion matrix [36] and the Kappa K index proposed by [39]. The Kappa index is expressed as the probability of correct classification on a scale of 0 to 1.

The Random Forest (RF) algorithm was used for mapping on the basis of over 300 training pixels, where classes were determined during the field survey. The validation of the classification was based on control points collected in the field. The Random Forest (RF) algorithm, developed by [30] was chosen for its good land use prediction capabilities [31] in the case of temporal analysis [32]. Several authors have shown that land cover classifications using RF outperform classifications using other types of algorithms, such as maximum likelihood classification [32]. The RF provides an algorithm for estimating missing values and the flexibility to perform several types of data analysis, including regression, classification, survival analysis, and unsupervised learning [33].

This is a non-parametric supervised classification algorithm that combines the decision tree algorithm with an aggregation technique. It is included in the "Random Forest" package of the "R" software. (version 4.3.1). The algorithm randomly selects a sample of observations and a sample of variables several times to produce a large number of small classification trees. These small trees are then grouped together and a majority voting rule is applied to determine the final category [30]. In order to maximize the band information and eliminate noise so that the discrimination of the classes studied can be improved, a PCA was calculated on the images composed of indices derived from the primary channels of the satellite images and the main bands used. The indices used included the Normalized Built-up Difference Index (NDBI), the Soil Adjustment Vegetation Index (SAVI), and the Normalized Moisture Difference Index (MNDWI) [55].

5. Conclusions

This study has highlighted the interest in using Landsat images to study the evolution of human habitats in urban and peri-urban areas in order to improve the understanding of their dynamics over time. A meticulous choice of satellite images and the method of classification enabled the obtainment of a clear and relevant rendering. The results showed that the dynamics of land use along the urban–rural gradient were characterized in 13 years (between 2007 and 2020) by a clear progression of buildings to the detriment of vegetation in the peri-urban zones. They provided a good understanding of the dynamics of these changes and indicated a strong dynamic in the landscape structure of Greater Lomé, marked by a rapid extension of built-up areas. Furthermore, the results of this study showed a marked extension in the peripheral areas of Greater Lomé, particularly towards the north and west, to the detriment of agricultural and wooded areas. In addition, towards the east, an evolution of the buildings was observed, but it was not continuous. The presence of the lower Zio valley, which constituted a green band on the images, caused the discontinuity of the evolution of the buildings in the east of the study area. However, the observation of the satellite images showed that the evolution of the buildings has narrowed this band of discontinuity, which was more stretched in 2020 than in 2016.

Finally, direct observations of the entire study area showed that the minimal conversions observed for built-up areas to other land use classes could be justified, for the most part, in places where houses have been washed away by floods. These places, which have become uninhabitable, were subsequently occupied by vegetation. It becomes necessary to carry out studies on the effects of the occupation of the lower Zio Valley on the dynamics of floods in Greater Lomé in order to better understand the problems and suggest solutions for mitigating their negative consequences, which have become more severe.

Author Contributions: Conceptualization, T.-H.B. and K.A.; methodology, K.A.; software, A.K.D.H.; validation, K.A., K.K. (Kossi Komi), and A.K.D.H.; formal analysis, K.A.; investigation, K.S.G.; resources, K.K. (Kossi Komi); data curation, A.K.D.H.; writing—original draft preparation, K.K. (Kossi Komi); writing—review and editing, J.-B.B.Z.; visualization, B.P.; supervision, K.K. (Kouami Kokou); project administration, K.A.; funding acquisition, K.A. All authors have read and agreed to the published version of the manuscript.

Funding: This study received financial support from The Regional Centre of Excellence on Sustainable Cities in Africa (CERViDA-DOUNEDON) through the funding of the project entitled "Opportunity Study for the Restoration of the Forest Landscape to Fight Against Urban Heat Islands (UHI) in the Context of Climate Change in the Greater Lome, Maritime Region"(grant N°5955 crédit IDA) and the WASCAL (West African Science Service Centre on Climate Change and Adapted Land Use) through the FURIFLOOD project.

Data Availability Statement: Data is contained within the article.

Acknowledgments: The authors are grateful to CERViDA_DOUNEDON, the Association of African Universities (AUA), and the World Bank as well as the WASCAL for funding this research.

Conflicts of Interest: The authors declare no conflicts of interest.

References

1. Fang, H.; Gu, Q.; Xiong, W.; Zhou, L.-A. Demystifying the Chinese housing boom. *NBER Macroecon. Annu.* **2016**, *30*, 105–166. [CrossRef]
2. Biakouye, K.A. Lomé Au-Delà de Lomé: Étalement Urbain et Territoires Dans une Capitale d'Afrique Sub-Saharienne. Ph.D. Thesis, University of Paris, Paris, France, 17 November 2014. Available online: https://www.theses.fr/2014PA100138 (accessed on 10 October 2023).
3. Benítez, A.; Prieto, M.; González, Y.; Aragón, G. Effects of tropical montane forest disturbance on epiphytic macrolichens. *Sci. Total Environ.* **2012**, *441*, 169–175. [CrossRef] [PubMed]
4. Fodé, S.; Kouami, K.; Mohamed, D.; Youssouf, C.; Sidiki, K.; Souleymane, K. Analyse Diachronique, Grâce Aux Images Landsat, de la Dynamique Spatiale des Forêts Sacrées du Haut Bassin du Niger en République de Guinée. *Rev. Fr. Photogramm. Télédétect.* **2022**, *223*, 250–266. [CrossRef]
5. Surya, B.; Salim, A.; Hernita, H.; Suriani, S.; Menne, F.; Rasyidi, E.S. Land Use Change, Urban Agglomeration, and Urban Sprawl: A Sustainable Development Perspective of Makassar City, Indonesia. *Land* **2021**, *10*, 556. [CrossRef]
6. Fashae, O.A.; Tijani, M.N.; Adekoya, A.E.; Tijani, S.A.; Adagbasa, E.G.; Aladejana, J.A. Comparative Assessment of the Changing Pattern of Land cover along the Southwestern Coast of Nigeria using GIS and Remote Sensing techniques. *Sci. Afr.* **2022**, *17*, e01286. [CrossRef]
7. Tiamgne, X.T.; Kalaba, F.K.; Nyirenda, V.R. Land use and cover change dynamics in Zambia's Solwezi copper mining district. *Sci. Afr.* **2021**, *14*, e01007. [CrossRef]
8. Nse, O.U.; Okolie, C.J.; Nse, V.O. Dynamics of land cover, land surface temperature and NDVI in Uyo City, Nigeria. *Sci. Afr.* **2020**, *10*, e00599. [CrossRef]
9. Coutard, O.; May, N.; Veltz, P. *La Ville Éclatée: Enjeux, Logistiques et Modalités d'une Régularisation Économique, Sociale et Territoriale*; Paris, France, 1996. Available online: https://medias.vie-publique.fr/data_storage_s3/rapport/pdf/974055800.pdf (accessed on 1 October 2023).
10. Förster, T.; Ammann, C. Les villes africaines et le casse-tête du développement. Acteurs et capacité d'agir dans la zone grise urbaine. *Int. De Polit. Dév.* **2018**, *10*, 1–23. [CrossRef]
11. Polorigni, B.; Radji, R.; Kokou, K. Perceptions, tendances et préférences en foresterie urbaine: Cas de la ville de Lomé au Togo. *Eur. Sci. J.* **2014**, *10*, 261–277.
12. Salomon, W.; Useni Sikuzani, Y.; Sambieni, K.R.; Kouakou, A.T.M.; Barima, Y.S.S.; Théodat, J.M.; Bogaert, J. Land Cover Dynamics along the Urban–Rural Gradient of the Port-au-Prince Agglomeration (Republic of Haiti) from 1986 to 2021. *Land* **2022**, *11*, 355. [CrossRef]
13. Wiwoho, B.S.; Phinn, S.; McIntyre, N. Two Decades of Land-Use Dynamics in an Urbanizing Tropical Watershed: Understanding the Patterns and Drivers. ISPRS. *Int. J. Geo-Inf.* **2023**, *12*, 92. [CrossRef]
14. Barros, J.L.; Tavares, O.A.; Santos, P.P. Land use and land cover dynamics in Leiria City: Relation between peri-urbanization processes and hydro-geomorphologic disasters. *Nat. Hazards* **2021**, *106*, 757–784. [CrossRef]
15. Awuh, M.E.; Officha, M.C.; Okolie, A.O.; Enete, I.C. Land-Use/Land-Cover Dynamics in Calabar Metropolis Using a Combined Approach of Remote Sensing and GIS. *J. Geogr. Inf. Syst.* **2018**, *10*, 398–414. [CrossRef]
16. Angel, S.; Parent, J.; Civco, D.L. The fragmentation of urban landscapes: Global evidence of a key attribute of the spatial structure of cities, 1990–2000. *Environ. Urban.* **2012**, *24*, 249–283. [CrossRef]
17. Herold, D.; Couclelis, H.; Clarke, K.C. The role of spatial metrics in the analysis and modeling of urban land use change. *Comput. Environ. Urban Syst.* **2005**, *29*, 369–399. [CrossRef]

18. Almeida, C.M.D.; Monteiro, A.M.V.; Câmara, G.; Soares-Filho, B.S.; Cerqueira, G.C.; Pennachin, C.L.; Batty, M. GIS and remote sensing as tools for the simulation of urban land-use change. *Int. J. Remote Sens.* **2005**, *26*, 759–774. [CrossRef]
19. Mohamed, M. *Le Centre-Ville de Lomé, Évolution de la Situation Foncière et de la Trame Urbaine*; ORSTOM: Lomé, Togo, 1983; 99p, Available online: https://www.documentation.ird.fr/hor/fdi:04087 (accessed on 11 February 2023).
20. Dziwonou, Y. Croissance Urbaine et Mécanismes Fonciers. Contribution à L'élaboration d'une Géomatique Cadastrale: Le Cas de la Ville de Lomé. Ph.D. Thesis, Géographie Urbaine et Aménagement, Université de Lomé, Lomé, Togo, 2000; 579p. Available online: https://www.theses.fr/1987TOU20006 (accessed on 1 March 2023).
21. Gondo, R.; Kolawole, O.D.; Mfundisi, K.B. Land use and land cover changes along the Boteti-Thamalakane River system in Ngamiland District, Botswan. *Sci. Afr.* **2023**, *20*, e01595.
22. Tian, P.; Li, J.; Gong, H.; Pu, R.; Cao, L.; Shao, S.; Shi, Z.; Feng, X.; Wang, L.; Liu, R. Research on Land Use Changes and Ecological Risk Assessment in Yongjiang River Basin in Zhejiang Province, China. *Sustainability* **2019**, *11*, 2817. [CrossRef]
23. Lambony, G. A Travers Images et Pratiques: Le Fait Citadin en Afrique Noire. Etude Comparée de Lomé (Togo) et de Harare (Zimbabwe). Ph.D. Thesis, EHESS, Paris, France, 1993; 592p.
24. Gbafa, K.S.; Tiem, S.; Kokou, K. Characterization of rainwater drainage infrastructure in the city of Lomé (Togo, West Africa). *Eur. Sci. J.* **2017**, *13*, 478–496. [CrossRef]
25. Guézéré, A. Oléyia (Taxi Moto) Acteurs Et Usagers d'un Mode de Transport Artisanal Récent à Lomé. Ph.D. Thesis, Université de Lomé, Lomé, Togo, 2008; 455p.
26. INSEED. Résultats Définitifs du RGPH-5 de Novembre 2022, République Togolaise. 2023. Available online: https://hcte-suisse.ch/wp-content/uploads/2023/04/Depliant-Resultats-Definitifs_RGPH5_02Avril2023.pdf (accessed on 9 September 2023).
27. Liu, J.; Heiskanen, J.; Aynekulu, E.; Pellikka, P.K.E. Seasonal variation of land cover classification accuracy of landsat 8 images in Burkina Faso. In Proceedings of the 36th International Symposium on Remote Sensing of Environment (ISRSE 2015), Berlin, Germany, 11–15 May 2015; Volume XL-7/W3.
28. Ozcan, O.; Aksu, G.A.; Erten, E.; Musaoglu, N.; Cetin, M. Degradation monitoring in Silvo-pastoral systems: A case study of the Mediterranean region of Turkey. *Adv. Space Res.* **2019**, *63*, 172–189. [CrossRef]
29. Gutman, G.; Byrnes, R.A.; Masek, J.; Covington, S.; Justice, C.; Franks, S.; Headley, R. Towards monitoring land-cover and land-use changes at a global scale: The global land survey 2005. *Photogramm. Eng. Remote Sens.* **2008**, *74*, 6–10.
30. Breiman, L. Random Forests. *Mach. Learn.* **2001**, *45*, 5–32. [CrossRef]
31. Gislason, P.O.; Benediktsson, J.A.; Sveinsson, J.R. Random forests for land cover classification. Pattern recognition. *Letters* **2006**, *27*, 294–300.
32. Schneider, A. Monitoring land cover change in urban and peri-urban areas using dense time stacks of Landsat satellite data and a data mining approach. *Remote Sens. Environ.* **2012**, *124*, 689–704. [CrossRef]
33. Grinand, C.; Rakotomalala, F.; Gond, V.; Vaudry, R.; Bernoux, M.; Vieilledent, G. Estimating deforestation in tropical humid and dry forests in Madagascar from 2000 to 2010 using multi-date Landsat satellite images and the random forests classifier. *Remote Sens. Environ.* **2013**, *139*, 68–80. [CrossRef]
34. Sun, H.; Forsythe, W.; Waters, N. Modeling urban land use change and urban sprawl: Calgary, Alberta, Canada. *Netw. Spat. Econ.* **2007**, *7*, 353–376. [CrossRef]
35. Gao, B.-C. NDWI—A normalized difference water index for remote sensing of vegetation liquid water from space. *Remote Sens. Environ.* **1996**, *58*, 257–266. [CrossRef]
36. Sari, I.L.; Weston, C.J.; Newnham, G.J.; Volkova, L. Assessing Accuracy of Land Cover Change Maps Derived from Automated Digital Processing and Visual Interpretation in Tropical Forests in Indonesia. *Remote Sens.* **2021**, *13*, 1446. [CrossRef]
37. Hansen, M.; Loveland, T. A Review of large area monitoring of land cover change using Landsat data. *Remote Sens. Environ.* **2012**, *122*, 66–74. [CrossRef]
38. Cabral, P. Délimitation d 'aires urbaines à partir d 'une image Landsat ETM+: Comparaison de méthodes de classification. *Can. J. Remote Sens.* **2007**, *33*, 422–430. [CrossRef]
39. De Sherbinin, A.; Carr, D.; Cassels, S.; Jiang, L. Population and environment. *Annu. Rev. Environ. Res.* **2007**, *32*, 345–373. [CrossRef] [PubMed]
40. Alam, A.; Bhat, M.; Maheen, M. Using Landsat satellite data for assessing the land use and land cover change in Kashmir valley. *GeoJournal* **2020**, *85*, 1529–1543. [CrossRef]
41. Foody, G.M. Status of land cover classification accuracy assessment. *Remote Sens. Environ.* **2020**, *80*, 185–201. [CrossRef]
42. Serra, P.; Pons, X.; Sauri, D. Post-classification change detection with data from different sensors: Some accuracy considerations. *Int. J. Remote Sens.* **2003**, *24*, 3311–3340. [CrossRef]
43. Munsi, M.; Malaviya, S.; Oinam, G.; Joshi, P. A landscape approach for quantifying land-use and land-cover change (1976–2006) in middle Himalaya. *Reg. Environ. Chang.* **2009**, *10*, 145–155. [CrossRef]
44. Shalaby, A.; Tateishi, R. Remote sensing and GIS for mapping and monitoring land cover and land-use changes in the Northwestern coastal zone of Egypt. *Appl. Geogr.* **2007**, *27*, 28–41. [CrossRef]
45. Baudot, Y. Télédétection Aérospatiale et Analyse Géographique de la Population des Villes dans les Pays en Développement. Ph.D. Thesis, Faculté des Sciences, Université Catholique de Louvain, Ottignies-Louvain-la-Neuve, Belgium, 1994; p. 233.

46. Louise, L.; Montpellier, H.; ESCAPE. *Analyse Diachronique de la Dynamique Paysagère sur le Bassin Supérieur de l'Ouémé (Bénin) à Partir de l'imagerie Landsat et MODIS: Cas d'étude du Communal de Djougou*; Rapport d'étude; Hydrosciences Montpellier: Montpellier, France, 2012.
47. Matlhodi, B.; Kenabatho, P.K.; Parida, B.P.; Maphanyane, J.G. Evaluating Land Use and Land Cover Change in the Gaborone Dam Catchment, Botswana, from 1984–2015 Using GIS and Remote Sensing. *Sustainability* **2019**, *11*, 5174. [CrossRef]
48. De Sherbinin, A.; Schiller, A.; Pulsipher, A. The Vulnerability of Global Cities to Climate Hazards. *Environ. Urban.* **2007**, *19*, 39–64. [CrossRef]
49. Gurmessa, F. Forest loss and climate change in Ethiopia. *Res. J. Agric. Environ. Manag.* **2015**, *4*, 216–224.
50. Emiru, T.; Naqvi, H.R.; Athick, M.A. Anthropogenic impact on land use land cover: Influence on weather and vegetation in Bambasi Wereda, Ethiopia. *Spat. Inf. Res.* **2018**, *26*, 427–436. [CrossRef]
51. Bamba, I.; Barima, Y.S.S.; Bogaert, J. Influence de la densité de la population sur la structure spatiale d'un paysage forestier dans le bassin du Congo en, R.D. Congo. *Trop. Conserv. Sci.* **2010**, *3*, 31–44. [CrossRef]
52. Useni, S.Y. Analyse Spatio-Temporelle des Dynamiques D'anthropisation Paysagère le long du Gradient Urbain-Rural à Lubumbashi (Haut-Katanga, République Démocratique du Congo. Ph.D. Thesis, Université de Lubumbashi, Lubumbashi, Democratic Republic of the Congo, 2017; 205p.
53. Mouzouri, M.; Irzi, Z.; Brahimi, A. *Etude de la Dynamique de L'occupation du sol da la Plaine Côtière de SAÏDIA (Littoral Méditerranéen du Nord-Est du Maroc) Durant la Période 2001–2009*; Colloque International des Utilisateurs du SIG: Oujda, Marocco, 2016; pp. 287–292.
54. Gracius, J.G. Vulnérabilités au Risque D'inondations et Aménagement du Territoire. Master's Thesis, Spécialisation en Gestion des Risques Naturels de l'Université de Liège, Cap-Haïtien, Haïti, 2016; 70p.
55. Rouse, J.W.; Haas, R.H.; Schell, J.A.; Deering, W.D. *Monitoring Vegetation Systems, the Great Plains with ERTS*; Nasa Special Publication; Nasa: Washington, DC, USA, 1974; Volume 351, p. 309.

land

MDPI

Article

Unveiling the Spatio-Temporal Evolution and Key Drivers for Urban Green High-Quality Development: A Comparative Analysis of China's Five Major Urban Agglomerations

Tonghui Yu [1], Xuan Huang [1], Shanshan Jia [1] and Xufeng Cui [2,*]

[1] School of Business, Xinyang Normal University, Xinyang 464000, China; yuth1987@xynu.edu.cn (T.Y.); huang1004@xynu.edu.cn (X.H.); js1218@xynu.edu.cn (S.J.)
[2] School of Business Administration, Zhongnan University of Economics and Law, Wuhan 430073, China
* Correspondence: cxf@zuel.edu.cn

Abstract: Faced with the dual challenges of ecological degradation and economic deceleration, promoting urban green high-quality development (UGHQD) is pivotal for achieving economic transformation, ecological restoration, and regional sustainable development. While the existing literature has delved into the theoretical dimensions of UGHQD, there remains a notable dearth of empirical studies that quantitatively assess its developmental levels, spatio-temporal evolution, and driving factors. This study examines 107 cities of China's five major urban agglomerations from 2003 to 2020, constructing a comprehensive evaluation indicator system for UGHQD. By employing methodologies, including the Dagum Gini coefficient, Kernel density estimation, Markov chain, and geographical detector, this study extensively assesses the spatial difference, dynamic evolution, and underlying driving forces of UGHQD in these urban agglomerations. The findings indicate: (1) The UGHQD level of the five major urban agglomerations has witnessed a consistent year-over-year growth trend, with coastal agglomerations like the Pearl River Delta (PRD) and Yangtze River Delta (YRD) outperforming others. (2) Pronounced regional differences exist in UGHQD levels across the urban agglomerations, with inter-regional differences primarily contributing to these differences. (3) The dynamic evolution of UGHQD distribution generally transitions from a centralized to a decentralized pattern, with a marked "club convergence" characteristic hindering cross-type leaps. (4) While a range of factors drive UGHQD in these agglomerations, technological innovation stands out as the principal factor inducing spatial differentiation. The comprehensive analysis and findings presented in this research not only contribute to academic knowledge but also hold practical implications for policymakers and practitioners striving for environmentally conscious land use planning and urban management.

Keywords: urban green high-quality development (UGHQD); spatio-temporal differences; Dagum Gini coefficient; Markov chain

1. Introduction

Currently, the world faces increasingly serious climate change and ecological degradation issues, which are progressively becoming major challenges to green economic recovery and sustainable social development [1]. In this context, there is a growing international consensus on the urgent need to foster green growth and cultivate a low-carbon, circular economy [2,3]. The United Nations World Commission on Environment and Development initially introduced the concept of sustainable development in 1987 through *Our Common Future*. Subsequently, David Pearce expanded upon this concept in his seminal work *Blueprint for a Green Economy*, which closely aligns with the central tenets of sustainable development and provides key guidelines for balancing economic growth with environmental preservation [4–6]. As the world's highest consumer of energy and a significant source of carbon emissions, China has undergone a phase of rapid economic growth marked by intensive resource use and elevated energy consumption [7,8]. Although China has successfully realized its aspirational objective of creating a comprehensively moderately prosperous society, this achievement has come at considerable environmental and resource costs [9,10]. Given the dual challenges of limited resources and ecological vulnerability, China's traditional development model, primarily reliant on increasing inputs of production factors, is no longer sustainable [11]. Consequently, China urgently needs to accelerate the transformation of its development model and shift its economy from being factor-driven to one focused on green, low-carbon development through "quality change", "efficiency change", and "driving force change" [12,13].

As key vehicles for industrialization and modernization, cities are experiencing significant expansion in size and population. Forecasts suggest that by 2050, nearly 67% of the global populace will inhabit urban areas. This heightened urban concentration promotes industrialization while simultaneously exacerbating environmental and resource-related challenges [14,15]. Using China as an illustrative case, data from the National Bureau of Statistics indicate that urban energy consumption constituted 85% of the national total in 2020 [16], thereby aggravating environmental degradation. Recently, urban agglomerations—advanced forms of spatial organization at mature stages of urban development—have emerged as strategic epicenters for regional economic and social advancement. China's 14th Five-Year Plan (2020–2025) emphasizes the role of such agglomerations in propelling innovations in intelligent manufacturing; the digital economy; and sustainable, low-carbon industries [17]. Notably, the five major urban agglomerations, the Yangtze River Delta (YRD), the Pearl River Delta (PRD), Beijing-Tianjin-Hebei (BTH), the Middle Reaches of the Yangtze River (MYR), and Chengdu-Chongqing (CC), are the most economically advanced and talent-attractive regions in China, serving as primary conduits linking China's economy with the global economic landscape [18]. However, due to the influence of various factors, such as population, geography, policies, educational resources, and development patterns, urban agglomerations differ greatly with respect to development potential, technological innovation capacity, and infrastructure development [19]. Thus, will these differences in development factors lead to serious spatial imbalances in urban green high-quality development (UGHQD) among urban agglomerations? Moreover, China is at a pivotal juncture in its urbanization trajectory, and high-pollution and high-energy-consumption industries still play an important role in pulling economic growth in development [20]. As crucial nodes of China's new urbanization agenda, these agglomerations are inevitably the hotspots for various environmental pollution issues [21,22]. Within this context, UGHQD, as an innovative development concept that is green-oriented and takes into account the synergistic progress of the economy, society, and technological innovation as well as the ecological environment, offers a strategic approach to solving the problem.

In view of this, this study focuses on a representative sample of 107 cities in China's five major urban agglomerations selected for their typicality in terms of economic scale, regional attributes, and policy environments. A multidimensional evaluation framework is constructed to assess UGHQD, incorporating factors related to economic development, social livelihood, ecological environment, and technological innovation. Initially, the entropy value method is employed to quantify composite scores for UGHQD across these urban agglomerations for the period 2003–2020. Subsequently, spatial differences and the sources of the differences are scrutinized using Dagum's Gini coefficient and its decomposition techniques. Kernel density estimation and Markov chain analyses are applied to systematically examine the spatio-temporal distribution and evolutionary dynamics of UGHQD within these urban agglomerations. Lastly, the geographical detector is employed to examine the driving factors behind spatial differences in UGHQD. The theoretical contributions of this study can be distilled into three main points. Firstly, this study delves deeper by shifting the empirical research focus to the city level, moving beyond the more commonly referenced provincial samples in the existing literature. Such an approach paves the way for more precise and detailed empirical conclusions. An in-depth examination of spatial differences and the dynamic evolution of UGHQD within China's five major urban agglomerations helps sharpen our understanding of strategies and pivotal aspects for the future promotion of UGHQD. Secondly, in contrast to existing studies that predominantly utilize green total factor productivity as the primary indicator, this paper constructs a comprehensive evaluation system for UGHQD. This system evaluates based on four dimensions, economic development, social livelihood, ecological environment, and technological innovation, thereby more closely aligning with the objectives of UGHQD. Thirdly, beyond simply examining the spatio-temporal evolution of UGHQD in China's five major urban agglomerations, this study employs appropriate methodologies to unveil the driving factors in spatial differences of UGHQD, offering empirical insights for crafting sustainable development strategies for contemporary urban agglomerations.

2. Literature Review

As a hot topic in recent years, high-quality development serves as a major strategic plan designed to address the shortcomings of previous development approaches, overcome technological bottlenecks, and reduce income inequality [23]. It signifies a strategic evolutionary path and an institutional innovation process that shifts economic growth from "scale expansion" and "factor-driven" to innovation-driven [24,25]. Integrating the concept of high-quality development into the framework of socialist modernization is pivotal as China's economy transitions to a new era [26]. In this context, "green" has emerged as an essential aspect of economic development. The new focus is on balancing economic efficiency with environmentally friendly development models, taking into full account resource utilization efficiency and environmental pollution emissions [27,28]. UGHQD represents a high degree of integration between "green development" and "high-quality development" [29]. Its connotations are multifaceted: it advocates for low-carbon, low-pollution, and high-efficiency production modes while sustaining economic growth [30]. These modes foster the advancement and deployment of clean energy, refine the composition of the energy matrix, alleviate ecological burdens, and improve resource utilization efficiency [31,32]. Furthermore, UGHQD places a heightened focus on social sustainability, emphasizing that economic growth should not solely prioritize material gains but that social benefits such as equity, sharing, and inclusivity should also be considered [33]. This necessitates addressing issues like job creation, narrowing income gaps, and improving education and healthcare conditions [34,35]. Lastly, UGHQD underscores the importance of innovation and technological progress [36,37]. This development paradigm, which places a premium on innovation, is positioned to markedly elevate production efficiency and expedite the transition and enhancement of industrial structures [38]. Overall, UGHQD aims to envision a "win-win-win" scenario for the economy, society, and the environment.

In the realm of assessing the UGHQD level, existing scholarly contributions predominantly fall into two categories: single-indicator and multi-indicator approaches. The single-indicator method primarily evaluates efficiency to determine the UGHQD level, typically using green total factor productivity as the metric. Earlier methodologies tended to rely on traditional approaches such as the Cobb–Douglas production function [39,40]. However, these traditional methods often overlooked the extensive impact of environmental costs and undesirable outputs on economic efficiency, revealing limitations in non-parametric estimations [41]. Consequently, they have evolved into more sophisticated methods, including data envelopment analysis (DEA) [42], super efficiency slacks-based measure (SBM) [43], and stochastic frontier analysis (SFA) [44–46]. Although a single total factor productivity measure cannot fully encapsulate all aspects of social economic operation and falls short of capturing the complex nuances of UGHQD, it does provide valuable insights for establishing a more comprehensive measurement system for UGHQD [47,48]. As for multi-indicator measurements, existing studies primarily focus on comprehensive assessments of high-quality developmental levels. Early studies often equated economic growth quality with efficiency, necessitating a more precise delineation of its broader implications [49]. In recent years, scholars have increasingly adopted a more comprehensive view of economic growth quality, arguing that it should possess a richer connotation to complement development speed. Consequently, factors such as green growth [50], industrial upgrading [51], and technological innovation [52] have been incorporated into assessments of economic growth quality. Existing literature mainly employs the five development concepts of innovation, coordination, greenness, openness, and sharing [53] or their derivatives as subsystems [54], or studies construct an indicator system from three subsystems: economic, social, and ecological [55]. Some scholars have also expanded the understanding of high-quality development, constructing indicator systems that consider dimensions like economic structure optimization, resource allocation efficiency, social services, and industrial recycling development [56,57].

From a regional perspective, factors such as geographical location, resource endowment, and environmental regulations collectively contribute to notable differences in green development and high-quality development levels across regions [58,59]. Numerous scholars have conducted in-depth analyses to elucidate these differences' characteristics and underlying causes, aiming to formulate more precise and targeted strategies for regional synergistic development [60,61]. In the study of regional differences in green development levels and convergence characteristics, the Theil index and β-convergence model are widely used [62]. However, the Theil index cannot describe the dynamic distribution of subgroup samples, nor can it decompose the sources of spatial differences in detail, limiting the accuracy of spatial difference analysis. Researchers have increasingly turned to the Dagum Gini coefficient to circumvent these issues and investigate the regional differences in urban green development along with their root causes [63]. The Dagum Gini coefficient addresses data overlap within sample sets and can effectively identify and trace the sources of regional differences. To describe the evolution of absolute differences in green development levels more accurately, scholars have employed Kernel density estimation methods to analyze the distributional dynamics of regional green development [64,65]. Building on this, some have introduced the Markov transition probability matrix to enhance data expressiveness and explanatory power [66]. While the spatial Durbin model [67] and the dynamic panel quantile model [68] are commonly applied to study the causes of regional differences, they fail to address endogeneity issues in regression analyses adequately. Researchers have adopted the geographical detector model to resolve this, allowing for a comprehensive assessment of interactions among multiple factors and thereby identifying the causes of differences [69]. The geographical detector model is gaining traction in economic, social, and environmental research, offering a valuable tool for scholars in these domains [70,71].

In surveying existing studies, it is evident that research on UGHQD has yielded certain results, establishing a theoretical foundation for this study while leaving room for further exploration. First, concerning the measurement of UGHQD levels, existing studies have predominantly utilized metrics such as green total factor productivity. Comprehensive evaluation frameworks based on an integrated set of indicators are notably scarce. This limitation often hampers the accurate and comprehensive depiction of UGHQD. Second, the majority of existing studies have exclusively explored the spatio-temporal differences in either green development or high-quality development, with few examining the spatio-temporal disparities in UGHQD and the driving factors that contribute to them. As a critical indicator for sustainable urban development, the study of these differences and their drivers holds practical significance; its absence hinders further in-depth research in this area. Third, the existing literature predominantly focuses on nations, provinces, and occasionally inter-provincial regions or watersheds as units of empirical analysis. In contrast, studies employing cities as the spatial dimension for investigation are relatively rare. This paper selects cities as the spatial research unit to more accurately capture the nuanced and precise dynamics of UGHQD.

3. Materials and Methods

3.1. Study Area

Guided by policy frameworks, such as *Outline of the Beijing-Tianjin-Hebei Cooperative Development Plan, Pearl River Delta Reform and Development Plan (2008–2020), Integrated Development Plan for the Yangtze River Delta Region, Development Plan for the Middle Reaches of the Yangtze River Urban Agglomeration*, and *Development Plan for the Chengdu-Chongqing Urban Agglomeration*, and drawing on existing research [72], this study systematically categorizes 107 cities within these five major urban agglomerations. The specific distribution of these cities is detailed in Figure 1.

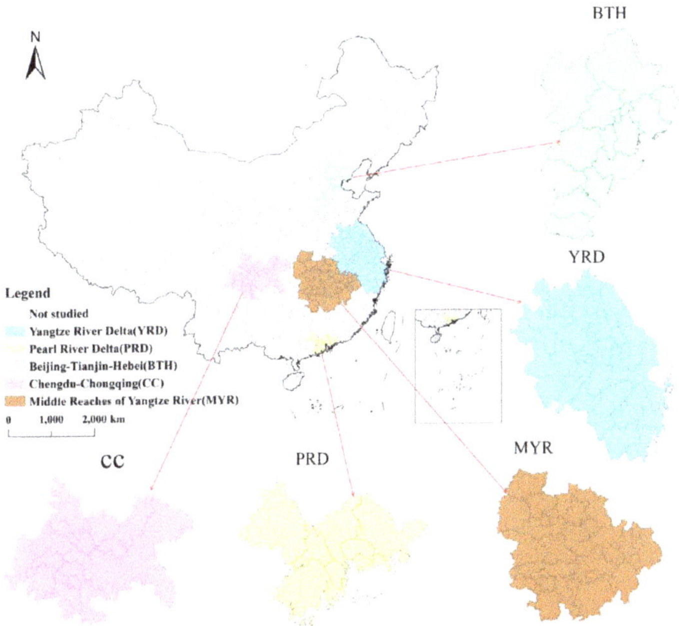

Figure 1. Spatial distribution of China's five major urban agglomerations.

3.2. Data Sources

This study focuses on China's five major urban agglomerations, covering the period from 2003 to 2020. The primary data sources included *China Statistical Yearbook, China Urban Statistical Yearbook, China Urban and Rural Construction Statistical Yearbook*, statistical yearbooks of various prefectures and cities, and the EPS database (https://www.epsnet.com.cn, accessed on 26 April 2023). The number of authorized green patents was obtained from the patent search database of China's State Intellectual Property Office (SIPO). For data gaps, interpolation methods were employed for imputation. Notably, the per capita real GDP used in this study was based on the GDP deflator of the province where the city is located and was adjusted to constant 2003 prices. Additionally, the urban population figures used in the calculations were based on the total urban population recorded at the end of each year.

3.3. Methods

3.3.1. Indicator System

Establishing a scientifically rigorous and practical evaluation indicator system is essential to deeply explore the UGHQD levels of China's five major urban agglomerations. Three primary considerations were emphasized in developing this system: policy documents, academic research findings, and data availability. Policy documents provided directive guidance for the construction of the indicator system. This study thoroughly reviewed numerous policy documents issued by the Chinese government related to UGHQD. Academic research offered a theoretical foundation for the system's design. We incorporated insights from scholars on the definition, characteristics, and potential indicator systems of high-quality development and green development, ensuring scientific integrity and the forward-thinking nature of our indicator choices [51–55]. Data availability ensured the operability of the indicator system. Meticulous review and selection processes were applied to each potential indicator's data sources, guaranteeing representativeness and reliability. The collected data were rigorously pre-processed and validated to mitigate potential biases, ensuring the accuracy and authenticity of the evaluation results. Integrating these considerations, this study proposes a UGHQD indicator system that encompasses four main dimensions, economic development (ED), social livelihood (SL), ecological environment (EE), and technological innovation (TI), featuring a total of 27 fundamental indicators. To ensure the objectivity and rationality of the weights for each indicator, this study calculates the weights of the fundamental indicators within the UGHQD indicator system using the entropy value method. Table 1 provides descriptions of the relevant indicators.

Economic development is a fundamental means to resolve major societal contradictions. To foster UGHQD and enhance both the quality and quantity of economic development, upgrading the driving forces behind economic development while maintaining stable economic growth is essential. Consequently, this study employed real GDP per capita, fixed-asset investment, logistics accessibility, intensity of social consumption, foreign trade dependence, foreign investment dependence, upgradation of the industrial structure, and rationalization of the industrial structure as the foundational indicators to assess the level of economic development.

The social livelihood dimension emphasizes the well-being of individuals, advocating for the continual improvement of people's livelihoods and social justice. This dimension focuses on elevating the populace's living standards, ensuring that every resident experiences tangible benefits, happiness, and security. This not only facilitates robust economic development but also lays the groundwork for social stability. Moreover, when people's livelihoods are adequately safeguarded, cities are more inclined to prioritize environmental sustainability, resulting in a symbiotic relationship between green development and the welfare of the populace. Consequently, this study employs the basic education level, transport infrastructure, public culture level, average wage level, health and medicine level, and internet penetration rate as the key indicators of the social livelihood dimension.

Regarding the ecological environment dimension, the growing global focus on environmental issues has positioned the ecological health of cities as a central evaluative dimension of UGHQD. A thriving ecological environment guarantees a healthy existence for citizens and underpins sustainable economic and social development. Consequently, this study employs green innovation achievements, intensity of science expenditure, intensity of education expenditure, cultivation of innovative talents, and technological innovation achievements as the key indicators of the ecological environment dimension.

The rationale for including technological innovation as a dimension in the UGHQD evaluation index system is derived from its pivotal role in facilitating economic transformation and enhancing resource efficiency. Technological innovation can drive the shift from traditional, polluting industries to low-carbon, clean industries and provide cutting-edge solutions to urban environmental challenges. Therefore, technological innovation is indispensable in steering cities toward green, high-quality development. Consequently, this study employs green innovation achievements, intensity of science expenditure, intensity of education expenditure, and cultivation of innovative talents as the core indicators of the technological innovation dimension.

3.3.2. Dagum Gini Coefficient

Dagum (1997) [73] decomposed the overall Gini coefficient into three components: the intra-regional differences G_w, the inter-regional differences G_{nb}, and the super-variable density G_t. Due to its capability to effectively decompose and elucidate regional differences, the Dagum Gini coefficient addresses the limitations of traditional measurement methods. In this study, the Dagum Gini coefficient and its decomposition method are employed to systematically investigate the sources of regional differences in the UGHQD level in China's five major urban agglomerations. This approach allows for the examination of sub-sample distributions that are unaffected by sample overlap. The formula for calculating the Dagum Gini coefficient is as follows:

$$G = \frac{1}{2n^2\mu}\sum_{j=1}^{k}\sum_{h=1}^{k}\sum_{i=1}^{n_j}\sum_{r=1}^{n_h}|y_{ji} - y_{hr}| \tag{1}$$

where G represents the overall Gini coefficient of the UGHQD level of the five major urban agglomerations, $y_{ji}(y_{hr})$ is the UGHQD level of any city within $j(h)$ urban agglomeration, and $n_j(n_h)$ is the number of cities within $j(h)$ urban agglomeration.

If only the Gini coefficient of the UGHQD level of each city within a region (e.g., within urban agglomeration j) is considered, the formula is as follows:

$$G_{jj} = \frac{1}{2n_j^2\mu_j}\sum_{i=1}^{n_j}\sum_{r=1}^{n_j}|y_{ji} - y_{jr}| \tag{2}$$

where μ_j is the expected value of the UGHQD level of each city in urban agglomeration j, $y_{ji}(y_{jr})$ is the UGHQD level of any city in urban agglomeration j, and n_j is the number of cities in urban agglomeration j.

The contribution of intra-regional differences is:

$$G_w = \sum_{j=1}^{k} G_{jj}p_js_j \tag{3}$$

where $p_j = n_j/n$, $s_j = n_j\mu_j/n\mu$, and G_w can be interpreted as a weighted average of the Gini coefficients within regions.

If the Gini coefficient of the UGHQD level of cities between regions (e.g., between urban agglomerations j and h) is considered, the formula is as follows:

$$G_{jh} = \frac{1}{n_jn_h(\mu_j + \mu_h)}\sum_{i=1}^{n_j}\sum_{r=1}^{n_h}|y_{ji} - y_{jr}| \tag{4}$$

where u_j (u_h) is the expected value of the UGHQD level of each city in the $j(h)$ urban agglomeration, y_{ji} (y_{jr}) is the UGHQD level of any city in the $j(h)$ urban agglomeration, and n_j (n_h) is the number of cities in the $j(h)$ urban agglomeration.

The contribution of the inter-regional differences and the contribution of the super-variable density are, respectively:

$$G_{nb} = \sum_{j=2}^{k}\sum_{h=1}^{j-1} G_{jh}(p_js_h + p_hs_j)D_{jh} \tag{5}$$

$$G_t = \sum_{j=2}^{k}\sum_{h=1}^{j-1} G_{jh}(p_js_h + p_hs_j)(1 - D_{jh}) \tag{6}$$

where D_{jh} is the relative impact of the UGHQD level among $j(h)$ urban agglomerations and D_{jh} is defined as shown in the following equation under the premise that $\mu_j > \mu_i$:

$$D_{jh} = \frac{d_{jh} - p_{jh}}{d_{jh} + p_{jh}} \tag{7}$$

where the formulas for d_{jh} and p_{jh} are shown below:

$$d_{jh} = \int_0^\infty dF_j(y) \int_0^y (y - x) dF_h(x) \tag{8}$$

$$p_{jh} = \int_0^\infty dF_h(y) \int_0^y (y - x) dF_j(x) \tag{9}$$

where $F_h(F_j)$ is the cumulative density distribution function for the $j(h)$ urban agglomerations.

Furthermore, utilizing the aforementioned Gini coefficient decomposition formula allows us to derive the decomposition equation for the overall Gini coefficient of the UGHQD levels across China's five major urban agglomerations:

$$G = G_w + G_{nb} + G_t \tag{10}$$

3.3.3. Kernel Density

Kernel density estimation serves as a method for analyzing data distribution characteristics drawing directly from the data itself. This method relaxes prior assumptions about data distribution, is not bound by specific functional expressions, and avoids logical inconsistencies arising from predetermined function forms. Given that the distribution of the UGHQD levels in China's five major urban agglomerations changes over time and exhibits significant uncertainty and complexity, the nonparametric Kernel density estimation method is well-suited for estimating its dynamic distribution trend. This study uses Kernel density estimation to systematically describe the distribution's location, shape, extensibility, and polarization characteristics for the UGHQD levels of the five major urban agglomerations throughout the sample period. Let the density function $f(c)$ for the random variable X at point c be formulated as [64,74]:

$$f(c) = \frac{1}{N\rho} \sum_{i=1}^{N} K\left(\frac{C_i - \bar{c}}{\rho}\right) \tag{11}$$

In Equation (11), N is the number of observations, C_i is the independent and equally distributed observations, $\bar{c}$ is the mean value, ρ is the bandwidth, and $K(\cdot)$ is the kernel function. This study chooses the more commonly used Gaussian kernel function to estimate the dynamic evolution trend of the distribution of the UGHQD level in the five major urban agglomerations. The kernel function expression is:

$$K(c) = \frac{1}{2\pi} \exp\left(-\frac{c^2}{2}\right) \tag{12}$$

3.3.4. Markov Chain

The Markov chain represents a discrete-time, discrete-state stochastic process. It captures the growth change of a variable by segmenting the data into λ categories and determining the growth change by calculating the probability distribution for each category and its evolutionary trend over time. One of the distinguishing features of the Markov chain is its "memory lessness"; the conditional distribution of state X_{t+1} is solely dependent on state X_t and is independent of prior states. In this study, we construct a transition probability matrix over a time span of d. When categorizing the UGHQD level into λ groups, one can establish a transition probability matrix of order $\lambda \times \lambda$. The transition probability formula is as follows:

$$P_{ij}^{t,t+d} = P\{X_{t+d} = j | X_t = i\} = \frac{\sum_{t=2003+d}^{2020} n_{ij}^{t,t+d}}{\sum_{t=2003}^{2020-d} n_i^t} \quad (13)$$

where $P_{ij}^{t,t+d}$ is the probability that the UGHQD level of a given urban agglomeration shifts from type i in year t to type j in year $t + d$, $n_{ij}^{t,t+d}$ denotes the number of regions belonging to type i in year t during the sample examination period that transition to type j after d years, and n represents the number of regions belonging to type i in year t.

The spatial Markov chain expands upon the traditional Markov chain by introducing the notion of "spatial lag". It transforms the two-dimensional transition probability matrix of $\lambda \times \lambda$ into a three-dimensional matrix of $\lambda \times \lambda \times \lambda$. This allows the spatial Markov chain to assess whether the UGHQD level of neighboring cities influences the transition probability of a city's UGHQD state.

3.3.5. Geographical Detector

The geographical detector is a statistical instrument designed to discern spatial disparities in geographic elements and investigate the driving factors underlying these differences. This methodology partitions the study area into several specific subregions and compares the overall variance within each region with the sum of variances across these subregions to determine if the geographic elements exhibit significant spatial heterogeneity. In this study, we use the q-value as a criterion to quantify the influence of various factors affecting the spatial differences in UGHQD levels among the five major urban agglomerations. The q-value ranges from 0 to 1; a q-value closer to 1 indicates a stronger explanatory power for the spatial differences in UGHQD. Specifically, a q-value of 1 or 0 suggests a high degree of consistency or lack of correlation, respectively, between the factor and the spatial differences in UGHQD.

Furthermore, this study introduces the concept of an interaction detector, which identifies the existence and strength of an interaction effect between bivariate variables. This is achieved by comparing the difference between the spatially superimposed q-values of two factors and the q-values of each individual factor [69].

$$q = 1 - \frac{1}{n\sigma^2} \sum_{h=1}^{m} n_h \bullet \sigma_h^2 \quad (14)$$

where $h = 1, 2, \ldots, m$ represents the partition of the independent or dependent variable, n is the total number of samples in the study area, σ^2 is the total discrete variance in the study area, n_h is the number of samples in partition h, and σ_h^2 is the discrete variance of the dependent variable in partition h.

Table 1. Indicator system of UGHQD.

Dimension Layer	Sub-Level	Explanation	Attributes	References
ED	Economic development level	Real GDP per capita	Positive	[26,49,57]
	Fixed-asset investment	Per capita investment in fixed assets	Positive	[26]
	Logistics accessibility	Per capita road freight volume	Positive	[55]
	Intensity of social consumption	Total retail sales of consumer goods per capita	Positive	[52]
	Foreign trade dependence	Ratio of total exports and imports to GDP	Positive	[49,55]
	Foreign investment dependence	Actual utilized of foreign capital to GDP	Positive	[55]
	Upgradation of industrial structure	Value added of tertiary industry to value added of secondary industry	Positive	[52,56]
	Rationalization of industrial structure	New Theil index	Reverse	[52,55,70]

Table 1. *Cont.*

Dimension Layer	Sub-Level	Explanation	Attributes	References
SL	Basic education level	Number of primary and secondary school teachers per student	Positive	[27]
	Transport infrastructure	Road mileage per unit area	Positive	[65]
	Public culture level	Number of library books per 10,000 people	Positive	[57]
	Average wage level	Average wage of urban employees	Positive	[66]
	Health and medicine level	Number of hospital beds per 10,000 people	Positive	[27,57]
	Internet penetration rate	Number of internet users per 100 people	Positive	[60]
EE	Urban wastewater treatment rate	Concentrated treatment rate of wastewater treatment plants	Positive	[55,56]
	Harmless treatment rate of garbage	Harmless treatment rate of household garbage	Positive	[55,56]
	Solid waste utilization rate	Comprehensive utilization rate of industrial solid waste	Positive	[52,55,56]
	Greening coverage in built-up areas	Area of landscaped green space to built-up area	Positive	[55–57]
	Air particulate pollution	Annual average of PM2.5	Reverse	[52,75]
	Industrial SO_2 emission	Industrial SO_2 emission to industrial output value	Reverse	[52,55]
	Industrial wastewater discharges	Industrial wastewater discharge to industrial output value	Reverse	[52,55,56]
	Industrial dust emission	Industrial smoke (dust) emission to industrial output value	Reverse	[55,56]
TI	Green innovation achievements	Number of green patent authorizations per 10,000 people	Positive	[52]
	Intensity of science expenditure	Per capita science expenditure	Positive	[55,56]
	Intensity of education expenditure	Per capita expenditure on education	Positive	[55]
	Cultivation of innovative talents	Number of university students per 10,000 people	Positive	[56]
	Technological innovation achievements	Number of patent authorizations per 10,000 people	Positive	[52]

4. Results

4.1. Level Measurement

This study employs the entropy method [76,77] to systematically evaluate the UGHQD levels of China's five major urban agglomerations, the Yangtze River Delta (YRD), the Pearl River Delta (PRD), Beijing-Tianjin-Hebei (BTH), the Middle Reaches of the Yangtze River (MYR), and Chengdu-Chongqing (CC), from 2003 to 2020. The study performs meanization processing on the urban sample data and, based on this, depicts the evolutionary trends of the UGHQD levels for these urban agglomerations, as shown in Figure 2. Drawing on existing studies [78,79] and using the "natural breaks" as a delineation criterion, this research identifies 2003, 2008, 2014, and 2020 as pivotal study years. The spatial distribution of UGHQD across the five major urban agglomerations is depicted in Figure 3, revealing the regional disparities in their development. As seen in Figure 2, two key features emerge: First, there was a clear upward trajectory in UGHQD levels across these urban agglomerations. However, the overall development level remained relatively low despite this progressive trend. The annual growth rate was modest, and noticeable differences among the different urban agglomerations were evident. A detailed analysis follows.

From 2003 to 2020, the average UGHQD level in these urban agglomerations increased from 0.0321 to 0.1216, marking a cumulative growth of 179.45%. From a spatiotemporal perspective, pronounced differences existed in the UGHQD levels across the five major urban agglomerations. The PRD and YRD exhibited average UGHQD levels that surpassed the collective mean. Notably, despite its relative strength, the PRD experienced a downturn in its development level in 2004 and again in 2018. Conversely, the YRD consistently showed a steady year-over-year ascent in its development. However, BTH, CC, and the MYR lagged in this metric. Specifically, the average UGHQD level in CC stood at just 0.0863 in 2020, consistently trailing other urban agglomerations throughout the sample period.

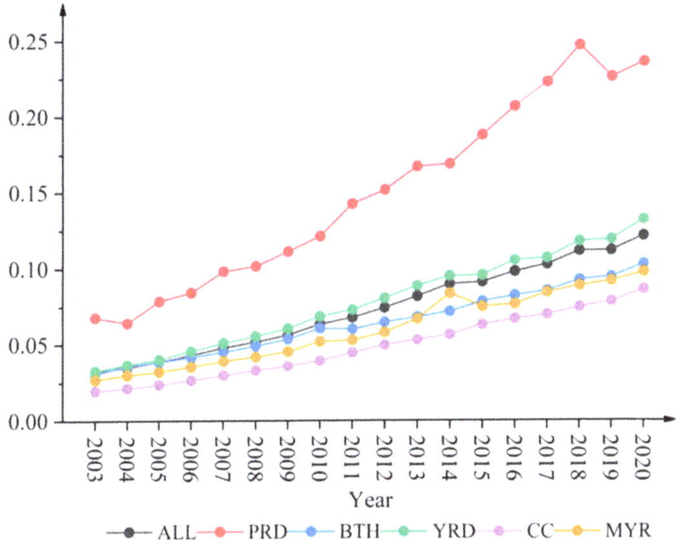

Figure 2. Evolutionary trend of the UGHQD level in China's five major urban agglomerations.

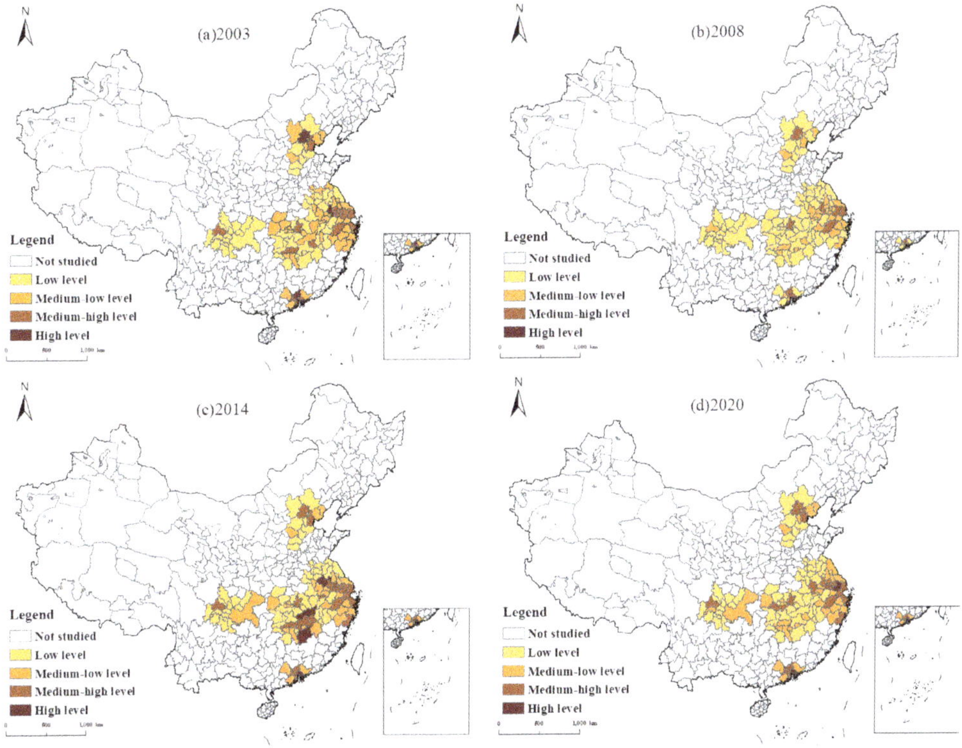

Figure 3. Spatial distribution of UGHQD level in China's five major urban agglomerations.

Analyzing the average annual growth rates, the combined rate for the five major urban agglomerations stood at 7.69%. Specifically, the YRD and CC registered growth rates of 8.06% and 8.56%, respectively, surpassing the aggregate average. This suggests that although the YRD currently leads in UGHQD, its potential for further growth is substantial. Conversely, CC, starting from a lower baseline, demonstrated the most robust annual growth, indicating potential for convergence with its peers in the future. Meanwhile, the growth rates for the PRD, BTH, and the MYR were 7.15%, 6.93%, and 7.42%, respectively, all falling below the aggregate average.

4.2. Spatial Differences

Utilizing the Dagum Gini coefficient and its decomposition methodology, this study quantitatively assesses the overall, intra-regional, and inter-regional differences and the super-variable density in the UGHQD level among China's five major urban agglomerations. The aim is to reveal the characteristics of spatial differences in UGHQD levels and their sources, as depicted in Figure 4.

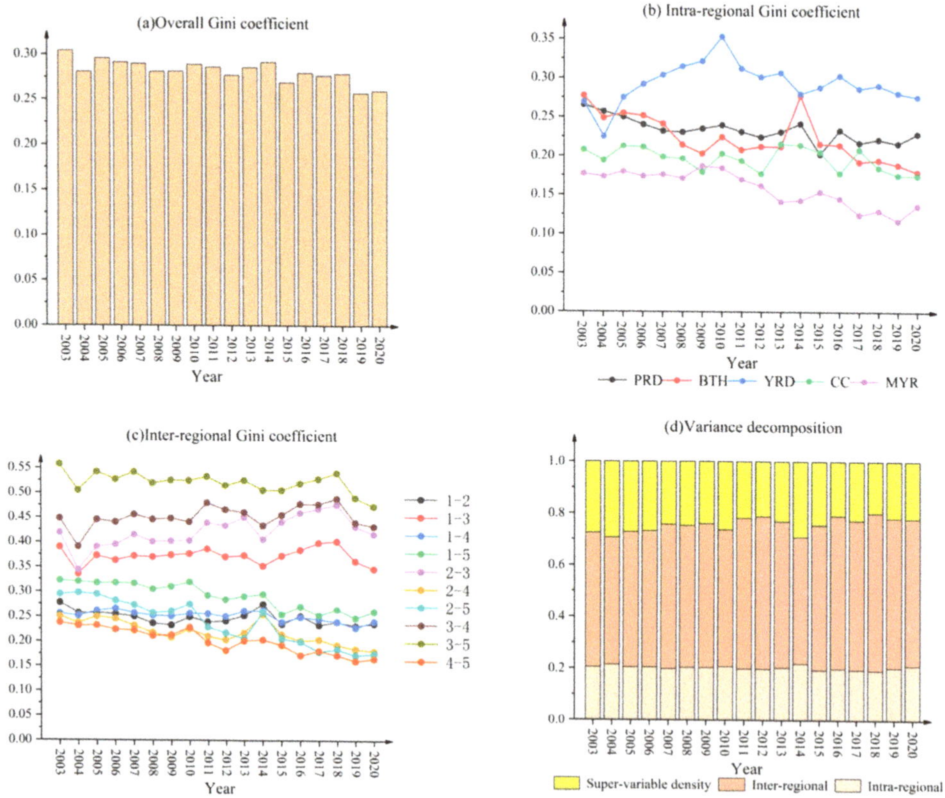

Figure 4. Gini coefficients and their decompositions in China's five major urban agglomerations.

4.2.1. Overall Differences and Evolutionary Trends

Figure 4a illustrates the evolutionary trends of the overall Gini coefficient for UGHQD across the five major urban agglomerations. From a static perspective, the Gini coefficient ranged from 0.2578 to 0.3043 during the sample period, with an average value of 0.2822, signifying a marked spatial imbalance. This difference largely stemmed from the persistently low development levels in CC and the MYR as opposed to the more developed

PRD and YRD. From a dynamic perspective, the Gini coefficient gradually declined from 0.3043 in 2003 to 0.2604 in 2020, with an annual average decrease rate of approximately 0.87%. Although this trend suggests that regional imbalances are slowly evolving toward regional synergy, the pace of this progress is relatively slow and inconsistent. The PRD and YRD showed limited growth rates, whereas CC and the MYR, starting from a lower baseline, are growing more rapidly, contributing to the observed fluctuations and narrowing of the Gini coefficient.

4.2.2. Intra-Regional Differences and Evolutionary Trends

Figure 4b illustrates the evolutionary trends of the intra-regional Gini coefficients for UGHQD across the five major urban agglomerations. From a static perspective, the Gini coefficients for the study period ranged from 0.1175 to 0.3535. The mean values for these urban agglomerations were as follows: the MYR (0.1585), CC (0.1964), BTH (0.2231), the PRD (0.2223), and the YRD (0.2954). These figures indicate significant differences in the levels of development within the urban agglomerations. Among them, the average value for the coastal YRD and PRD exceeded 0.2000, which is significantly higher than for other urban agglomerations. From a dynamic perspective, except for the YRD, all other urban agglomerations showed a declining trend in spatial differences. The average annual decrease rates were: the PRD (0.83%), CC (0.96%), the MYR (1.44%), and BTH (2.42%). In contrast, the YRD's intra-regional differences have expanded gradually by approximately 0.15% annually. CC, interestingly, exhibited considerable short-term fluctuations—for example, sharply rising from 0.1775 in 2012 to 0.2143 in 2014 before dropping to 0.1782 in 2016. This volatility may be attributed to its lower initial level of UGHQD, making it more vulnerable to unstable economic conditions. In conclusion, a higher UGHQD level correlates with more significant spatial disequilibrium within a region and vice versa.

4.2.3. Inter-Regional Differences and Evolutionary Trends

Figure 4c illustrates the evolutionary trends of the inter-regional Gini coefficients for UGHQD among the five major urban agglomerations. From a static perspective, the inter-regional Gini coefficients during the sample period ranged from 0.1608 to 0.5577, and the average values exceeded 0.2000. These data highlight pronounced spatial imbalances in UGHQD across these urban agglomerations. The most significant difference was observed between the YRD and MYR, registering an average Gini coefficient of 0.5212. In contrast, more modest disparities were evident between the MYR and CC, with average Gini coefficients of 0.2017. From a dynamic perspective, the spatial differences between urban agglomerations exhibited fluctuations during the study period, with most inter-regional differences showing a declining trend, although a few still displayed a minor increase. For instance, in the case of the PRD and YRD, the difference narrowed from 0.5577 in 2003 to 0.5052 in 2004, rebounded to 0.5424 in 2007, fell to 0.5052 in 2015, rebounded to 0.5401 in 2018, and then declined rapidly again. The evolutionary patterns of the Gini coefficients among the other urban agglomerations are broadly similar, albeit with some variations in specific fluctuation nodes and magnitudes.

4.2.4. Sources of Differences in the UGHQD Level and Their Contribution

Figure 4d illustrates the sources of differences in the UGHQD level among the five major urban agglomerations and their respective contributions. From a static perspective, inter-regional differences dominated the overall differences, with contribution rates ranging from 48.91% to 60.87% and an average of 55.34%. In contrast, intra-regional differences accounted for between 19.13% and 21.45% of the total differences, averaging 20.26%. Furthermore, the fluctuating contribution of super-variable density fell between 20.01% and 29.26%, with an average of 24.40%. This evidence suggests that efforts to mitigate the prevalent spatial differences in UGHQD should focus on bridging the development gap between urban agglomerations, particularly those between the coastal and central-western regions. From a dynamic perspective, the contribution rate of inter-regional differences followed a "decreasing-then-increasing" pattern throughout the study period, with comparable magnitudes of decline and ascent. The contribution rate of super-variable density oscillated more frequently, albeit within a narrower range. Meanwhile, the contribution rate of intra-regional differences remained relatively stable, showing only minor fluctuations between 2012 and 2018. Collectively, these trends indicated an anticipated stabilization in the respective contribution rates of the differences mentioned above.

4.3. Dynamic Evolution

Figure 5 illustrates the Kernel density estimation for the overall and individual UGHQD levels across the five major urban agglomerations. Subsequent sections delve deeper, analyzing the Kernel density curves with a focus on attributes such as location, shape, ductility, and polarization features.

From the perspective of spatial distribution, the Kernel density distribution curves for the five major urban agglomerations—both collectively and individually—manifested discernible rightward shifts over time. This finding corroborates the substantial improvements in UGHQD levels, as elaborated in preceding sections. CC and the YRD exhibited the most pronounced rightward shifts, signifying faster UGHQD improvement rates. In contrast, the weaker shift in the PRD suggests a relatively sluggish pace of improvement.

From the perspective of distributional shape, a decreasing peak height and widening curve in Kernel density distribution indicate a transition from a centralized to a more dispersed spatial pattern among the five agglomerations. A more granular analysis uncovers that the majority of urban agglomerations in the PRD, the YRD, BTH, and the MYR exhibited expanding development profiles. CC, however, demonstrated a unique pattern: the height of the main peak decreased before increasing, followed by a narrowing of the peak's width, indicating a more concentrated form of UGHQD.

From the perspective of distributional ductility, all urban agglomerations displayed varying degrees of the "right-tail" phenomenon, meaning that some cities within each agglomeration were leading in UGHQD levels. Specifically, Guangzhou and Shenzhen in the PRD, Beijing and Tianjin in BTH, Shanghai and Hangzhou in the YRD, Chengdu and Chongqing in CC, and Wuhan and Changsha in the MYR, were all leading in terms of UGHQD. Furthermore, some urban agglomerations exhibited annual extension trends, indicative of a growing difference between leading cities and others, potentially due to policy preferences and resource allocation advantages.

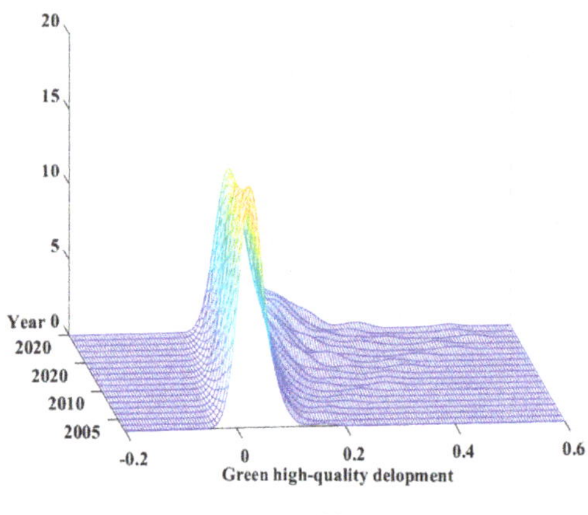

(a) ALL

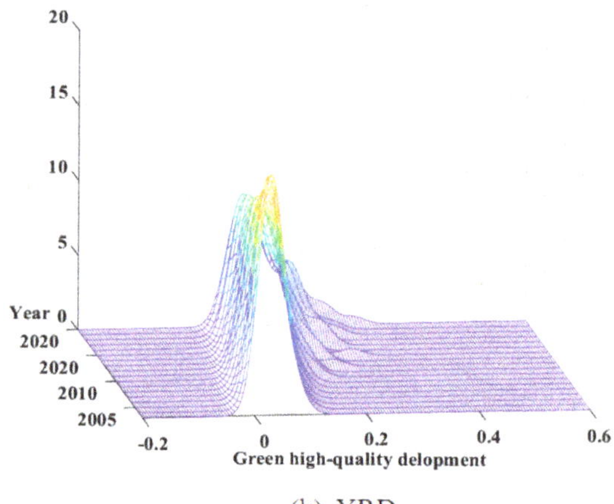

(b) YRD

Figure 5. *Cont.*

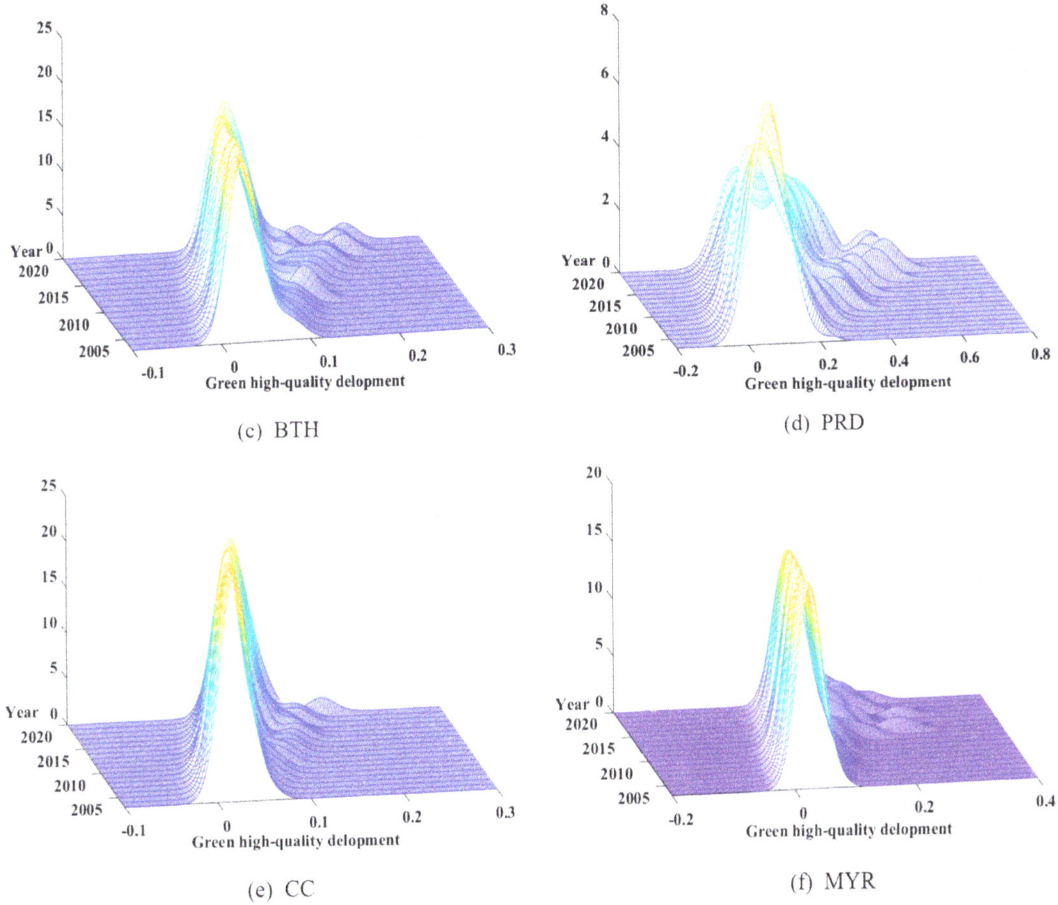

Figure 5. Evolution of the distribution of UGHQD in China's five major urban agglomerations.

From the perspective of distributional polarization, all five major urban agglomerations generally displayed characteristics of a "single main peak with multiple side peaks", showing an evolutionary trend towards multi-level differentiation. Further analysis of each urban agglomeration substantiates an orderly transition from single-peak to multiple-peak distributions, highlighting the increasing feature of multi-polarization in UGHQD.

4.4. Transfer Probabilities

By leveraging the quartile method, our study categorizes the UGHQD scores of 107 cities in China's five major urban agglomerations into four distinct classes: low, medium-low, medium-high, and high. To ascertain the probability and direction of categorical transitions for each city, we employ Markov chain analysis using Matlab 2022b software, resulting in transition probability matrices as depicted in Tables 2 and 3.

Table 2. Traditional Markov transfer probability matrix for UGHQD.

		Span of 1 Year				Span of 2 Years				Span of 3 Years			
Region	**Type**	**L**	**ML**	**MH**	**H**	**L**	**ML**	**MH**	**H**	**L**	**ML**	**MH**	**H**
ALL	L	0.817	0.180	0.000	0.002	0.668	0.328	0.000	0.004	0.521	0.467	0.006	0.006
	ML	0.008	0.773	0.204	0.015	0.000	0.600	0.387	0.013	0.000	0.431	0.552	0.017
	MH	0.000	0.009	0.830	0.161	0.000	0.005	0.693	0.302	0.000	0.003	0.531	0.466
	H	0.002	0.010	0.019	0.969	0.000	0.008	0.017	0.975	0.000	0.003	0.023	0.974
YRD	L	0.822	0.173	0.005	0.000	0.649	0.346	0.005	0.000	0.503	0.486	0.011	0.000
	ML	0.000	0.785	0.210	0.006	0.000	0.593	0.401	0.006	0.000	0.393	0.595	0.012
	MH	0.000	0.017	0.821	0.162	0.000	0.019	0.658	0.323	0.000	0.021	0.483	0.497
	H	0.000	0.006	0.013	0.981	0.000	0.000	0.029	0.971	0.000	0.000	0.035	0.965
PRD	L	0.829	0.171	0.000	0.000	0.707	0.268	0.024	0.000	0.585	0.366	0.049	0.000
	ML	0.026	0.789	0.184	0.000	0.028	0.639	0.333	0.000	0.029	0.471	0.471	0.029
	MH	0.000	0.026	0.821	0.154	0.000	0.000	0.676	0.324	0.000	0.000	0.543	0.457
	H	0.000	0.000	0.029	0.971	0.000	0.000	0.067	0.933	0.000	0.000	0.080	0.920
BTH	L	0.814	0.186	0.000	0.000	0.644	0.356	0.000	0.000	0.508	0.492	0.000	0.000
	ML	0.000	0.793	0.207	0.000	0.000	0.603	0.397	0.000	0.000	0.414	0.586	0.000
	MH	0.000	0.000	0.855	0.145	0.000	0.000	0.729	0.271	0.000	0.000	0.561	0.439
	H	0.000	0.000	0.000	1.000	0.000	0.000	0.000	1.000	0.000	0.000	0.000	1.000
CC	L	0.764	0.236	0.000	0.000	0.583	0.417	0.000	0.000	0.403	0.583	0.014	0.000
	ML	0.028	0.764	0.208	0.000	0.014	0.569	0.403	0.014	0.000	0.389	0.569	0.042
	MH	0.000	0.000	0.764	0.236	0.000	0.000	0.571	0.429	0.000	0.000	0.413	0.587
	H	0.000	0.000	0.018	0.982	0.000	0.000	0.024	0.976	0.000	0.000	0.030	0.970
MYR	L	0.802	0.198	0.000	0.000	0.619	0.365	0.008	0.008	0.444	0.532	0.016	0.008
	ML	0.008	0.778	0.175	0.040	0.000	0.603	0.357	0.040	0.000	0.421	0.532	0.048
	MH	0.000	0.000	0.793	0.207	0.000	0.000	0.645	0.355	0.000	0.000	0.485	0.515
	H	0.000	0.010	0.058	0.932	0.000	0.012	0.047	0.942	0.000	0.000	0.056	0.944

Note: "L, ML, MH, H" stand for low, medium-low, medium-high, and high levels, respectively.

4.4.1. Traditional Markov Chain Analysis

Table 2 provides the transition probability matrices for UGHQD levels in China's five major urban agglomerations across one-, two-, and three-year time horizons. The diagonal entries in these matrices signify the likelihood of UGHQD levels remaining constant within the respective time horizons, while the off-diagonal entries indicate transition probabilities. During the study period, diagonal probabilities ranged from 43.11% to 97.41%, substantially exceeding off-diagonal values. This suggests that UGHQD levels are more likely to remain stable within 1–3 years, displaying a clear pattern of "club convergence". Importantly, areas with higher UGHQD levels demonstrated greater stability, while less-developed regions exhibited increased fluidity. In terms of stability, the ranking is as follows: medium-low < low < medium-high < high. Examining specific urban agglomerations, such as the YRD, the PRD, BTH, and the MYR, largely corroborates these overall trends, pointing to more entrenched developmental patterns. In contrast, CC demonstrated lower stability but a higher proclivity for upward movement, suggesting a self-reinforcing growth trajectory. Further analysis reveals that lower-tiered regions are expected to experience diminishing stability and a growing likelihood of upward mobility within the next 1–3 years as we extend the time horizon. Although these less-developed areas face increased odds of improvement, transitioning across categorical boundaries at this stage remains a formidable challenge.

4.4.2. Spatial Markov Chain Analysis

Traditional Markov chains depict the trend of type transitions in the UGHQD level within the five major urban agglomerations. However, it is essential to recognize that cities exhibit specific spatial correlations in this development. Given these urban agglomerations' geographic dispersion and their limited mutual influence, this study predominantly examines the interaction and impact of UGHQD within each urban agglomeration. Consequently, using the first-order neighboring spatial weight matrix, this study constructs the spatial Markov transition probability matrix for these urban agglomerations' UGHQD, as presented in Table 3.

Table 3. Spatial Markov transfer probability matrix for UGHQD.

Region	Neighborhood	Type	L	ML	MH	H	Region	Neighborhood	Type	L	ML	MH	H
YRD	I	L	0.897	0.103	0.000	0.000	BTH	I	L	0.893	0.107	0.000	0.000
		ML	0.000	0.913	0.087	0.000			ML	0.000	0.727	0.273	0.000
		MH	0.000	0.000	0.833	0.167			MH	0.000	0.000	1.000	0.000
		H	0.000	0.000	0.000	1.000			H	0.000	0.000	0.000	1.000
	II	L	0.703	0.281	0.016	0.000		II	L	0.769	0.231	0.000	0.000
		ML	0.000	0.833	0.167	0.000			ML	0.000	0.917	0.083	0.000
		MH	0.000	0.047	0.814	0.140			MH	0.000	0.000	0.700	0.300
		H	0.000	0.000	0.000	1.000			H	0.000	0.000	0.000	1.000
	III	L	0.500	0.500	0.000	0.000		III	L	0.600	0.400	0.000	0.000
		ML	0.000	0.721	0.265	0.015			ML	0.000	0.808	0.192	0.000
		MH	0.000	0.014	0.913	0.072			MH	0.000	0.000	1.000	0.000
		H	0.000	0.022	0.044	0.933			H	0.000	0.000	0.000	1.000
	IV	L	0.000	0.000	0.000	0.000		IV	L	0.000	0.000	0.000	0.000
		ML	0.000	0.500	0.500	0.000			ML	0.000	0.667	0.333	0.000
		MH	0.000	0.000	0.706	0.294			MH	0.000	0.000	0.844	0.156
		H	0.000	0.000	0.000	1.000			H	0.000	0.000	0.000	1.000
PRD	I	L	0.923	0.077	0.000	0.000	CC	I	L	0.907	0.093	0.000	0.000
		ML	0.000	0.778	0.222	0.000			ML	0.125	0.750	0.125	0.000
		MH	0.000	0.250	0.750	0.000			MH	0.000	0.000	0.800	0.200
		H	0.000	0.000	0.000	0.000			H	0.000	0.000	0.000	0.000
	II	L	0.800	0.200	0.000	0.000		II	L	0.552	0.448	0.000	0.000
		ML	0.083	0.583	0.333	0.000			ML	0.026	0.872	0.103	0.000
		MH	0.000	0.000	0.818	0.182			MH	0.000	0.000	0.667	0.333
		H	0.000	0.000	0.000	1.000			H	0.000	0.000	0.000	1.000
	III	L	0.818	0.182	0.000	0.000		III	L	0.000	0.000	0.000	0.000
		ML	0.000	1.000	0.000	0.000			ML	0.000	0.625	0.375	0.000
		MH	0.000	0.000	0.810	0.190			MH	0.000	0.000	0.857	0.143
		H	0.000	0.000	0.000	1.000			H	0.000	0.000	0.000	1.000
	IV	L	0.500	0.500	0.000	0.000		IV	L	0.000	0.000	0.000	0.000
		ML	0.000	0.900	0.100	0.000			ML	0.000	0.000	1.000	0.000
		MH	0.000	0.000	1.000	0.000			MH	0.000	0.000	0.655	0.345
		H	0.000	0.000	0.077	0.923			H	0.000	0.000	0.031	0.969
MYR	I	L	0.881	0.119	0.000	0.000	MYR	III	L	0.625	0.375	0.000	0.000
		ML	0.083	0.750	0.167	0.000			ML	0.000	0.788	0.135	0.077
		MH	0.000	0.000	0.625	0.375			MH	0.000	0.000	0.714	0.286
		H	0.000	0.000	0.000	1.000			H	0.000	0.000	0.036	0.964
	II	L	0.746	0.254	0.000	0.000		IV	L	0.000	0.000	0.000	0.000
		ML	0.000	0.848	0.130	0.022			ML	0.000	0.563	0.438	0.000
		MH	0.000	0.000	0.923	0.077			MH	0.000	0.000	0.819	0.181
		H	0.000	0.056	0.000	0.944			H	0.000	0.000	0.100	0.900

Note: I, II, III, and IV represent the neighborhood levels of low, medium-low, medium-high, and high, respectively.

Table 3 reveals that, after accounting for spatial lag, the values on the diagonal significantly exceeded those off-diagonal. This indicates that even when considering spatial lag, the UGHQD of the five major urban agglomerations maintained a "club convergence" feature. Transitions mainly occurred between adjacent types, and inter-type transition probabilities were nearly negligible. As spatial lag increased, the developmental stability of low and medium-low regions declined, accompanied by a rise in their upward transition probability. In contrast, the stability of medium-high and high regions showed an increasing trend. For instance, in the YRD, the diagonal values in the transition probability matrix ranged between 50.00% and 100.00%, underscoring the distinct "club convergence" feature. As spatial lag increased, the likelihood of retaining the status quo in the low-level regions decreased from 89.70% to 50.00%, while the probability of upward transition jumps from 10.30% to 50.00%. The medium-low region's probability of maintaining the status quo decreased from 91.30% to 50.00%, while its likelihood for upward transition increased from 8.70% to 26.50%. The probabilities for the medium-high and high-level regions varied between 70.60% and 100.00%. While other urban agglomerations showed different transition probabilities, their characteristics largely align with those of the YRD.

4.5. Driving Factors

To elucidate systematically and scientifically the intrinsic drivers behind the spatial and temporal differences in UGHQD across the five major urban agglomerations, this study draws on methodologies from the existing literature [80,81], with a focus on four dimensions of driving forces for the spatial difference of UGHQD levels in 2020: economic development (X_1), social livelihood (X_2), ecological environment (X_3), and technological innovation (X_4). Initially, with the assistance of ArcGIS, "natural breaks" were used to convert each detector factor into a type variable. Subsequently, the geographical detector model was employed to quantify each factor's influence on the spatial differences in development quality, as reflected by their respective q-values; higher q-values indicate stronger influence. Detailed findings are presented in Table 4.

Table 4. Power of factors to drive UGHQD in 2020.

Urban Agglomeration	Factor	q-Value	Urban Agglomeration	Factor	q-Value
ALL	X_1	0.6550 ***	BTH	X_1	0.9021 ***
	X_2	0.6513 ***		X_2	0.8048
	X_3	0.1834 *		X_3	0.1760
	X_4	0.8074 ***		X_4	0.9232 ***
YRD	X_1	0.7501 ***	CC	X_1	0.6885 **
	X_2	0.8055 ***		X_2	0.4418
	X_3	0.0218		X_3	0.0527
	X_4	0.8290 ***		X_4	0.3474 *
PRD	X_1	0.8148 *	MYR	X_1	0.8020 ***
	X_2	0.6784		X_2	0.7495 **
	X_3	0.3574		X_3	0.0463
	X_4	0.8515 **		X_4	0.8581 ***

Note: "*, **, ***" indicate significant at 10%, 5%, and 1% level, respectively.

4.5.1. Factor Detection Analysis

The results in Table 4 reveal that each factor has a differentiated impact on the spatial difference of UGHQD across the five major urban agglomerations. Viewed as a whole, the driving effects of the factors are ranked as follows: technological innovation (0.8074) > economic development (0.6550) > social livelihood (0.6513) > ecological environment (0.1834). Notably, technological innovation significantly outperformed the other factors, establishing it as the principal driver of spatial difference in UGHQD. This underscores the increasing global reliance on innovative and knowledge-based economies. Concurrently, economic development and social livelihood were secondary drivers, providing essential material and economic foundations. However, in CC, economic development surpassed technological innovation as the primary driver, registering a value below 0.5000, thus indicating untapped potential for technological innovation. Intriguingly, for all five major urban agglomerations, the ecological environment factor was the weakest driver and failed to achieve statistical significance, highlighting pressing challenges in resource utilization and ecological governance as critical bottlenecks in their path toward UGHQD.

4.5.2. Interactive Detection Analysis

By using interaction detectors, this study quantitatively evaluates the potential additive effects among the driving factors. Figure 6 shows the outcomes of interaction detection for spatial differences in UGHQD across the five major urban agglomerations. As evidenced by Figure 6, the influence of bi-factor interactions consistently exceeded that of single-factor interactions, exhibiting a "bi-variable enhancement". Specifically, the joint influence of two interacting drivers surpassed the contribution of any single factor toward UGHQD yet fell short of the sum of the influences of the two individual factors. Notably, the q-value arising from the interaction of technological innovation (TI) with the other factors was prominent. In contrast, the q-value for economic development's interaction with other factors was less marked. These findings highlight the crucial role of TI in guiding the spatio-temporal evolution of UGHQD within these urban agglomerations, providing invaluable insights for policymaking. Notably, the synergistic effect of technological innovation's interaction with the ecological environment substantially outweighs their individual impacts, underscoring the importance of STI in enhancing urban resource utilization and ecological governance.

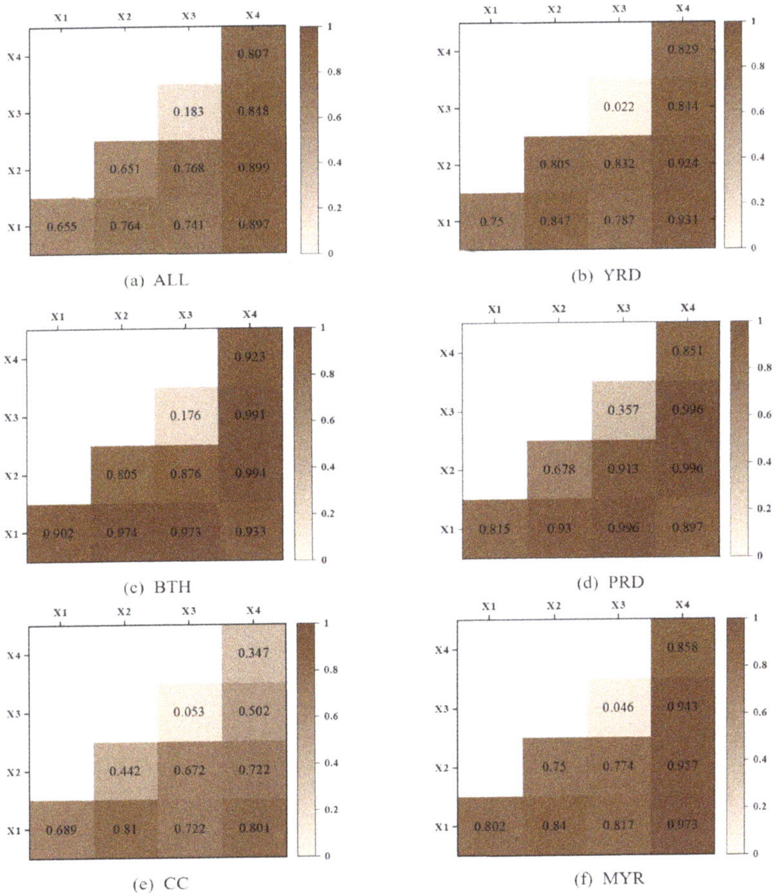

Figure 6. Results of interaction detection.

5. Discussion

Given the pressing challenges posed by global climate change and ecological degradation, the significant threats to societal and economic activities are increasingly garnering widespread attention. Specifically, establishing a nexus between "ecological welfare" and "economic prosperity"—ensuring ecological sustainability while concurrently fostering economic growth—has become a pivotal discourse among scholars and policymakers. As China transitions from high-speed to high-quality growth, it becomes imperative to deepen our understanding of the theories underpinning UGHQD. Although existing literature has engaged with the theoretical aspects of UGHQD, there is a significant lack of empirical studies quantitatively assessing its developmental levels, spatio-temporal evolution, and driving factors. Departing from previous research that employed green total factor productivity to gauge UGHQD [31,47,48], this study introduces a comprehensive evaluative framework incorporating four key dimensions: economic development (ED), social livelihood (SL), ecological environment (EE), and technological innovation (TI). Unlike prior works focusing on China's provincial-level administrative regions [52,56], this paper analyzes 107 cities across China's five major urban agglomerations, offering an in-depth exploration of UGHQD's spatio-temporal evolution characteristics and its key drivers. Our research furnishes a robust empirical methodology for the nuanced and exhaustive measurement of UGHQD and establishes valuable yardsticks for mapping the trajectory of sustainable urban development in this transformative era. Furthermore, the synergistic advancement of UGHQD is pivotal not just for China's contemporary socio-economic progress but also serves as an innovative blueprint for sustainable development that can be considered by countries globally.

The empirical findings of this study indicate that the five major urban agglomerations exhibited a gradual improvement in UGHQD. However, the spatial and temporal differences during the sample period were marked by a significant gap, particularly between the coastal urban agglomerations and those in the central and western regions. Due partly to the early implementation of open economy policies and unique resource endowments, coastal areas have attracted substantial foreign investment and high-quality talent. They have also introduced advanced technologies, collectively contributing to the optimization and transformation of their economic structures and lending robust support to UGHQD. Taking the YRD as an example, 2021 data indicate that the GDP of eight cities within the urban agglomeration—including Shanghai, Suzhou, and Hangzhou—has exceeded one trillion dollars. These cities maintain an unassailable lead in economic strength among the five major urban agglomerations. Benefiting from geographic advantages and supportive macro policies, the YRD had already constructed a more mature industrial and service system by the 1990s. Its highly advanced industrial structure provides abundant employment opportunities and economic growth points, thereby offering a solid foundation for UGHQD. Like the YRD, the nine cities in the PRD have also succeeded in attracting considerable talent and capital driven by reform and opening up policies. These cities excel in several industry sectors, including advanced manufacturing, high-tech industries, and financial services. By contrast, the urban agglomerations in central and western China manifest apparent deficiencies in these areas, characterized by relatively weak industrial and service systems and an over-reliance on low value-added and resource-intensive traditional industries. This not only constrains their rate of economic growth but also imposes significant ecological and environmental pressure. In aspects such as foreign investment, technological innovation, and government incentives, urban agglomerations in the central and western regions also display evident shortcomings. These multilevel and multidimensional factors lead to the relative lag in UGHQD observed in central and western urban agglomerations.

The empirical findings of this study further reveal that technological innovation is the core driver of the spatio-temporal difference in UGHQD across the five major urban agglomerations. The possible explanations for this conclusion are as follows: Firstly, technological innovation catalyzes the transformation and upgrading of economic structures, enabling a shift from low-value-added, pollutant-intensive traditional industries to high-value-added, low-carbon, and eco-friendly emerging sectors. For example, the YRD has successfully pivoted from traditional manufacturing to advanced manufacturing and services, particularly in integrated circuits, biomedicine, and electric vehicles. Secondly, technological innovation often enhances resource utilization efficiency while minimizing carbon emissions and environmental degradation. For instance, innovative cleaner production technologies in the PRD have led many manufacturing enterprises to reduce carbon emissions and environmental pollution while improving productivity significantly. Additionally, urban agglomerations with strong innovation capabilities tend to attract skilled talent and foreign investment, creating a positive feedback loop that further fuels technological advancement. The problem of brain drain in some cities in the MYR is partly due to the relatively weak capacity of scientific and technological innovation, which is unable to attract and retain high-level talent. Lastly, technological innovation also offers cost-effective and efficient solutions for environmental governance, thus reinforcing its role as a prerequisite for sustainable development. Collectively, technological innovation is the key to economic transformation and talent attraction as well as a necessary condition for environmental governance and sustainable development, which explains why technological innovation is the chief factor causing spatial differences in UGHQD in the five major urban agglomerations.

In light of the findings presented in this paper, we propose the following policy recommendations. Firstly, there is an urgent need to reassess the traditional focus on GDP-centric development, promoting instead a coordinated approach between economic growth and ecological sustainability. Local governments should prioritize environmentally efficient and people-centered development strategies. Secondly, regional characteristics should be acknowledged, and barriers should be broken to optimize resource allocation, restructure industries, and foster technological innovation. Lastly, given the critical role of technological advancement, sustained governmental support in R&D investment, tax incentives, and talent development are imperative for continuous technological and economic progress.

The limitations of this study are twofold. On the one hand, the present study investigates the UGHQD of 107 cities in China's five major urban agglomerations, a focus that holds unique value and significance from a geographical standpoint. However, this scope means that the role of other small and medium-sized cities, as well as rural areas, is not adequately addressed in this paper. In an era of accelerated globalization, the study also needs a comparative analysis of UGHQD with international cities, thereby missing an international perspective. Future research might consider expanding the geospatial sample and conducting more systematic comparative studies across different spatial and temporal scales. On the other hand, the analysis in this paper covers the driving factors of economic development, social livelihood, ecological environment, and technological innovation, providing a diversified perspective for understanding the complex phenomenon of UGHQD. Nonetheless, additional soft factors such as cultural influences, educational systems, public environmental awareness, and external variables like policy landscapes and global trends require further exploration for their potential impact on UGHQD. Therefore, future research should analyze external drivers, including the policy environment, in greater depth to better understand how they influence the process of UGHQD in urban agglomerations.

6. Conclusions

This study took 107 cities of China's five major urban agglomerations as research objects, constructed a comprehensive evaluation indicator system of UGHQD level based on four dimensions (economic development, social livelihood, ecological environment, and technological innovation), and empirically examined the spatio-temporal difference in the UGHQD level of the five major urban agglomerations and the trend of dynamic evolution by using the analytical methods, such as the Dagum Gini coefficient, the Kernel density estimation, and the Markov chain. Further, the driving factors behind the spatial differences in the UGHQD level were analyzed in depth using geographic detectors. The research revealed several key findings:

(1) The UGHQD levels of the five major urban agglomerations demonstrated a consistent upward trajectory during the period of 2003–2020. Coastal regions, specifically the PRD and YRD, consistently outperformed inland agglomerations. The Chengdu-Chongqing urban agglomeration, however, persistently lagged behind its counterparts, necessitating close scrutiny.

(2) Significant spatial differences existed in the UGHQD levels among the urban agglomerations. Inter-regional differences among the sub-clusters primarily drove these differences. Coastal urban agglomerations consistently led in terms of UGHQD, creating a pronounced development gap when compared to the central and western urban agglomerations.

(3) The spatio-temporal evolution of UGHQD levels in 107 cities within these five major urban agglomerations exhibited a trend of moving from a concentrated to a dispersed pattern. The dynamism of this distribution varied among different urban agglomerations. Additionally, the phenomenon of "club convergence" was observed in UGHQD levels, making it challenging to achieve a leap across different types. Upon accounting for spatial lag effects, the potential for upward mobility was more prominent in low-level areas.

(4) Diverse driving factors underpinned UGHQD levels in the five major urban agglomerations. Among them, the impact of technological innovation (TI) notably surpassed other factors. Interactive detection analysis further revealed a prominent synergistic effect between technological innovation (TI) and other driving factors, affirming that technological innovation served as the primary driver behind the spatial differentiation of UGHQD.

Author Contributions: Conceptualization, T.Y. and X.H.; methodology, X.H. and S.J.; writing—original draft preparation, T.Y. and X.H.; writing—review and editing, T.Y. and X.C.; funding acquisition, T.Y. and X.C. All authors have read and agreed to the published version of the manuscript.

Funding: National Social Science Foundation of China: 22BZZ039; Major Project of Philosophy and Social Sciences in Higher Education Institutions in Hubei Province of China: 22ZD015.

Data Availability Statement: All data generated or analyzed during this study are included in this published article.

Conflicts of Interest: The authors declare no conflict of interest.

References

1. Chen, H.; Qi, S.; Tan, X. Decomposition and prediction of China's carbon emission intensity towards carbon neutrality: From perspectives of national, regional and sectoral level. *Sci. Total Environ.* **2022**, *825*, 153839. [CrossRef] [PubMed]
2. Wei, L.; Liu, Z. Spatial heterogeneity of demographic structure effects on urban carbon emissions. *Environ. Impact Assess. Rev.* **2022**, *95*, 106790. [CrossRef]
3. Andiappan, V.; Foo, D.C.Y.; Tan, R.R. Process-to-Policy (P2Pol): Using carbon emission pinch analysis (CEPA) tools for policy-making in the energy sector. *Clean Technol. Environ. Policy* **2019**, *21*, 1383–1388. [CrossRef]
4. Huang, M.; Ye, Q. The Marxist green development concept and green development in contemporary China: Comment on incompatibility theory between environment and development. *Econ. Res. J.* **2017**, *52*, 17–30.
5. Xu, Y.; Wen, S.; Tao, C.Q. Impact of environmental tax on pollution control: A sustainable development perspective. *Econ. Anal. Policy* **2023**, *79*, 89–106. [CrossRef]
6. Jansen, L. The challenge of sustainable development. *J. Clean. Prod.* **2003**, *11*, 231–245. [CrossRef]
7. Wang, Y.; Guo, C.; Du, C.; Chen, X.; Jia, L.; Guo, X.; Chen, R.; Zhang, M.; Chen, Z.; Wang, H. Carbon peak and carbon neutrality in China: Goals, implementation path and prospects. *China Geol.* **2021**, *4*, 720–746. [CrossRef]
8. Iftikhar, Y.; Wang, Z.; Zhang, B.; Wang, B. Energy and CO_2 emissions efficiency of major economies: A network DEA approach. *Energy* **2018**, *147*, 197–207. [CrossRef]
9. Zhu, X. Understanding China's growth: Past, present, and future. *J. Econ. Perspect.* **2012**, *26*, 103–124. [CrossRef]
10. Cui, X.; Cai, T.; Deng, W.; Zheng, R.; Jiang, Y.; Bao, H. Indicators for Evaluating High-Quality Agricultural Development: Empirical Study from Yangtze River Economic Belt, China. *Soc. Indic. Res.* **2022**, *164*, 1101–1127. [CrossRef]

11. Kumar, R.R.; Stauvermann, P.J.; Patel, A. Exploring the link between research and economic growth: An empirical study of China and USA. *Qual. Quant.* **2016**, *50*, 1073–1091. [CrossRef]
12. Wang, J. Revive China's green GDP programme. *Nature* **2016**, *534*, 37. [CrossRef] [PubMed]
13. Färe, R.; Grosskopf, S.; Norris, M. Productivity growth, technical progress, and efficiency change in industrialized countries: Reply. *Am. Econ. Rev.* **1997**, *87*, 1040–1044.
14. Wang, X.; Wang, K.; Zhang, Y.; Gao, J.; Xiong, Y. Impact of Climate on the Carbon Sink Capacity of Ecological Spaces: A Case Study from the Beijing–Tianjin–Hebei Urban Agglomeration. *Land* **2023**, *12*, 1619. [CrossRef]
15. Rahman, M.M.; Alam, K. Impact of industrialization and non-renewable energy on environmental pollution in Australia: Do renewable energy and financial development play a mitigating role? *Renew. Energy* **2022**, *195*, 203–213. [CrossRef]
16. Peng, Y.; Chen, Z.; Lee, J. Dynamic convergence of green total factor productivity in Chinese cities. *Sustainability* **2020**, *12*, 4883. [CrossRef]
17. Fang, C.; Yu, D. Urban agglomeration: An evolving concept of an emerging phenomenon. *Landsc. Urban Plan.* **2017**, *162*, 126–136. [CrossRef]
18. Zhang, Q.; Xiao, Y.; Tang, X.; Huang, H. The spatial-temporal evolution and influencing factors of eco-efficiency in the five major urban agglomerations of China. *Econ. Geogr.* **2022**, *42*, 54–63. [CrossRef]
19. Tian, Y.; Wang, R.; Liu, L.; Ren, Y. A spatial effect study on financial agglomeration promoting the green development of urban agglomerations. *Sustain. Cities Soc.* **2021**, *70*, 102900. [CrossRef]
20. Li, L.; Hu, J. Ecological total-factor energy efficiency of regions in China. *Energy Policy* **2012**, *46*, 216–224. [CrossRef]
21. Wang, Y.; Yao, L.; Xu, Y.; Sun, S.; Li, T. Potential heterogeneity in the relationship between urbanization and air pollution, from the perspective of urban agglomeration. *J. Clean. Prod.* **2021**, *298*, 126822. [CrossRef]
22. Wang, Z.; Li, J.; Liang, L. Spatio-temporal evolution of ozone pollution and its influencing factors in the Beijing-Tianjin-Hebei Urban Agglomeration. *Environ. Pollut.* **2020**, *256*, 113419. [CrossRef] [PubMed]
23. Zhibiao, L.; Yonghui, L. Structural transformation, TFP and high-quality development. *China Econ.* **2022**, *17*, 70–82. [CrossRef]
24. He, X.; Shen, K. Modernized economic system, total factor productivity and high quality development. *Shanghai J. Econ.* **2018**, *6*, 25–34. [CrossRef]
25. Chen, Y.; Zhu, M.; Lu, J.; Zhou, Q.; Ma, W. Evaluation of ecological city and analysis of obstacle factors under the background of high-quality development: Taking cities in the Yellow River Basin as examples. *Ecol. Indic.* **2020**, *118*, 106771. [CrossRef]
26. Wang, R.; Wang, F. Exploring the role of green finance and energy development towards high-quality economic development: Application of spatial Durbin model and intermediary effect model. *Int. J. Environ. Res. Public Health* **2022**, *19*, 8875. [CrossRef]
27. Zhu, K.; Song, D.; Zhang, L.; He, Y.; Zhang, S.; Liu, Y.; Tian, X. Evolving Trends and Influencing Factors of the Rural Green Development Level in Chongqing. *Land* **2023**, *12*, 1333. [CrossRef]
28. Hickel, J.; Kallis, G. Is green growth possible? *New Political Econ.* **2020**, *25*, 469–486. [CrossRef]
29. Zhou, G.; Zhang, Z.; Fei, Y. How to evaluate the green and high-quality development path? An FsQCA approach on the China pilot free trade zone. *Int. J. Environ. Res. Public Health* **2022**, *19*, 547. [CrossRef]
30. Shi, X.; Li, L. Green total factor productivity and its decomposition of Chinese manufacturing based on the MML index: 2003–2015. *J. Clean. Prod.* **2019**, *222*, 998–1008. [CrossRef]
31. Ma, D.; Zhu, Q. Innovation in emerging economies: Research on the digital economy driving high-quality green development. *J. Bus. Res.* **2022**, *145*, 801–813. [CrossRef]
32. Schandl, H.; West, J. Resource use and resource efficiency in the Asia–Pacific region. *Glob. Environ. Chang.* **2010**, *20*, 636–647. [CrossRef]
33. Barro, R.J. *Quantity and Quality of Economic Growth*; Banco Central de Chile: Santiago, Chile, 2002.
34. Yamazaki, A. Jobs and climate policy: Evidence from British Columbia's revenue-neutral carbon tax. *J. Environ. Econ. Manag.* **2017**, *83*, 197–216. [CrossRef]
35. He, S.; Fang, B.; Xie, X. Temporal and Spatial Evolution and Driving Mechanism of Urban Ecological Welfare Performance from the Perspective of High-Quality Development: A Case Study of Jiangsu Province, China. *Land* **2022**, *11*, 1607. [CrossRef]
36. Li, J.; Dong, K.; Dong, X. Green energy as a new determinant of green growth in China: The role of green technological innovation. *Energy Econ.* **2022**, *114*, 106260. [CrossRef]
37. Melander, L.; Arvidsson, A. Green innovation networks: A research agenda. *J. Clean. Prod.* **2022**, *357*, 131926. [CrossRef]
38. Ge, T.; Cai, X.; Song, X. How does renewable energy technology innovation affect the upgrading of industrial structure? The moderating effect of green finance. *Renew. Energy* **2022**, *197*, 1106–1114. [CrossRef]
39. Michaelides, P.; Milios, J. TFP change, output gap and inflation in the Russian Federation (1994–2006). *J. Econ. Bus.* **2009**, *61*, 339–352. [CrossRef]
40. Jefferson, G.H.; Rawski, T.G.; Zhang, Y. Productivity growth and convergence across China's industrial economy. *J. Chin. Econ. Bus. Stud.* **2008**, *6*, 121–140. [CrossRef]
41. Emrouznejad, A.; Yang, G. A survey and analysis of the first 40 years of scholarly literature in DEA: 1978–2016. *Socio-Econ. Plan. Sci.* **2018**, *61*, 4–8. [CrossRef]
42. Wu, D.; Wang, Y.; Qian, W. Efficiency evaluation and dynamic evolution of China's regional green economy: A method based on the Super-PEBM model and DEA window analysis. *J. Clean. Prod.* **2020**, *264*, 121630. [CrossRef]

43. Guo, A.; Liu, P.; Zhong, F.; Yang, C.; Luo, X. Borrowing Size and Urban Green Development Efficiency in the City Network of China: Impact Measures and Size Thresholds. *Land* **2022**, *11*, 493. [CrossRef]
44. Coto-Millán, P.; de la Fuente, M.; Fernández, X.L. Determinants of the European electricity companies efficiency: 2005–2014. *Energy Strategy Rev.* **2018**, *21*, 149–156. [CrossRef]
45. Sun, P.; Liu, L.; Qayyum, M. Energy efficiency comparison amongst service industry in Chinese provinces from the perspective of heterogeneous resource endowment: Analysis using undesirable super efficiency SBM-ML model. *J. Clean. Prod.* **2021**, *328*, 129535. [CrossRef]
46. Zhang, H.; Qin, Y.; Xu, J.; Ren, W. Analysis of the Evolution Characteristics and Impact Factors of Green Production Efficiency of Grain in China. *Land* **2023**, *12*, 852. [CrossRef]
47. Jiang, H.; Jiang, P.; Wang, D.; Wu, J. Can smart city construction facilitate green total factor productivity? A quasi-natural experiment based on China's pilot smart city. *Sustain. Cities Soc.* **2021**, *69*, 102809. [CrossRef]
48. Zhao, X.; Nakonieczny, J.; Jabeen, F.; Shahzad, U.; Jia, W. Does green innovation induce green total factor productivity? Novel findings from Chinese city level data. *Technol. Forecast. Soc. Chang.* **2022**, *185*, 122021. [CrossRef]
49. Chao, X.; Ren, B. The fluctuation and regional difference of quality of economic growth in China. *Econ. Res. J.* **2011**, *46*, 26–40.
50. Wu, H.; Li, Y.; Hao, Y.; Ren, S.; Zhang, P. Environmental decentralization, local government competition, and regional green development: Evidence from China. *Sci. Total Environ.* **2020**, *708*, 135085. [CrossRef]
51. Zhong, Z.; Chen, Z. Business environment, technological innovation and government intervention: Influences on high-quality economic development. *Manag. Decis.* **2023**, *61*, 2413–2441. [CrossRef]
52. Liu, L.; Ding, T.; Wang, H. Digital economy, technological innovation and green high-quality development of industry: A study case of China. *Sustainability* **2022**, *14*, 11078. [CrossRef]
53. Ma, S.; Huang, J. Analysis of the spatio-temporal coupling coordination mechanism supporting economic resilience and high-quality economic development in the urban agglomeration in the middle reaches of the Yangtze River. *PLoS ONE* **2023**, *18*, e0281643. [CrossRef] [PubMed]
54. Pan, W.; Wang, J.; Lu, Z.; Liu, Y.; Li, Y. High-quality development in China: Measurement system, spatial pattern, and improvement paths. *Habitat Int.* **2021**, *118*, 102458. [CrossRef]
55. Ren, P.; Liu, J. Theoretical connotation, evaluation criteria and path to realization of high quality green development. *Inn. Mong. Soc. Sci.* **2019**, *40*, 123. [CrossRef]
56. Zheng, W.; Zhang, L.; Hu, J. Green credit, carbon emission and high quality development of green economy in China. *Energy Rep.* **2022**, *8*, 12215–12226. [CrossRef]
57. Weng, Q.; Qin, Q.; Li, L. A comprehensive evaluation paradigm for regional green development based on "Five-Circle Model": A case study from Beijing-Tianjin-Hebei. *J. Clean. Prod.* **2020**, *277*, 124076. [CrossRef]
58. Xie, R.; Yuan, Y.; Huang, J. Different types of environmental regulations and heterogeneous influence on "green" productivity: Evidence from China. *Ecol. Econ.* **2017**, *132*, 104–112. [CrossRef]
59. Chen, Y.; Tian, W.; Zhou, Q.; Shi, T. Spatiotemporal and driving forces of Ecological Carrying Capacity for high-quality development of 286 cities in China. *J. Clean. Prod.* **2021**, *293*, 126186. [CrossRef]
60. Liu, L.; Gu, T.; Wang, H. The Coupling Coordination between Digital Economy and Industrial Green High-Quality Development: Spatio-Temporal Characteristics, Differences and Convergence. *Sustainability* **2022**, *14*, 16260. [CrossRef]
61. Li, Z.; Yang, W.; Wang, C.; Zhang, Y.; Yuan, X. Guided High-Quality Development, Resources, and Environmental Forcing in China's Green Development. *Sustainability* **2019**, *11*, 1936. [CrossRef]
62. Zhou, X.; Wu, W. The measurement and analysis of the inclusive green growth in China. *J. Quant. Tech. Econ.* **2018**, *8*, 3–20. [CrossRef]
63. Ke, N.; Lu, X.; Kuang, B.; Zhang, X. Regional disparities and evolution trend of city-level carbon emission intensity in China. *Sustain. Cities Soc.* **2023**, *88*, 104288. [CrossRef]
64. Yang, J.; Zou, R.; Cheng, J.; Geng, Z.; Li, Q. Environmental technical efficiency and its dynamic evolution in China's industry: A resource endowment perspective. *Resour. Policy* **2023**, *82*, 103451. [CrossRef]
65. Tan, S.; Hu, B.; Kuang, B.; Zhou, M. Regional differences and dynamic evolution of urban land green use efficiency within the Yangtze River Delta, China. *Land Use Policy* **2021**, *106*, 105449. [CrossRef]
66. Liu, Z.; Wakasi, R.A. A research of the regional disparities and distributional dynamic evolution of high-quality agricultural development in China. *Quant. Tech. Econ.* **2021**, *6*, 28–44. [CrossRef]
67. Song, M.; Wu, J.; Song, M.; Zhang, L.; Zhu, Y. Spatiotemporal regularity and spillover effects of carbon emission intensity in China's Bohai Economic Rim. *Sci. Total Environ.* **2020**, *740*, 140184. [CrossRef]
68. Huang, Y.; Zhu, H.; Zhang, Z. The heterogeneous effect of driving factors on carbon emission intensity in the Chinese transport sector: Evidence from dynamic panel quantile regression. *Sci. Total Environ.* **2020**, *727*, 138578. [CrossRef]
69. Wang, J.; Xu, C. Geodetector: Principle and prospective. *Acta Geogr. Sin.* **2017**, *72*, 116–134.
70. Peng, W.; Kuang, T.; Tao, S. Quantifying influences of natural factors on vegetation NDVI changes based on geographical detector in Sichuan, western China. *J. Clean. Prod.* **2019**, *233*, 353–367. [CrossRef]
71. Wang, G.; Peng, W.; Zhang, L. Estimate of population density and diagnosis of main factors of spatial heterogeneity in the metropolitan scale, western China. *Heliyon* **2023**, *9*, e16285. [CrossRef]

72. Liu, X.; Zhang, P.; Shi, X. Industrial Agglomeration, Technological Innovation and High-quality Economic Development: Empirical Research based on China's Five Major Urban Agglomerations. *Reform* **2022**, *36*, 68–87.
73. Dagum, C. A New Approach to the Decomposition of the Gini Income Inequality Ratio. *Empir. Econ.* **1997**, *22*, 515–531. [CrossRef]
74. Cui, X.; Zhang, J.; Huang, W.; Liu, C.; Shan, L.; Jiang, Y. Spatial Pattern and Mechanism of the Life Service Industry in Polycentric Cities: Experience from Wuhan, China. *J. Urban Plan. Dev.* **2023**, *149*, 05023015. [CrossRef]
75. Cui, X.; Yang, S.; Zhang, G.; Liang, B.; Li, F. An Exploration of a Synthetic Construction Land Use Quality Evaluation Based on Economic-Social-Ecological Coupling Perspective: A Case Study in Major Chinese Cities. *Int. J. Environ. Res. Public Health* **2020**, *17*, 3663. [CrossRef]
76. Zou, Z.; Yi, Y.; Sun, J. Entropy method for determination of weight of evaluating indicators in fuzzy synthetic evaluation for water quality assessment. *J. Environ. Sci.* **2006**, *18*, 1020–1023. [CrossRef]
77. Li, D.; Yang, J.; Hu, T.; Wang, G.; Cushman, S.A.; Wang, X.; László, K.; Su, R.; Yuan, L.; Li, B.; et al. The seeds of ecological recovery in urbanization–Spatiotemporal evolution of ecological resiliency of Dianchi Lake Basin, China. *Ecol. Indic.* **2023**, *153*, 110431. [CrossRef]
78. Zhang, S.; Wu, Z.; Wang, Y.; Hao, Y. Fostering green development with green finance: An empirical study on the environmental effect of green credit policy in China. *J. Environ. Manag.* **2021**, *296*, 113159. [CrossRef] [PubMed]
79. Wang, F.; Wang, R.; He, Z. The impact of environmental pollution and green finance on the high-quality development of energy based on spatial Dubin model. *Resour. Policy* **2021**, *74*, 102451. [CrossRef]
80. Ye, D.; Yang, L.; Zhou, M. Spatiotemporal Variation in Ecosystem Health and Its Driving Factors in Guizhou Province. *Land* **2023**, *12*, 1439. [CrossRef]
81. Bai, L.; Jiang, L.; Yang, D.; Liu, Y. Quantifying the spatial heterogeneity influences of natural and socioeconomic factors and their interactions on air pollution using the geographical detector method: A case study of the Yangtze River Economic Belt, China. *J. Clean. Prod.* **2019**, *232*, 692–704. [CrossRef]

land

MDPI

Article

Identifying the Spatial Patterns and Influencing Factors of Leisure and Tourism in Xi'an Based on Point of Interest (POI) Data

Xiaoshuang Qu [1,*], Gaoyang Xu [1], Jinghui Qi [2] and Hongjie Bao [3]

1 School of Business, Zhengzhou University of Aeronautics, Zhengzhou 450064, China; sxyxgy@zua.edu.cn
2 School of Civil Engineering and Architecture, Zhengzhou University of Aeronautics, Zhengzhou 450064, China; qijh2018@zua.edu.cn
3 School of Management, Northwest Minzu University, Lanzhou 730030, China; hongjie@xbmu.edu.cn
* Correspondence: qxs20150906@zua.edu.cn

Abstract: Leisure and tourism spaces are shared by both residents and tourists seeking a higher quality of life. Most of the literature focuses only on the study of a particular type of leisure or tourism space in cities and lacks an overall exploration of the distribution patterns of urban leisure and tourism formats. Based on the leisure and tourism point of interest (POI) data of 11 districts in Xi'an, this paper uses geospatial analysis to examine the spatial patterns of leisure and tourism facilities and their influencing factors in Xi'an. It is found in this study that the distributions overall and the various types of leisure and tourism spaces in Xi'an show the characteristics of central urban agglomeration and sparse dispersion in the surrounding urban areas. Different types of leisure and tourism patterns have obvious spatial scale dependence, but there are differences in the scope of spatial selection. In general, the core agglomeration area has limited radiation and driving effects on the peripheral areas, and there is a prominent phenomenon of imbalance in the distribution of leisure and tourism facilities following a single industrial structure. The formation of the spatial patterns of leisure and tourism is the result of a combination of dominant factors, driving factors, safeguarding factors, and other triggering factors. Urban leisure and tourism spaces are intertwined, and the spatial balance and industrial diversification of leisure and tourism can be promoted through scientific spatial planning. This study aims to provide services for urban land planning and policy-making by revealing the spatial distribution principles of leisure and tourism sites in tourist cities as represented by Xi'an.

Keywords: leisure; tourism; spatial pattern; influence factor; POI; Xi'an City

1. Introduction

Although it is widely recognized that leisure and tourism are closely related in terms of motivation, experience, and behavior [1], existing research has generally viewed leisure and tourism as two separate fields, and direct theoretical exchanges between the two have been relatively limited [2]. There are two representative views on the relationship between "leisure" and "tourism". The first viewpoint suggests that leisure includes tourism, which is a form of off-site leisure different from local leisure [3]. The second view is that not all tourist activities belong to leisure. According to the classification system established by the World Tourism Organization, tourism includes not only leisure but also non-leisure tourism, such as business tourism. Therefore, there is an intersection between leisure and tourism, but not the relationship of including and being included [4]. For example, some scholars have also paid attention to leisure-oriented tourism activities, such as leisure travel [5,6], as well as the continuity of daily leisure activities with tourism activities [7,8].

Ryan and Kinder (1996) argued that the separation of leisure and tourism research is due to the significant difference between rich tourism experiences and leisure activities that occur outside of vacations [9]. Tourism differs from leisure by space, as tourism occurs in destinations different from the place of origin [10]. The commonly accepted assumption is that leisure is for local residents and tourism serves visitors [11]. However, is there a clear boundary between leisure and tourism space in a city? According to Kaplan (1960), social development and lifestyle changes have inevitably led to the integration of leisure and tourism. How to integrate the two is an important challenge for geographic research, and it is also a concern for government agencies and planning departments [12]. Generally, cities with perfect leisure functions are more likely to serve tourists well and become more attractive tourist destinations for them. In recent years, an increasing number of cities in China have begun to pay attention to the cultivation of the leisure and tourism industry. These cities aim to promote the improvement of urban functions through the development of leisure and tourism.

Although there are certain differences between daily leisure activities near the home and leisure travel activities, the daily leisure space of residents may also be the travel space for tourists to a large extent. Leisure and tourism facilities are generic for both local residents and travelers [13]. Urban leisure and tourism spaces are not only recreational spaces for local residents but also key nodes for tourist activities [14]. Urban leisure and tourism spaces thus have integration and intersectionality, and the boundary between serving local residents and tourists will be increasingly blurred. The spatial pattern is the foundation of industrial planning and construction and is of great significance for revealing the characteristics of spatial structure and the mechanisms of spatial differentiation [15]. Therefore, studying the characteristics of urban leisure and tourism space and the factors that influence their creation is of great significance for building the urban leisure and tourism industry, optimizing the layout of urban leisure and tourism space, and better meeting the needs of residents and tourists for a better life. It also helps urban managers formulate scientific and reasonable leisure and tourism development strategies.

Given the independence of "leisure" and "tourism" research, there are also differences in their research perspectives and content. Leisure research focuses on public entertainment and park management, highlighting the government's public service function with a welfarism orientation, while tourism research tends to have a commercial orientation [11]. Research on urban leisure space mainly focuses on a particular type of public urban leisure space, such as parks [16], green spaces [17], sports and recreation sites [18], cultural venues [19], and coastal areas [20]. Most of the literature focuses on the planning and management of urban leisure space [21], public access to recreational spaces [22], health perceptions [23], and service experiences [24]. Research on urban tourism space mainly focuses on a certain type of tourism resource, such as hotels [25,26], homestay [27], scenic spots [28], and cultural heritage sites [29], etc., and studies on the spatial patterns of urban tourism [30,31], tourist spatial behavior [32], urban tourism spatial planning [33], and the evolution of the spatial structure of urban tourism [34]. However, in general, there are few studies that combine leisure and tourism space from the overall urban perspective.

Traditionally, urban leisure and tourism spatial data were mostly obtained through case studies [20], participatory observation [21], in-depth interviews [35], and other methods. With the advent of the big data era, some open platforms now provide big data, such as digital footprints [36], mobile positioning data [37], geotagged photos [38], and point of interest (POI) data [39], which have changed the original paradigms around geospatial data acquisition and research methods [40]. It also provides an opportunity for the accurate measurement of urban leisure and tourism space. Among them, POI data are favored in urban functional space research [41,42] due to their advantages, such as easy access, large data volume, and precise positioning [43].

It has become a trend to use information technology and big data to study urban leisure and tourism space, but at present, big data analysis and geographic information technology have not been widely used, and the existing research mostly focuses on the single format of urban leisure or tourism. There is a lack of academic attention to the overall large-scale spatial pattern of leisure and tourism and its influence mechanisms. Based on this scenario, this paper applies POI data mining, ArcGIS spatial analysis, and geodetector technology to Xi'an, a premiere tourist city in China, to analyze the spatial distribution characteristics and influencing factors on Xi'an's various types of leisure and tourism industries. This paper reveals the principles of spatial distribution of leisure and tourism sites in Xi'an to further optimize the spatial layout and planning of leisure and tourism in Xi'an and provide a reference for other domestic cities committed to the development of leisure and tourism industries.

2. Materials and Methods

2.1. Study Area

Xi'an as the capital of Shaanxi Province as well as the ancient capital of 13 dynasties in China, is an important central city in Western China, and is also a "World Historic City" designated by the United Nations Educational, Scientific, and Cultural Organization (UNESCO). This paper studies the 11 municipal districts of Xi'an, with a total area of approximately 5145.67 square kilometers. Xi'an is rich in tourism resources, including many historic relics, such as Emperor Qinshihuang's Mausoleum site and museum, Weiyang Palace, Daming Palace, and many other historical sites, as well as nationally recognized areas for nighttime cultural and tourism consumption, such as Beiyuanmen Historical and Cultural Block and Datang Night City. The cultural and tourism industry is the pillar industry of Xi'an. The "14th Five-Year Plan" for National Economic and Social Development and the Outline of Vision Goals for 2035 list "promoting the construction of national cultural and tourism consumption pilot cities and building national-level tourism and leisure blocks with distinctive cultural characteristics" as important tasks. The leisure and tourism industry in Xi'an has great development potential for the future. Moreover, Xi'an has a profound historical and cultural heritage, with a prominent image as an ancient capital, and is a representative of Chinese tourist cities. Therefore, this paper studies the leisure and tourism spatial layout in Xi'an and its influencing factors, which has substantial theoretical and practical significance for the study of spatial characteristics of leisure and tourism sites and industrial planning for tourist cities.

2.2. Data Source

POI data are point element data representing real geographic entities, containing the geographical location and attribute information for all kinds of spaces and facilities related to human production habits and life. These data have the characteristics of accurate geographic information and rich volume [43]. In this study, the POI data of leisure and tourism formats in 11 districts of Xi'an were acquired through the Application Programming Interface (API) provided by Amap, which is one of the map platforms that currently provides POI data for free. By applying for a key to the API of Amap and searching for keywords such as POI classification and city, users can obtain batch access to all POI data of a certain type in their area. A total of 75,323 leisure and tourism POI data points were finally retained after data deduplication, merging, cleaning, screening, and bias correction. Each POI data point contains attribute information such as name, latitude and longitude, address, and type.

Combining the classification of urban leisure and tourism in the literature [44] and the classification of POI data from Amap, this paper classifies the leisure and tourism industry in Xi'an into five categories, including catering services, accommodation services, shopping services, sports and entertainment, and scenic spots. The POI data obtained for each type of location in the leisure and tourism industry are shown in Table 1.

Table 1. Classification and Statistics of Leisure and Tourism POI Data in Xi'an City.

Category	POI Type	POI Number	Percentage
Catering services	Chinese restaurants, specialty restaurants, general cuisine restaurants, cafes, tea houses, etc.	50,485	67.02
Accommodation services	Star-rated hotels, guesthouses, homestays, etc.	13,295	17.65
Shopping services	Shopping malls, supermarkets, shopping centres, specialty markets, etc.	2881	3.83
Sports and entertainment	Sports venues, entertainment venues, fitness centres, nursing homes, etc.	8337	11.07
Scenic spots	Lakes, mountains, wetlands, parks and squares, historical sites, temples and Taoist temples, etc.	325	0.43

2.3. Methods

ArcGIS 10.3 software includes a variety of spatial analysis tools, of which the nearest neighbor index method, kernel density estimation, and Ripley's K function are common methods for analyzing spatial distribution features. These three spatial analysis methods can be used to clearly portray the spatial distribution characteristics and laws of leisure and tourism in Xi'an.

2.3.1. Nearest Neighbor Index Method

The nearest neighbor index (NNI) method is a spatial measurement method that quantitatively describes the proximity of spatial point elements to describe the spatial distribution pattern. The spatial distribution trends of point data are judged by calculating the NNI, which is the ratio between the average observed distance of geographical element points and the expected average distance in a random distribution. The formulas are as follows [45]:

$$\mathrm{NNI} = \mathrm{d(NN)}/\mathrm{d(ran)} \tag{1}$$

$$\mathrm{d(ran)} = 0.5\sqrt{\mathrm{A}/\mathrm{N}} \tag{2}$$

In Equation (1), d(NN) denotes the average observed distance between point elements and their nearest neighbors in space, and d(ran) denotes the expected average distance between elements in a random state. In Equation (2), A represents the area of the study area, and N represents the number of sample points within the study area. If NNI > 1, it indicates that the sample points tend to be uniformly distributed. If NNI < 1, it indicates that the sample points are clustered, and the smaller the value, the higher the degree of clustering. If NNI = 1, it indicates that the sample points are randomly distributed.

2.3.2. Kernel Density Estimation

Kernel density estimation is a method to calculate the density of point elements in their neighborhood, which directly reflects the spatial distribution pattern of elements [46]. Its calculation formula is as follows [45]:

$$\mathrm{f(x)} = \sum_{\mathrm{i=1}}^{\mathrm{n}} \frac{1}{\pi \mathrm{r}^2} \mathrm{k}\left(\frac{\mathrm{d_{ix}}}{\mathrm{r}}\right) \tag{3}$$

In Equation (3), f(x) is the kernel density value at x, n is the total number of samples, r is the search radius, k represents the distance weight, and d_{ix} is the distance from POI point i to x.

2.3.3. Ripley's K Function

Ripley's K function is a point density distance function that counts the number of points within a search range established by a certain radius and can determine whether the elements have statistically significant clustering within a certain range [45]. This is because although the nearest neighbor index can portray the overall spatial distribution of point data, the same point data may present different spatial characteristics at different scales. Therefore, the use of Ripley's K function can analyze the spatial clustering characteristics of various leisure and tourism modes in Xi'an at multiple scales. The formula is as follows [45]:

$$L(d) = \sqrt{\frac{A\sum_{i=1}^{n}\sum_{j=1}^{n}\frac{W_{ij}(d)}{n^2}}{\pi}} - d \tag{4}$$

In Equation (4), A represents the area of the study area, and n represents the number of various leisure and tourist spots. W_{ij} (d) represents the distance between the leisure and tourism points i and j within the range of distance d. L (d) represents the degree of agglomeration of leisure and tourism points within the range of distance d. The maximum and minimum values obtained by the test are defined as the upper and lower envelope values. If L (d) > 0 and the simulation results are above the upper envelope, it indicates that this type of leisure and tourism point is agglomerated. If L (d) < 0 are below the lower envelope, it means that this type of leisure and tourism spot is evenly distributed. If L (d) = 0 and the simulation results are between the upper and lower envelopes, it indicates that this type of leisure and tourism site is randomly distributed.

2.3.4. Geographic Detector

Geodetection is a spatial analysis method used to detect spatial differentiation and reveal the driving force behind it. The basic assumption is that if the independent variable has an influence on the spatial differentiation of the dependent variable, then the independent variable and the dependent variable have similar spatial distributions [45]. The extent to which a Factor X is able to explain the spatial divergence of Y is measured by the q-value. By calculating the q-value of a single factor and the q-value of two factors superimposed on each other, geodetectors can not only analyze the degree of influence of a particular independent variable on the dependent variable but also determine whether there is an interaction between the two factors and the strength of the interaction. The formula for calculating the influence q is as follows [45]:

$$q_x = 1 - \frac{1}{N\sigma^2}\sum_{i=1}^{k} N_i{\sigma_i}^2 \tag{5}$$

In Equation (5), q_x is the influence of the independent variable X on the density of the leisure and tourism industries in Xi'an. ${\sigma_i}^2$ and σ^2 are the discrete variances of the leisure and tourism densities in layers and regions. N_i and N are the number of units in the layer and region, respectively. For $q \in [0, 1]$, the larger the value, the greater the influence of the independent variable on the spatial differentiation of leisure and tourism, and vice versa.

3. Leisure and Tourism Spatial Characteristics

3.1. Spatial Agglomeration Characteristics

The collected POI data of leisure and tourism in Xi'an were imported into ArcGIS for nearest neighbor analysis, and the nearest neighbor index (Table 2) was calculated to characterize the various types of leisure and tourism spatial agglomerations in Xi'an both as smaller units and overall. As shown in Table 2, the average observed distance for the overall leisure and tourism space is 29.62 m and the expected average distance is 166.77 m. The NNI is 0.178 and the Z value is −431.79. Therefore, the overall leisure and tourism space in Xi'an has a significant tendency toward agglomeration. The NNIs of the five types of leisure and tourism spaces, namely, catering services, accommodation services, shopping services, sports and entertainment, and scenic spots, are 0.151, 0.189, 0.359, 0.288, and 0.649, respectively. The NNIs are all less than 1, and the Z scores are all less than

−2.58 with a P of 0. All of them passed the significance test at the 1% level. The results indicate that the five types of leisure and tourism spaces in Xi'an are all significantly clustered, but the clustering characteristics of different types of leisure and tourism spaces show some differences. The degree of agglomeration of the identified types of leisure and tourism spaces, including the overall assessment, in descending order is as follows: catering services > overall leisure and tourism > accommodation services > sports and entertainment > shopping services > scenic spots.

Table 2. Nearest Neighbor Distance for Leisure and Tourism Spaces in Xi'an City.

Category	The Average Observed Distance (m)	The Expected Average Distance (m)	NNI	Z	P	Spatial Distribution Type
Overall	29.62	166.77	0.178	−431.79	0	Significant agglomeration
Catering services	28.57	189.19	0.151	−364.94	0	Significant agglomeration
Accommodation services	69.62	370.54	0.189	−179.14	0	Significant agglomeration
Shopping services	264.75	736.44	0.359	−65.77	0	Significant agglomeration
Sports and entertainment	141.97	493.35	0.288	−124.41	0	Significant agglomeration
Scenic spots	1323.26	2039.63	0.649	−12.11	0	Significant agglomeration

Specifically, catering service points are the most numerous and most spatially clustered, indicating that the strongest demand of local residents and foreign tourists is for catering and leisure service spaces in Xi'an. The agglomeration degree of accommodation services is only second to that of catering services, and the NNI values of these two types of leisure and tourism are relatively similar overall, reflecting to some extent that the overall development of leisure and tourism in Xi'an is highly synchronized with the catering and accommodation industries. The degree of concentration of sports and entertainment and shopping services is at the medium level, and there is still a gap between the number of leisure spots in the periphery of the city and that in the center of the city. The NNI of scenic spots is 0.649, whereby although it also shows significant agglomeration, the degree of agglomeration is relatively weak compared to the overall distribution and the distributions of other types. The scenic type is influenced by the natural environment and history, and its spatial layout is subject to a low degree of human intervention. The scenic spot category in Amap includes various types and volumes of scenic spots, such as tourist attractions, wetlands, and ruins, which are included in POI data crawling. The map can only obtain point data spatially and does not record social attributes such as scenic spot size, resource grade, and popularity. Due to the scattered distribution of differently graded scenic spots in various districts of Xi'an, the concentration of scenic leisure and tourism spots is relatively low.

3.2. Spatial Distribution Characteristics

The kernel density estimation method is used to calculate the distribution densities overall and for various leisure and tourism types in Xi'an, and the natural fracture method is used to divide the kernel density of various leisure and tourism spots into three grades: low-density areas, medium-density areas, and high-density areas. The spatial distributions overall and for various types of leisure and tourism spaces in Xi'an are shown in Figure 1. On the whole, the overall distribution and all types of leisure and tourism spaces in Xi'an show the characteristics of central urban agglomeration and sparse dispersion in the surrounding urban areas.

As seen in Figure 1a, the overall distribution of leisure and tourism sites shows the characteristics of "one center, multiple points, and two axes". Leisure and tourism in Xi'an are centered on the Bell and Drum Tower. The north–south direction takes Weiyang Road, North Street, and Changan South Road as its axis, which is basically consistent with Metro Line 2. The east–west direction takes Fengqing Road and Huancheng South Road as its axis, which mostly coincides with the nodes of Metro Lines 2 and 6. The core area spans four districts, including Lianhu, Beilin, Xincheng, and Yanta, forming a high-density hierarchical agglomeration. The medium-density area is distributed around the core area, in line with the core-edge development pattern. All other districts have multiple tourism hotspots that are scattered in distribution. Leisure and tourism hotspots in Weiyang District are distributed near the physical location of the municipal government and Xi'an City Sports Park. The hot spots of Yanta District are located in and around the Shaanxi Provincial Historical Museum and Tang Paradise. Leisure and tourism hotspots in Chang'an District are concentrated in Chang'an District's university town and near Chang'an Park. Baqiao District is focused on commercial centers such as Huayang City and Xi'an Sky City Shopping Centre. Leisure and tourism hotspots in Lintong District are concentrated in areas such as Emperor Qinshihuang's Mausoleum site and museum, Mount Li, and Huaqing Pool. The distribution of high-level and medium-level agglomerations reflects the characteristics of proximity to transport, scenic spots, and business districts.

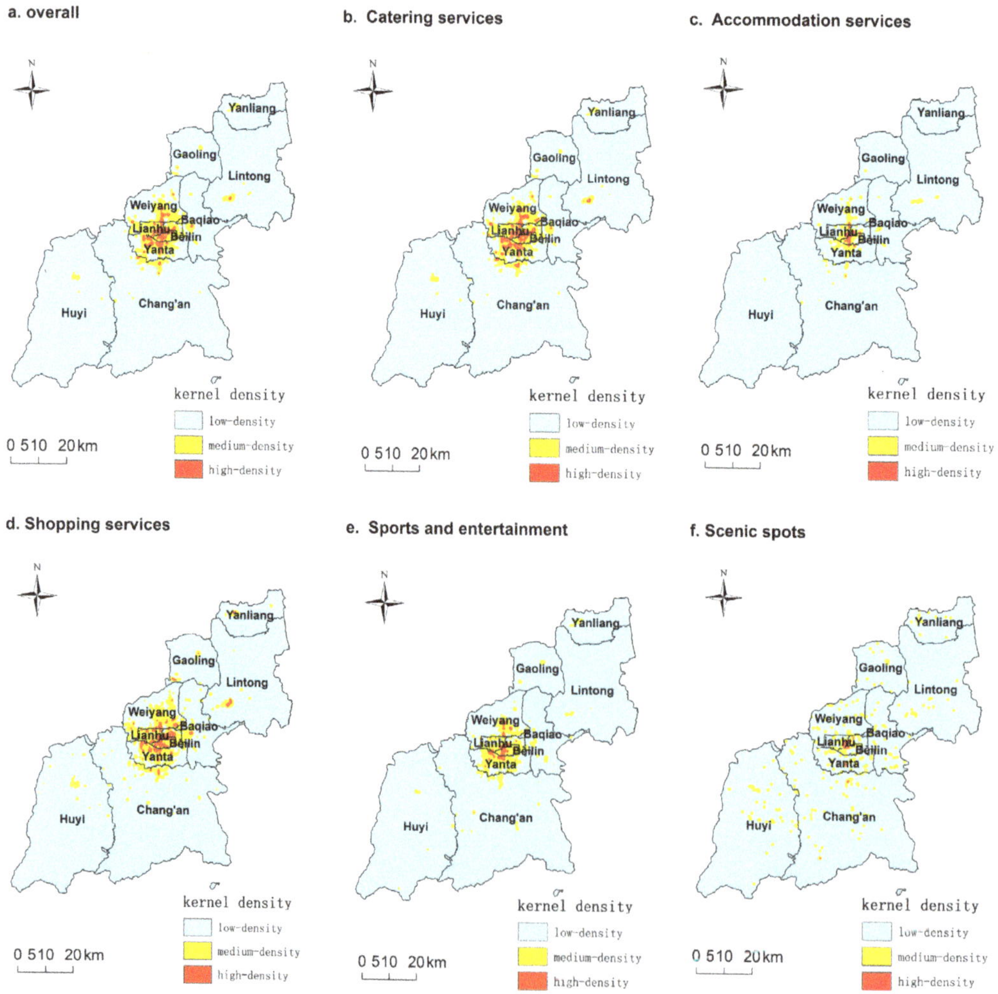

Figure 1. Kernel Density Analysis of the Spatial Distribution of Leisure and Tourism in Xi'an.

The catering category (Figure 1b) accounts for the largest proportion of the overall leisure and tourism POI data, and its distribution pattern and form are closest to the overall leisure and tourism spatial characteristics. The most densely clustered area is basically within the Ring Road, and its spatial scope is largely consistent with the historical urban area defined in the Conservation Plan for Xi'an Historical and Cultural City (2020–2035), which is the area with the most historical and cultural heritage in Xi'an. The middle-grade agglomeration area is outside the core area but within the area bounded by the Ring Road.

The accommodation category (Figure 1c) is characterized by "small agglomerations and small dispersion". The core area of high-level agglomeration is centered on the Bell and Drum Tower, which is mainly located at the junction of the three districts of Lianhu, Xincheng, and Beilin and consists of the areas of North Street, Tieta Temple, South Street, and East Street. Intermediate agglomerations are distributed on the periphery of high agglomerations, and in addition to the three districts in which the high agglomerations are located, they are distributed in contiguous or point-like forms in the districts of Weiyang, Yanta, Lintong, and Chang'an.

The shopping category (Figure 1d) shows a spatial distribution pattern of "small aggregations, multiple points, and large dispersion". The high-level agglomeration area is centered in the Bell and Drum Tower area, spreading out in a planar shape towards the surrounding area. For example, the shopping leisure and tourism spots in Yanta District are concentrated in the main blocks of Chang'an Road, Yanta Road, Furong South Road, Qujiang Road, Tangyan Road, etc. Medium-level clustering areas are distributed in all districts and are relatively scattered.

Sports and entertainment (Figure 1e) are characterized by a "linear and widely dispersed" pattern. High-density cluster areas are mainly distributed in the four districts of Weiyang, Lianhu, Xincheng, Beilin, and Yanta. They start from Weiyang Road and Beijie Street in the north and end at Chang'an South Road in the south, with a concentrated distribution along the axis. Other areas have scattered medium-density clusters.

Scenic spots (Figure 1f) show a spatial distribution pattern of "small aggregations and large dispersion". The high-density agglomerations are the Bell and Drum Tower scenic area, the Xi'an City Wall area, the Daming Palace National Heritage Park area, Dayan Pagoda, and the Tang Paradise area. Medium-density agglomerations are more dispersed, with multiple distribution points in each district.

3.3. Spatial Scale Characteristics

Ripley's K function is used to judge the significance of multiscale agglomeration for the spatial distributions overall and in various types of leisure and tourism spaces in Xi'an. In Figure 2, the distributions overall and for various types of leisure and tourism spaces in Xi'an city follow the principles of significant agglomeration at different spatial scales. Based on the distance at which the peak occurs, overall leisure and tourism distribution reaches peak agglomeration at 19.3 km, which is higher than the peak agglomeration for any of the individual leisure and tourism types. This indicates that the five types of leisure and tourism spaces work together to enhance the agglomeration strength of the distribution of leisure and tourism spaces overall.

The catering service distribution is highly similar to the overall distribution for leisure and tourism, and catering services reach peak agglomeration at 19.1 km, which is higher than other types of leisure and tourism. This suggests that catering service points show agglomeration characteristics at larger spatial scales, with a wide range of spatial choices for location. This is followed by accommodation services and sports and entertainment locations, both of which reach their peak agglomeration at 18.7 km, with a slightly lower location selection ability. Scenic spots and shopping services reach peak agglomeration at 17.6 km and 17.0 km, respectively. Since scenic spots are more influenced by history and geography and shopping and commercial facilities are more influenced by business districts and location, these two categories of leisure and tourism have a lower level of ability to choose their locations and range among all categories. In conclusion, leisure and tourism facilities in Xi'an show obvious spatial scale dependence, but different types of leisure and tourism facilities have different degrees of spatial scale dependence. This difference is related to factors such as the nature of leisure and tourism facilities, location, environment, and market demand.

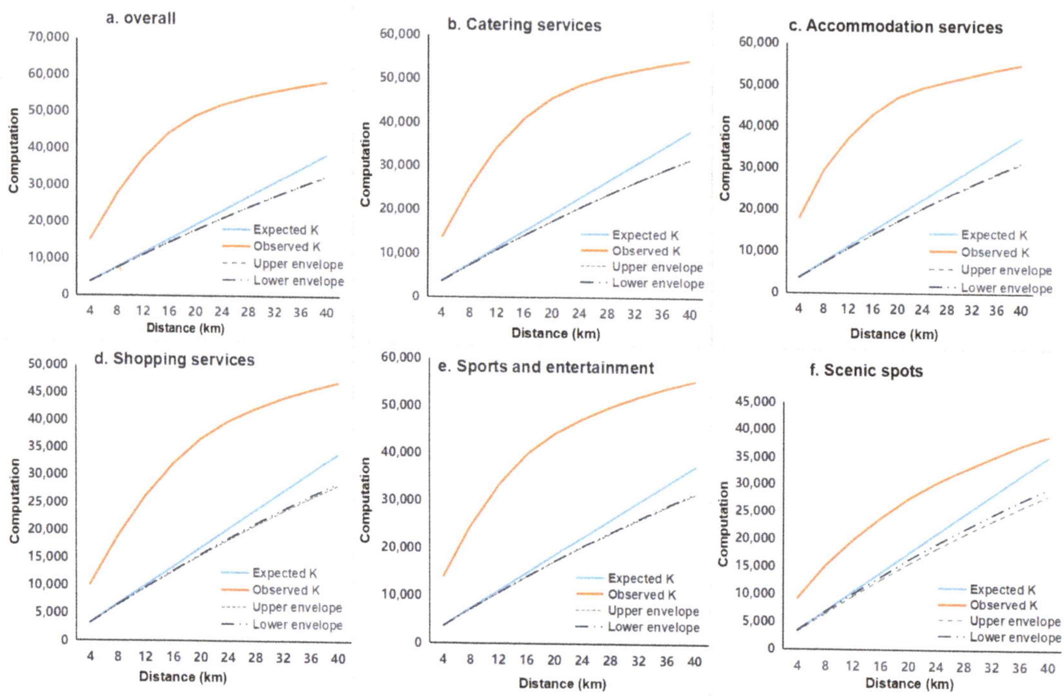

Figure 2. Ripley's K Function Analysis of Leisure and Tourism Spaces in Xi'an City.

4. Factors Influencing the Spatial Distribution of Leisure and Tourism

4.1. Impact Indicators and Analysis

The formation of the spatial pattern of urban leisure and tourism is the result of a combination of multiple factors. It has been argued that the economy, population, transport, infrastructure, and policy are the main factors influencing the distribution of urban leisure and tourism patterns [30,47]. The previous results of the spatial distribution of leisure and tourism in Xi'an also show that the leisure and tourism sites are characterized by near traffic and densely populated business districts. In addition to these aspects mainly considered by previous scholars, this paper argues that industrial structure is also a main factor affecting the distribution of leisure and tourism. Generally speaking, areas with a developed economy and a high proportion of the service industry have a dense distribution of commercial and leisure facilities, which provide an adequate supply for the development of leisure and tourism. Where the population is concentrated, people's demand for leisure and tourism is strong, and commercial and leisure activities are more frequent. Transport and infrastructure are the guarantee for leisure and tourism activities and provide convenience for people's leisure and tourism activities. In addition, factors such as policy and history also influence the spatial distribution of urban leisure and tourism, but these aspects are often difficult to quantify and are not considered in quantitative analyses.

Therefore, an indicator system is constructed for influencing factors, which includes seven representative indicators in five dimensions: economic level, demographic factors, industrial structure, traffic conditions, and infrastructure. The level of economic development is composed of the average house price at level X_1 and the per capita disposable income of urban residents, X_2. The demographic factor is represented by the population density of permanent residents, X_3. The industrial structure consists of the proportion of the service industry to GDP, X_4, and the density of cultural service enterprises, X_5. The traffic conditions are measured by the X_6 indicator of road network density within the region. Generally, the more developed the evening economy is, the more perfect the facilities are, and the more it contributes to the development of leisure and tourism. Therefore, the night light intensity, X_7, is used to reflect the comprehensive level of infrastructure development in all urban areas.

Geodetector technology is used to calculate the q-value of the influence of each factor on the spatial differentiation of leisure and tourism in Xi'an, in which the dependent variable is the spatial density both overall and for the various types of leisure and tourism sites in 11 districts of Xi'an, and the independent variables are the seven indicators in the above index system. Geodetector analysis (Table 3) shows that all seven indicators are significant at the 1% level, indicating that the seven selected factors have a significant influence on the spatial differentiation of leisure and tourism in Xi'an, and the intensity of the effect is, in descending order, population density X_3 > density of cultural service enterprises X_5 > per capita disposable income of urban residents X_2 > night light intensity X_7 > the proportion of the service industry in GDP X_4 > road network density X_6 > average house price level X_1.

Table 3. Geographical Detection Analysis of Factors Influencing the Spatial Distribution of Leisure and Tourism in Xi'an City.

Indicator Dimension	Detection Factors	q Ranking	Overall	Catering Services	Accommodation Services	Shopping Services	Sports and Entertainment	Scenic Spots
Economic level	Average house price level, X_1	7	0.137	0.230	0.097	0.137	0.261	0.123
	Per capita disposable income of urban residents, X_2	3	0.679	0.760	0.580	0.679	0.738	0.664
Demographic factors	Population density, X_3	1	0.866	0.935	0.877	0.866	0.873	0.839
Industrial structure	The proportion of the service industry to GDP, X_4	5	0.434	0.489	0.374	0.434	0.594	0.411
	Density of cultural service enterprises, X_5	2	0.811	0.872	0.636	0.811	0.773	0.744
Traffic conditions	Road network density, X_6	6	0.430	0.486	0.304	0.430	0.410	0.329
Infrastructure	Night light intensity, X_7	4	0.663	0.715	0.571	0.663	0.726	0.538

Population density has the strongest explanatory power for the spatial distribution of leisure and tourism in Xi'an, which suggests that leisure and tourism in Xi'an is distributed with a strong population dependency. In the dimension of economic development level, the average housing price in the region has the lowest influence on the spatial distribution of leisure and tourism sites, while the per capita disposable income of urban residents has the third highest explanatory power, which is relatively strong in explaining the spatial differentiation of leisure and tourism. The income level of local residents largely influences the consumption demand for leisure and tourism; the higher the income level, the stronger the demand of residents for leisure and tourism. The leisure and tourism industry not only carries commercial attributes but also has public and welfare characteristics, so the average housing price level may not be a key factor affecting the spatial distribution of leisure and tourism. In terms of industrial structure, the proportion of the service industry to GDP and the density of cultural service enterprises rank fifth and second, respectively, in their explanatory power for the spatial differentiation of leisure and tourism in Xi'an. This is related to the fact that the leisure and tourism industry has cultural attributes and belongs to the service sector, which is also in line with Xi'an's urban image as a historical and cultural ancient capital. The intensity of nighttime lighting and the road network both have strong explanatory power for the spatial differentiation of leisure and tourism sites in Xi'an, ranking fourth and sixth, respectively, which indicates that the development of the leisure and tourism industry is more closely related to the city's high-quality facilities and traffic conditions.

4.2. Interaction Detection of Influencing Factors

The influence value of q reflects the single-factor explanatory force of each factor on spatial differentiation, but the combined effect of different factors may have different effects on spatial differentiation. The interaction detector can be used to analyze whether the combined effect of any two factors on the dependent variable Y is independent, as well as the change in explanatory power. The results of analysis of the interactive influence of factors in the spatial differentiation of leisure and tourism sites in Xi'an (Table 4) show that the interaction of every two factors manifests a nonlinear enhancement (NE) or a bifactor enhancement (BE). That is, the interactions of every two factors are all greater than the explanatory power of each factor alone, and the influence of each factor on the spatial differentiation of leisure and tourism in Xi'an is interrelated and thus not an isolated effect. On the whole, the interaction between the factors is dominated by BE, i.e., the influence of the interaction of two factors is greater than the maximum influence of either factor alone. The interaction between housing price and other factors will show more NE results; that is, the influence of housing price interacting with other factors is greater than the sum of the influences of single factors. Although the explanatory power of housing prices in single factor impact analysis is not strong, when interacting with other factors, it also has a strong explanatory power on the spatial differentiation of leisure and tourism in Xi'an, with all calculations exceeding 0.8. The interaction of the other six factors also enhances the explanatory power of spatial differentiation.

Table 4. Interactive Analysis of Factors Influencing the Spatial Distribution of Leisure and Tourism Sites in Xi'an City.

A∩B	Overall	Catering Services	Accommodation Services	Shopping Services	Sports and Entertainment	Scenic Spots
$X_1 \cap X_2$	0.889 (NE)	0.874 (BE)	0.828 (NE)	0.889 (NE)	0.813 (BE)	0.927 (NE)
$X_1 \cap X_3$	0.951 (BE)	0.953 (BE)	0.904 (BE)	0.951 (BE)	0.879 (BE)	0.939 (BE)
$X_1 \cap X_4$	0.976 (NE)	0.984 (NE)	0.871 (NE)	0.976 (NE)	0.864 (NE)	0.945 (NE)
$X_1 \cap X_5$	0.972 (NE)	0.943 (BE)	0.977 (NE)	0.972 (NE)	0.985 (BE)	0.988 (NE)
$X_1 \cap X_6$	0.997 (NE)	0.989 (NE)	0.994 (NE)	0.997 (NE)	0.999 (NE)	0.991 (NE)
$X_1 \cap X_7$	0.966 (NE)	0.995 (BE)	0.919 (NE)	0.966 (NE)	0.888 (BE)	0.936 (NE)
$X_2 \cap X_3$	0.868 (BE)	0.937 (BE)	0.877 (BE)	0.868 (BE)	0.878 (BE)	0.844 (BE)
$X_2 \cap X_4$	0.799 (BE)	0.862 (BE)	0.630 (BE)	0.799 (BE)	0.762 (BE)	0.747 (BE)
$X_2 \cap X_5$	0.813 (BE)	0.874 (BE)	0.637 (BE)	0.813 (BE)	0.777 (BE)	0.749 (BE)
$X_2 \cap X_6$	0.975 (BE)	0.984 (BE)	0.870 (BE)	0.975 (BE)	0.864 (BE)	0.943 (BE)
$X_2 \cap X_7$	0.797 (BE)	0.870 (BE)	0.623 (BE)	0.797 (BE)	0.785 (BE)	0.723 (BE)
$X_3 \cap X_4$	0.997 (BE)	0.998 (BE)	0.999 (BE)	0.997 (BE)	0.996 (BE)	0.994 (BE)
$X_3 \cap X_5$	0.973 (BE)	0.952 (BE)	0.982 (BE)	0.973 (BE)	0.982 (BE)	0.992 (BE)
$X_3 \cap X_6$	0.975 (BE)	0.999 (BE)	0.921 (BE)	0.975 (BE)	0.894 (BE)	0.944 (BE)
$X_3 \cap X_7$	0.974 (BE)	0.999 (BE)	0.921 (BE)	0.974 (BE)	0.894 (BE)	0.942 (BE)
$X_4 \cap X_5$	0.836 (BE)	0.918 (BE)	0.653 (BE)	0.836 (BE)	0.788 (BE)	0.747 (BE)
$X_4 \cap X_6$	0.674 (BE)	0.670 (BE)	0.757 (NE)	0.674 (BE)	0.865 (BE)	0.680 (BE)
$X_4 \cap X_7$	0.839 (BE)	0.930 (BE)	0.660 (BE)	0.839 (BE)	0.791 (BE)	0.753 (BE)
$X_5 \cap X_6$	0.996 (BE)	0.989 (BE)	0.993 (NE)	0.996 (BE)	1.000 (BE)	0.993 (BE)
$X_5 \cap X_7$	0.839 (BE)	0.930 (BE)	0.660 (BE)	0.839 (BE)	0.791 (BE)	0.753 (BE)
$X_6 \cap X_7$	0.975 (BE)	0.999 (BE)	0.921 (NE)	0.975 (BE)	0.894 (BE)	0.945 (NE)

4.3. Influencing Mechanisms

The formation of the spatial patterns of urban leisure and tourism sites is a complex process influenced by many factors. Based on the analysis of geographic detectors in combination with findings from relevant studies [48–51], the impact mechanisms affecting the spatial pattern of leisure and tourism sites in Xi'an are further analyzed from four aspects: dominant factors, driving factors, guarantee factors, and other triggering factors (Figure 3).

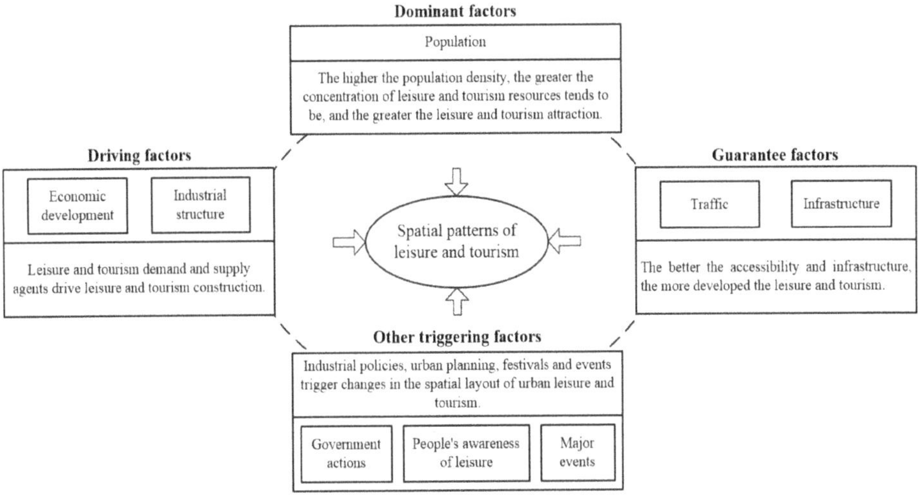

Figure 3. Impact Mechanisms of Leisure and Tourism Spatial Patterns.

(1) Dominant factors

Demographic factors are the dominant factors influencing the formation of the spatial pattern of leisure and tourism sites in Xi'an. Population is regarded as a source of tourism growth [49]. The purpose of leisure and tourism is to provide good facilities, leisure environments, and services and to meet the leisure and tourism needs of residents and tourists. The higher the degree of population concentration, the greater the market demand for leisure and tourism. In general, population density is closely related to the location, resources, and attractiveness of the area. The three districts with high population density in Xi'an are Beilin District, Lianhu District, and Xincheng District, which are the center of Xi'an. These districts have the highest concentration of leisure and tourism resources, which are the richest in historical and cultural connotations for the region and the most expressive of the charms of ancient capital culture. Allocentric tourists are more likely to be attracted by the cultural heritage of a given destination [50]. These districts are an important activity area for foreign tourists to experience the original lifestyle of the local residents.

(2) Driving factors

The level of economic development and industrial structure are important driving factors influencing the spatial layout of leisure and tourism sites in Xi'an. The higher the level of economic development in the region, the higher the disposable income of the residents, and the stronger the demand for leisure and tourism consumption [51]. Leisure and tourism includes catering, accommodations, entertainment, tourism, and other forms of business belonging to the scope of the service industry, and the degree of development of the city's service industry has a direct impact on the development of leisure and tourism. Culture and tourism are inextricably linked, especially in a city such as Xi'an, where history and culture are deeply rooted, and cultural tourism resources are an important carrier of leisure and tourism activities. As the main operators in the leisure and tourism market, cultural enterprises provide important support for the development of high-quality leisure and tourism products and the optimization and upgrading of the leisure and tourism industry in Xi'an. These enterprises drive the innovation and development of the leisure and tourism industry.

(3) Guarantee factors

Transport facilities and infrastructure play an important role in tourism development [51,52] and are important guarantees for the development of leisure and tourism. From the spatial distribution characteristics of leisure and tourism in Xi'an, there is a notably strong dependence on traffic, and the middle-level and high-level agglomerations are mostly distributed along the metro lines, main roads, ring roads, and other axes and circuits. The areas with strong lighting at night are Beilin, Lianhu, Yanta, and Xincheng, including the Bell and Drum Tower business district and the emerging cultural and tourism consumption agglomeration areas, such as Datang Night City. The supporting tourism service facilities and various infrastructure in these areas are relatively mature, providing an important guarantee for the functional improvement of leisure and tourism clusters.

(4) Other triggering factors

The formation and evolution of urban leisure and tourism distributions are also influenced by other factors that are difficult to quantify, such as government actions (policies and regulations, urban positioning, land planning, industrial planning, etc.), public awareness of leisure and paid vacation systems, and major events (conferences and exhibitions, sports events, festivals, web celebrity events, etc.). In the future, with the continuous enhancement of public awareness of leisure, the service functions of cities for residents and tourists will become increasingly perfected. The emergence of new technologies and new forms of businesses will also further influence the formation of the spatial patterns of leisure and tourism services.

5. Discussion

Previous studies have mostly focused on the spatial distribution of a certain type of leisure or tourism type in the city. This paper reveals the characteristics of the spatial distributions of leisure and tourism types in Xi'an, a nationally recognized urban center of excellent tourism and their influencing mechanisms from the overall perspective of leisure and tourism space in the city. This study argues that urban leisure and tourism spaces are spaces shared by hosts and guests. Urban leisure and tourism can be divided into five types: catering services, accommodation services, sports and entertainment, shopping services, and scenic spots.

The reason for studying urban leisure and tourism spaces together is not only that they are intertwined and difficult to distinguish, but also that leisure and tourism can enhance each other's development. Rising living standards have awakened people's sense of leisure, and a city with good leisure facilities and environment is equally attractive to tourists and can also enrich their experience. As seen from the above analysis, the distribution of scenic spot sites is strongly influenced by the environment and history, and its ability of location selection is limited. In contrast, recreational facilities have a strong location selection ability and can complement traditional attractions. Quality leisure facilities not only serve residents well but also provide tourists with alternatives to attractions. For example, many night-time specialty snacks and shopping bazaars, night-time cultural performances, and other leisure projects in Xi'an have attracted a large number of tourists and become new tourist hotspots, promoting the development of night-time tourism in Xi'an.

Theoretically, this study deepens the exploration of the spatial law of urban leisure and tourism and enriches the study of the relationship between leisure and tourism. At the practical level, it is aligned with the law of urban development and practical requirements in China. China's National Tourism and Leisure Development Program (2022–2030) and the 14th Five-Year Plan for Tourism Development both explicitly propose promoting the development of tourism and leisure and creating a number of nationally designated tourism and leisure cities with distinctive cultural characteristics. Consequently, the exploration of the spatial principles guiding leisure and tourism development will help Chinese tourist cities plan the layout of tourism and leisure facilities more scientifically.

The results of the nearest-neighbor distance and kernel density analyses show that the distributions overall and for various types of leisure and tourism spaces in Xi'an present the characteristics of concentration in the central urban area and sparse dispersion in the surrounding urban areas, but there are differences in the degree of concentration and the patterns among different types of services. This finding verifies the conclusion of some scholars that the distribution of urban leisure tourism resources has an obvious core-periphery structure [30,31]. The high-density distributions overall and for various types of leisure and tourism sites in Xi'an exhibit a proximity to transportation, scenic areas, and commercial districts. Generally, the most concentrated core area of the distributions both overall and for various types of leisure and tourism services is centered on the Bell and Drum Tower scenic spot, within the scope of the Ring Road, which is the most richly preserved area of historical remains relating to Xi'an's history as an ancient capital and was delineated in Xi'an's early urban planning as the urban center. This conclusion is in line with the views of some studies. In the case of tourist historic cities, tourism is closely intertwined with the daily lives of local people [53]. The best value hotels are most likely to be located in and around the central districts of urban tourist destinations where population and economic activities are denser [25]. However, the distribution of medium-sized agglomerations in Xi'an shows significant differences, with a distribution of multiple points or axes. The spatial concentration of tourism investment should be shifted from the spot approach to the axis approach while those axes are equipped as comprehensive spatial strategic in the regional tourism plans [54]. Therefore, attention should be paid to these points and axes outside the center of the city, which have a high potential for future development of the leisure and tourism industry.

The spatial distribution of leisure and tourism in Xi'an is to some extent representative of the distinctive characteristics of leisure and tourism spaces in cities within the historical capital category, i.e., the "star effect" of the historical core area is obvious, but the peripheral areas of the core area may have "image masking". The image of the city as a tourist destination is not set by the will of tourism developers but by the perception of the tourism market [55]. Tourists are influenced by a variety of factors when choosing a destination, including its perceived image. Invisible competition exists within the same region and destinations with a distinctive image and high visibility are often more popular with tourists. Xi'an, as the ancient capital for thirteen dynasties in China, has a distinctive urban image as a historical capital. The ancient city wall, the Bell and Drum Towers, and the Emperor Qinshihuang's Mausoleum Site and Museum are the most attractive cultural symbols of the ancient capital to tourists, and therefore, the layout of the city's leisure and tourism industry is centered on these attractions, with a highly concentrated industrial layout. The number of leisure and tourism facilities in areas outside the central city is relatively few and scattered, and the development of leisure and tourism resources that are not closely related to the cultural image of Xi'an's ancient capital, such as landscape resources, parks, public spaces, etc., have received relatively low-key development efforts.

Although this paper analyses the spatial distribution of the leisure and tourism industry in Xi'an only from the perspective of leisure and tourism market supply and does not analyze the spatial flow of residents and tourists on the demand side, it can be indirectly speculated that the flow of residents and tourists in Xi'an will also be very concentrated, forming distinctive "hot spots" and "cold spots" of leisure and tourism. This spatial imbalance will inevitably affect the overall development of the industry. Local residents outside the central city do not have equitable access to the city's leisure resources, and foreign tourists are influenced by the misimpression that their activities in Xi'an are limited to the central city. The sparse distribution and poor accessibility of leisure and tourism resources outside the central city further constrains the mobility of residents and tourists outside the core area. In the long run, the gap between the leisure and tourism industries in the center and the surrounding areas will increase day by day, which would be detrimental to the overall economic and social development of the city. It should be noted that the POI data used in this paper are the data of all leisure and tourism sites, which can only reflect the distribution of leisure and tourism facilities as a whole, and which cannot be used to clearly distinguish the respective activity spaces of residents and tourists.

Through the study of the influencing factors on the spatial distribution of leisure and tourism in Xi'an, it is found that the spatial distribution of leisure and tourism patterns in Xi'an is related to factors such as population density, density of cultural service enterprises, per capita disposable income of residents, infrastructure, and transport. This conclusion is basically consistent with the findings of push and pull studies of tourism activity [51]. However, compared with general tourist cities, Xi'an specializes in cultural tourism products; therefore, when considering the factors affecting the spatial distribution of leisure and tourism in Xi'an, this paper includes the density of cultural service enterprises and the proportion of the service industry in GDP, which are two factors reflecting the industrial structure, in the system of influencing factors. The data show that the influence of these two factors, especially the density of cultural service enterprises, is relatively strong.

In summary, leisure and tourism are important drivers of the economic and social development of cities. The rationality of the spatial layout of leisure and tourism in Xi'an and the development of leisure and tourism have a positive impact on the construction of Xi'an as a national central city and are also inspirational to the development of leisure and tourism in other cities across the country. The high-quality development of leisure and tourism in Xi'an should include spatial balance and industrial diversification. Based on this, this study proposes the following suggestions:

(1) The radiation effect of the core agglomeration should be enhanced, and new growth poles should be cultivated. The layout of the leisure and tourism industry in Xi'an is relatively concentrated, with various types of leisure and tourism resources mainly concentrated in the central urban areas of Beilin, Xincheng, and Lianhu. The "one core" pattern has basically formed, but the radiation and driving effect of the core agglomeration area on the outer region is limited. In the future, efforts should be made to further optimize resource allocation, promote the transfer of resource elements, strengthen regional industrial cooperation, promote the extension of transportation networks, and cultivate new growth poles to promote the integration and coordinated development of the main urban area and other sections. Furthermore, efforts should aim to improve the rationality of the overall layout of leisure and tourism formats in the region of Xi'an, as well as the ability to provide leisure and tourism services. The noncentral urban areas should increase their marketing and publicity efforts to fully demonstrate the unique charms of their respective areas and promote the reasonable flow of tourists between the central urban areas and the peripheral areas.

(2) The advantageous resources of each district should be fully tapped to form a diversified leisure and tourism system. At present, cultural tourism in Xi'an city center, Yanta District, and Lintong District has formed a certain influence due to their unique historical and cultural resources. However, from the overall scope of the Xi'an urban area, the situation of whole-area tourism remains undeveloped, and important leisure and tourism resources such as the landscape resources represented by the Qinling Mountains and the Weihe River and the city parks represented by the Xi'an City Sports Park have not yet been fully tapped. In the future, while successfully building the historical and cultural core area, we can fully tap into the advantageous leisure and tourism resources of Baqiao, Chang'an, and Gaoling districts to create diversified leisure and tourism experiences, such as cultural performances, ecological leisure and holidays, festivals and recreation, athletics, business meetings, parks and outdoor recreation, and other leisure and tourism modes.

6. Conclusions

Urban leisure and tourism space is a space for local residents and tourists to share a better life, as well as a space to show the unique charm of the city. Exploring the spatial pattern of urban leisure and tourism and the influencing mechanisms is of guiding significance for urban planning and industrial layout. Based on the leisure and tourism POI data in Xi'an, this paper classifies the leisure and tourism sites into five types: catering services, accommodation services, shopping services, sports and recreation, and scenic spots, and analyzes the spatial distributions of leisure and tourism overall and for different types and their influencing factors. The results of this study show that the leisure and tourism space in Xi'an is significantly clustered, which is in line with the spatial development pattern of "core-periphery". Using geodetectors, we analyze the influences and interactions of population, industrial structure, economic development level, traffic conditions, and infrastructure on the spatial differentiation of leisure and tourism types in Xi'an. This paper then explores the influencing mechanisms of the spatial patterns driving leisure and tourism sites in Xi'an from four aspects: dominant factors (population), driving factors (economic level, industrial structure), guaranteeing factors (infrastructure, traffic condition), and other triggering factors (governmental behaviors, consumption orientation, major events).

This study finds that in Xi'an, a famous historical capital of China, despite the mature development of the leisure and tourism industry, the distribution of the industry is highly concentrated in the central urban area, and the spillover and driving effect of the central urban area on the peripheral areas is relatively limited. The distinctive image of Xi'an as an ancient capital is a "double-edged sword", which exacerbates to a certain extent the imbalance in the distribution of the leisure and tourism industry between the central city and the peripheral regions. This phenomenon should be taken into account by government policy and the relevant planning and management departments of the city. The layout of urban leisure facilities should consider the balance within the urban area and safeguard the common leisure rights of urban residents. The development of the tourism industry also fully exploits the advantageous leisure and tourism resources of various districts in the city, and some cultural and tourism integration and innovative leisure and tourism projects can be considered for layout in other areas outside the central urban area. Combined with the historical and cultural projects in the central urban area, these services will provide tourists with diversified and high-quality leisure and tourism products.

The limitations of this study are as follows: First, limited by the characteristics of the POI data itself, the acquired data for leisure and tourism sites in Xi'an are spatial point data, which fail to reflect the difference in the volume of the leisure and tourism industry, and there is a certain amount of error in the spatial distance and kernel density analyses. Second, this paper only examines the distribution of leisure and tourism facilities in Xi'an from the perspective of supply, without considering the respective activity spaces and views of residents and tourists from the perspective of demand. Third, there are many factors affecting the spatial distribution of urban leisure and tourism facilities, but it is difficult to obtain data at the district level and the street level. Therefore, when analyzing the factors affecting the spatial differentiation of leisure and tourism types, only seven main indicators were selected for analysis, and some factors could not be taken into account due to the difficulty of obtaining or quantifying data. Future research is required to connect POI big data with qualitative surveys, focusing not only on the spatial layout of leisure and tourism patterns but also on human mobility and the views of residents and tourists. Through a variety of data sources, the rationality and regularity of the spatial distribution of urban leisure and tourism can be explored more deeply, with a view to providing better guidance for the development of the urban leisure and tourism industry and urban planning.

Author Contributions: Conceptualization, X.Q. and H.B.; methodology, X.Q., G.X. and J.Q.; software, X.Q. and J.Q.; validation, X.Q., G.X. and H.B.; formal analysis, X.Q.; investigation, X.Q. and H.B.; resources, X.Q. and G.X.; data curation, X.Q. and G.X.; writing—original draft preparation, X.Q., J.Q. and G.X.; writing—review and editing, X.Q. and G.X.; visualization, X.Q., J.Q. and H.B.; supervision, J.Q. and H.B.; project administration, X.Q.; funding acquisition, X.Q. All authors have read and agreed to the published version of the manuscript.

Funding: The National Natural Science Foundation of China: 72102216.

Data Availability Statement: The data presented in this study are available on request from the first author.

Conflicts of Interest: The authors declare no conflict of interest.

References

1. Cohen, E. Who is a tourist? A conceptual clarification. *Sociol. Rev.* **1974**, *22*, 527–555. [CrossRef]
2. Smith, S.; Godbey, G. Leisure, recreation and tourism. *Ann. Tour. Res.* **1991**, *18*, 85–100. [CrossRef]
3. Veal, A.J. *Leisure and Tourism Police and Planning*, 2nd ed.; CABI Publishing: Oxon, UK, 2002; p. 3.
4. Chen, Y.C.; Guo, J.; Xu, H. Leisure tourism: Research status, difference and connotation exploration. *Geogr. Geo-Inf. Sci.* **2014**, *30*, 94–98.
5. Slak Valek, N.; Fotiadis, A. Is tourism really an escape from everyday life? Everyday leisure activities vs leisure travel activities of expats and emirati nationals living in the UAE. *Int. J. Cult. Tour. Hosp. Res.* **2019**, *12*, 238–254. [CrossRef]
6. Kuhn, F.; Kock, F.; Lohmann, M. Personal prestige through travel? Developing and testing the personal prestige inventory in a tourism context. *Consum. Behav. Tour. Hosp.* **2023**, *18*, 1–16. [CrossRef]
7. Carr, N. The tourism-leisure behavioral continuum. *Ann. Tour. Res.* **2022**, *29*, 972–986. [CrossRef]
8. Chang, S.; Gibson, H.J. The relationships between four concepts (involvement, commitment, loyalty, and habit) and consistency in behavior across leisure and tourism. *Tour. Manag. Perspect.* **2015**, *13*, 41–50. [CrossRef]
9. Ryan, C.; Kinder, R. Sex, tourism and sex tourism: Fulfilling similar needs? *Tour. Manag.* **1996**, *17*, 507–518. [CrossRef]
10. Leiper, N. The Framework of tourism: Towards a definition of tourism, tourist, and the tourist industry. *Ann. Tour. Res.* **1979**, *6*, 390–407. [CrossRef]
11. Godbey, G.; Shim, J. The development of leisure studies in North America: Implications for China. *J. Zhejiang Univ. Humanit. Soc. Sci.* **2008**, *38*, 22–29.
12. Kaplan, M. *Leisure in America: A Social Inquiry*; John Wiley & Sons: New York, NY, USA, 1960.
13. Honggen, X.; Huyton, J.R. Tourism and leisure: An integrative case in China. *Int. J. Contemp. Hosp. Manag.* **1996**, *8*, 18–24. [CrossRef]
14. Cai, M.Y.; Shi, J.Z.; He, X.R. From tourism to leisure: The evolving direction for the tourism cities of China. *Econ. Geogr.* **2022**, *42*, 225–231.
15. Fu, B.J. The integrated studies of geography: Coupling of patterns and processes. *Acta Geogr. Sin.* **2014**, *69*, 1052–1059.
16. Siu, K.W.M. Accessible park environments and facilities for the visually impaired. *Facilities* **2013**, *31*, 590–609. [CrossRef]
17. le Brasseur, R. Citizen sensing within urban greenspaces: Exploring human wellbeing interactions in deprived communities of Glasgow. *Land* **2023**, *12*, 1391. [CrossRef]
18. Friedman, M.T.; Beissel, A.S. Beyond "who pays?": Stadium development and urban governance. *Int. J. Sports Mark. Spons.* **2021**, *22*, 107–125. [CrossRef]
19. Li, S.; Liu, S.; Ding, X. Exploring the spatial distribution pattern and influencing factors of Shanghai's cultural functional elements based on the point of interest data. *Open House Int.* **2022**, *10*, 504–520. [CrossRef]
20. Gon, M.; Osti, L.; Pechlaner, H. Leisure boat tourism: Residents' attitudes towards nautical tourism development. *Tour. Rev.* **2016**, *71*, 180–191. [CrossRef]
21. Ikudayisi, A.E.; Taiwo, A.A. Accessibility and inclusive use of public spaces within the city-centre of Ibadan, Nigeria. *J. Place Manag. Dev.* **2022**, *15*, 316–335. [CrossRef]
22. Meshram, K.; O'Cass, A. Empowering senior citizens via third places: Research driven model development of seniors' empowerment and social engagement in social places. *J. Serv. Mark.* **2013**, *27*, 141–154. [CrossRef]
23. Ishak, S.A.; Hussein, H.; Jamaludin, A.A. Neighbourhood Parks as a potential stress reliever: Review on literature. *Open House Int.* **2018**, *43*, 52–64. [CrossRef]
24. Erdoğan, H.H.; Enginkaya, E. Exploring servicescape experiences across museum types. *J. Serv. Mark.* **2023**, *37*, 706–718. [CrossRef]
25. Al-Saad, S.A.; Jawarneh, R.N.; Aloudat, A.S. Spatiotemporal cluster analysis of reputable tourist accommodation in Greater Amman Municipality, Jordan. *J. Hosp. Tour. Technol.* **2023**, *14*, 579–597. [CrossRef]
26. Qin, Y.; Qin, J.; Liu, C. Spatial-temporal evolution patterns of hotels in China: 1978–2018. *Int. J. Contemp. Hosp. Manag.* **2021**, *33*, 2194–2218.

27. Li, S.; Liu, S.; Ding, X. Spatial agglomeration pattern of homestay inn and influencing factors based on the comparison of Hangzhou, Huzhou, and Enshi cities. *Prog. Geogr.* **2023**, *39*, 1698–1707.
28. Liang, H.; Zhang, Q. Do social media data indicate visits to tourist attractions? A case study of Shanghai, China. *Open House Int.* **2022**, *47*, 17–35. [CrossRef]
29. Brokalaki, Z.; Patsiaouras, G. Commodifying ancient cultural heritage: The market evolution of the Parthenon temple. *J. Hist. Res. Mark.* **2022**, *14*, 4–23. [CrossRef]
30. Dongdong, X.; Zhenfang, H.; Huangping, S.; Xueying, S.; Huan, L.; Linjiao, T. The spatial characteristics and its influencing factors of leisure tourism resources in Nanjing. *J. Nanjing Norm. Univ. Nat. Sci. Ed.* **2017**, *40*, 127–133.
31. Yu, C.; Lian, T.; Geng, H.; Li, S. Analyzing the structure of tourism destination network based on digital footprints: Taking Guilin, China as a case. *Data Technol. Appl.* **2023**, *57*, 56–83. [CrossRef]
32. Md Khairi, N.D.; Ismail, H.N.; Syed Jaafar, S.M.R. Knowledge of tourist spatial behaviour to improve Melaka world heritage site management. *Int. J. Tour. Cities* **2022**, *8*, 88–106. [CrossRef]
33. Gkoumas, A.; D'Orazio, F. Public-space tactical intervention as urban tourist allure. *Int. J. Tour. Cities* **2020**, *6*, 711–730. [CrossRef]
34. Cocola-Gant, A.; Gago, A.; Jover, J. Tourism, gentrification and neighbourhood change: An analytical framework– reflections from Southern European Cities. In *The Overtourism Debate*; Oskam, J.A., Ed.; Emerald Publishing Limited: Bingley, UK, 2020; pp. 121–135.
35. Moleiro, D.F.; Carneiro, M.J.; Breda, Z. Assessment of residents' perceptions and attitudes towards the appropriation of public spaces by tourists: The case of Aveiro. *Int. J. Tour. Cities* **2021**, *7*, 922–942. [CrossRef]
36. Go, H.; Kang, M.; Nam, Y. The traces of ecotourism in a digital world: Spatial and trend analysis of geotagged photographs on social media and Google search data for sustainable development. *J. Hosp. Tour. Tecnol.* **2020**, *11*, 183–202. [CrossRef]
37. Kovács, Z.; Smith, M.; Teleubay, Z.; Kovalcsik, T. Measuring Visitor Flows Using Mobile Positioning Data in Three Hungarian Second-Tier Cities. *Int. J. Tour. Cities* 2023, *ahead-of-print*. [CrossRef]
38. Wang, F.; Hu, W.; Zhu, Y.; Jiang, C. The locality of Beijing historic areas from a dynamic perspective based on geo-tagged photos. *Int. J. Tour. Cities* **2019**, *5*, 75–89. [CrossRef]
39. Tache, A.V.; Popescu, O.C.; Petrișor, A.I. Conceptual model for integrating the Green-Blue Infrastructure in planning using geospatial tools: Case study of Bucharest, Romania Metropolitan Area. *Land* **2023**, *12*, 1432. [CrossRef]
40. Yang, Z.S.; Long, Y.; Nicolas, D. Opportunities and limitations of big data applications to human and economic geography: The state of the art. *Prog. Geogr.* **2015**, *34*, 410–417.
41. Höpken, W.; Müller, M.; Fuchs, M.; Lexhagen, M. Flickr data for analysing tourists' spatial behaviour and movement patterns: A comparison of clustering techniques, *J. Hosp. Tour. Technol.* **2020**, *11*, 69–82. [CrossRef]
42. Jing, Z.; Luo, Y.; Li, X.; Xu, X. A multi-dimensional city data embedding model for improving predictive analytics and urban operations. *Ind. Manag. Data Syst.* **2022**, *122*, 2199–2216. [CrossRef]
43. Ivan, I.; Singleton, A.; Horák, J.; Inspektor, T. *The Rise of Big Spatial Data*; Springer: London, UK, 2017.
44. Long, L.C.; Qin, Z.B.; Mo, X.Y. Leisure tourism spatial characteristics and influencing factors in tourism cities: A case study of Guilin. *Soc. Sci.* **2022**, *301*, 30–37.
45. Wang, J.F.; Li, X.H.; Christakos, G.; Liao, Y.L.; Zhang, T.; Gu, X.; Zheng, X.Y. Geographical detectors-based health risk assessment and its application in the neural tube defects study of the Heshun region, China. *Int. J. Geogr. Inf. Sci.* **2010**, *24*, 107–127. [CrossRef]
46. Cui, X.; Zhang, J.; Huang, W.; Liu, C.; Shan, L.; Jiang, Y. Spatial Pattern and Mechanism of the Life Service Industry in Polycentric Cities: Experience from Wuhan, China. *J. Urban Plann. Dev.* **2023**, *149*, 05023015. [CrossRef]
47. Li, W.W.; Cui, T.; Ma, X.L.; Zhang, X.Y. A study on the spatial patterns and causes of the attractiveness of scenic sports in Hangzhou. *Tour. Sci.* **2023**, *37*, 19–39.
48. Li, L.; Hou, G.L.; Xia, S.Y.; Huang, Z.F. Spatial distribution characteristics and influencing factors of leisure tourism resources in Chengdu. *J. Nat. Resour.* **2020**, *35*, 683–697. [CrossRef]
49. Baum, T. Human resources in tourism: Still waiting for change. *Tour. Manag.* **2007**, *28*, 1383–1399. [CrossRef]
50. Sinclair, T.M.; Stabler. *The Economics of Tourism*; Routledge: London, UK, 1997.
51. Brida, J.G.; Pulina, M. A literature review on the Tourism-Led-Growth Hypothesis. *Work. Pap. Crenos* **2010**, *17*, 1–26.
52. Williams, A.M.; Baláž, V. Low-cost carriers, economies of flows and regional externalities. *Reg. Stud.* **2008**, *43*, 677–691. [CrossRef]
53. Terhorst, P.; Erkuş-Öztürk, H. Urban tourism and spatial segmentation in the field of restaurants: The case of Amsterdam. *Int. J. Cult. Tour. Hosp. Res.* **2015**, *9*, 85–102. [CrossRef]
54. Pazhuhan, M.; Shiri, N. Regional tourism axes identification using GIS and TOPSIS model (Case study: Hormozgan Province, Iran). *J. Tour. Anal. Rev. Análisis Turístico* **2020**, *27*, 119–141. [CrossRef]
55. Leisen, B. Image segmentation: The case of a tourism destination. *J. Serv. Mark.* **2001**, *15*, 49–66. [CrossRef]

land

MDPI

Article

Spatiotemporal Variations of Production–Living–Ecological Space under Various, Changing Climate and Land Use Scenarios in the Upper Reaches of Hanjiang River Basin, China

Pengtao Wang [1], Xupu Li [2,*], Liwei Zhang [2], Zhuangzhuang Wang [3], Jiangtao Bai [4], Yongyong Song [2], Hongzhu Han [1], Ting Zhao [1], Guan Huang [1] and Junping Yan [2]

1 School of Tourism & Research Institute of Human Geography, Xi'an International Studies University, Xi'an 710128, China; wnpengtao@xisu.edu.cn (P.W.); hhz@xisu.edu.cn (H.H.); 107242021100052@xisu.edu.cn (T.Z.); hguan703@xisu.edu.cn (G.H.)
2 School of Geography and Tourism, Shaanxi Normal University, Xi'an 710119, China; zlw@snnu.edu.cn (L.Z.); syy2016@snnu.edu.cn (Y.S.); yanjp@snnu.edu.cn (J.Y.)
3 State Key Laboratory of Urban and Regional Ecology, Research Center for Eco-Environmental Sciences, Chinese Academy of Sciences, Beijing 100085, China; zzwang_st@rcees.ac.cn
4 School of History and Archives, Yunnan University, Kunming 650091, China; geodata@snnu.edu.cn
* Correspondence: xupuli@snnu.edu.cn; Tel.: +86-29-8531-0525

Abstract: Land is an important resource that supports the production, life, and ecological development of human society. The current research on production–living–ecological space (PLES) is mainly focusing on the identification of single and dominant functions of land space, and the comprehensive spatial function measurement index of PLES (PLESI) is less known in the effective quantitative evaluation of multifunctionality of different land use categories. Integrating the CMIP6 (Coupled Model Intercomparison Project phase 6) scenario data and the future land use simulation model (FLUS), this research took the upper reaches of the Hanjiang River (URHR) as an example to explore the temporal and spatial variations in land use, PLES, and PLESIs during 2000–2020, and in the SSP2-4.5 and SSP5-8.5 scenarios from 2021 to 2100. The findings were as follows: (1) Forest land is the most widely distributed type of land; correspondingly, ecological space has the widest distribution area in PLES, followed by production space. (2) The area of dry land and building land increased between 2000 and 2010, accompanied by the increase in living space. From 2010 to 2020, the growth rate of building land tended to slow down while forest land increased, and the conflict of PLES eased. (3) The transfer between forest land and dry land is projected to intensify under the SSP2-4.5 scenario, while it is projected to occur between forest land and grassland under the SSP5-8.5 scenario. As for the changes in PLES, the SSP2-4.5 scenario has a greater impact than the SSP5-8.5 scenario. Spatially, several sub-basins in the northern URHR are the main areas of land use and PLES change. (4) PLESI presents a significant downward trend from 2000 to 2020 while trending upward under the SSP5-8.5 scenario and trending downward slightly under the SSP2-4.5 scenario between 2020 and 2100. Combining climate scenarios and the future land use simulation, this research would support the effective utilization of regional land resources and ecosystem management decision-making.

Keywords: production–living–ecological spaces; climate scenarios; land multifunctionality index; FLUS; Hanjiang River

1. Introduction

Land is a key and scarce resource to support human social development in production, life, and ecology [1,2]. It is closely related to food security, ecosystem health, and social sustainable development [3–5]. Over the past decades, the world has experienced the rapid expansion of urbanization, human activity, and extreme climate events, which comprehensively affected the regional soil environment, hydrological cycle, biodiversity, climate, etc., with a great influence on the ecological functions of land in urbanized areas [6]. In turn, the relationship between humans and land has become increasingly tense [7–9], and conflict between production, life, and ecological land is growing more frequent in urbanized areas [10,11]. China has proposed comprehensive requirements for building a "production-living-ecological space, PLES" with "promoting the high efficient and compositive production space, livable and moderate living space and ecological space with picturesque scenery" [12]. PLES has become an important way to ameliorate rural–urban disparity and bolster harmonious and sustainable development.

In recent years, great efforts have been made in research on PLES: First, in terms of classification and identification of PLES, most researchers divide different lands into three types of ES, LS, and PS on the basis of the predominant function of the land categories and have made numerous investigations about the spatiotemporal patterns of PLES at the national scale [13], provincial scale [14], city and county scale [15,16], township scale [17], economic belt and urban agglomeration [18–20], watershed and basin scale [21]. Simultaneously, temporal and spatial changes are inevitably accompanied by the process of the mutual encroachment or tradeoff of PLES, such as the spatial coordination and conflict of PLES [22,23]. As for the temporal and spatial change, the influence mechanism of natural environment, social, and economic development are crucial factors for the assessment of PLES. Current studies on multi-year variations in PLES focus on historical periods. However, under the far-reaching impact of global changes and human activities [24,25], great uncertainties of the coordination and conflict of PLES exist regarding the changes in future land use. Based on the systematic research on the driving forces behind PLES, quantitative models and methods have been applied to deduce the processes of land use change and to generate predictions of PLES in the future [18].

Multiple models of land change evolution and prediction have been produced [26–28]. The future land use simulation model, FLUS, was created with consideration of the comprehensive impact of climatic variations and human activities on land utilization changes [29,30]. With support from multi-source data, this model can explicitly simulate land use at a global scale or regional scale and obtain high-precision and multi-scenario land use prediction results, laying the foundation for predictions of future PLES scenarios and clearly revealing the long-term patterns of evolution of PLES [31].

Furthermore, current research on PLES is mainly concerned with the identification of dominant functions of land utilization, ignoring that land use has compound land functions [21]. For instance, cultivated land can be used as agricultural land to fulfil food production functions, while it can also be ecological land, fulfilling diverse ecosystem functions, including climate regulation, carbon sequestration, flood mitigation, nutrient cycling, etc. [32–34]. Some research has been conducted with the aim of classifying and evaluating the multifunctionality of different land utilization types. Specifically, the PLES indicator system (PLESI) and the four-level scoring method were employed to achieve the multi-functional measurement of production function, living function, and ecological function for different land categories [21,35]. Existing classifications of PLES mostly consist of PS, ES, and LS based on land use types and it lacks multi-functional and comprehensive classification methods in PLESI evaluation.

The upper reaches of the Hanjiang River (URHR), the core area of the Qinling Mountains, is the key water conservation area of the South-to-North Water Diversion Project [36,37]. As an extremely crucial national ecological security barrier, its ecological function has an irreplaceable role to play in the construction of ecological civilization in China. However, little research has explored PLES in the URHR and Qinling Mountains. In this context, this study attempts to measure the evolutionary characteristics of the PLES and land multifunctionality through the PLESI model, with the integration of CMIP6 future climate and socio-economic data and the FLUS model. Specifically, the objectives of this study were to (1) analyze the patterns of temporal and spatial development of PLES between 2000 and 2100 under two climate scenarios in the URHR; (2) investigate the transfer evolution characteristics of land use and PLES under two different scenarios; and (3) discuss the implication of scenarios analysis and the PLESI model on land resources management. This is an effective exploration of comprehensive PLES research of long-term time series on the typical ecological reserve, aiming to provide scientific and sufficient support for the regional land resource management and social sustainable development.

2. Materials and Methods

2.1. Study Area

The Hanjiang River is the largest branch of the Yangtze River, Asia's longest river. It originates in the Ningqiang County in the southwest of Shaanxi Province, and joins the Yangtze River at Wuhan City, Hubei Province. The Upper Reaches of the Hanjiang River is located in the middle of mainland China and in the geographic demarcation line between southern and northern China [38]. In Shaanxi Province, the Hanjiang River spans 652 km, covering three main prefecture-level cities, including Hanzhong, Ankang and Shangluo, and Taibai County and Feng County of Baoji City, with a basin area of 62,384 km^2 [37] (Figure 1).

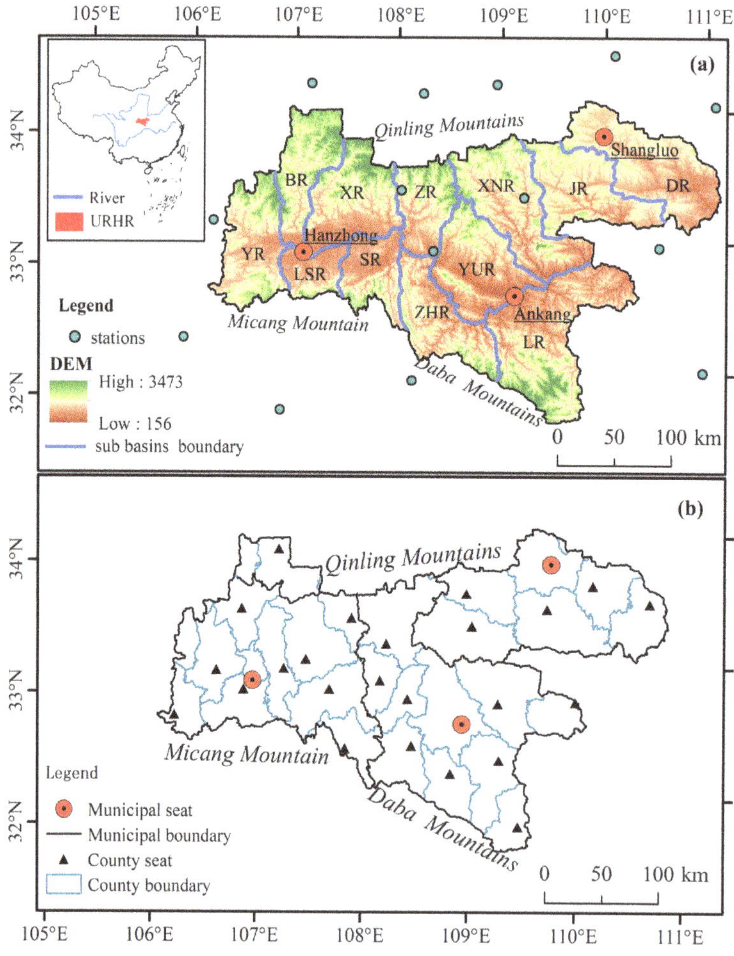

Figure 1. Geography situation of the URHR: (**a**) location, elevation and sub basins distribution of the URHR in China; (**b**) location of administrative cities.

Topographically, it is structured with three mountains and two basins. The Qinling Mountains constitute the northern boundary, Daba Mountain and Micang Mountain constitute the southern boundary, and Hanjiang River flows through the canyon areas from the west to the east, forming Ankang Basin and Hanzhong Basin. Based on regional river zoning characteristics and related studies, the URHR in the ownership of numerous tributaries consists of 12 sub-basins through the hydrological analysis module in GIS [39]. The mainstream area is determined by the flow route of Hanjiang River in Shaanxi Province. YR, BR, LSR, XR and SR mainly belong to Hanzhong Basin, ZR, ZHR, XNR, YUR and LR to Ankang Basin, and JR and DR in Shangluo City are tributaries of Hanjiang.

The URHR is the core area in the south of the water distribution line of the Qinling Mountains. It is in the 0 °C isotherm line, 800 mm iso-precipitation line and 2000 h sunshine hours isochron line in January in mainland China, with unique natural environmental characteristics [36,37]. The region is mainly dominated by the warm temperate continental climate, characterized with warmth, rain and moisture, which has the annual average precipitation ranging from 653 mm to 1183 mm and annual average temperature ranging from 12 °C to 18 °C [39]. Known as China's Central Water Tower, the study area contains abundant water resources and is an important water conservation area of the Yellow River, the Yangtze River, and China's South-to-North Water Diversion project. The ecological environment is diverse with rich biological resources and extremely important ecosystem service functions, named as "natural gene bank" of biodiversity in China. The URHR plays an irreplaceable role in the local and national ecological security and social development.

However, due to the terrain conditions and the extreme rainfall events in summer, the flood disaster in the study area is frequent, triggering significant threats to the lives and safety of local residents [37]. In terms of social and economic development, the Qinling Mountains region belongs to the largest centralized contiguously poor area in China. The large-scale social production and urban industrial development have been restricted due to its functioning as the ecological protection area. Therefore, how to weight the relationship between economic growth and environmental care effectively, on the premise of strengthening the ecological location and ecological function of the Qinling Mountains, has become an important issue for local sustainable development.

2.2. Research Framework

This study was conducted in three phases: collection and processing of basic data and climate data, simulation of future land use, and evaluation of spatio-temporal variations of PLES and PLESI (Figure 2).

First, site-scale climate scenario data of CMIP6 model were obtained with statistical downscaling method, then spatial interpolation was employed to generate future climate scenario raster data of the URHR. In the second step, with the FLUS model, the future land scenario data of URHR was produced using the historical and future data of driving factors of land use. In the third step, the temporal and spatial evolution of future PLES and PLESI scenarios in the URHR were analyzed through the classifications of land functions and the multifunctional measurement model of land types in order to provide specific advice on the regional land use management and territorial space optimization, etc.

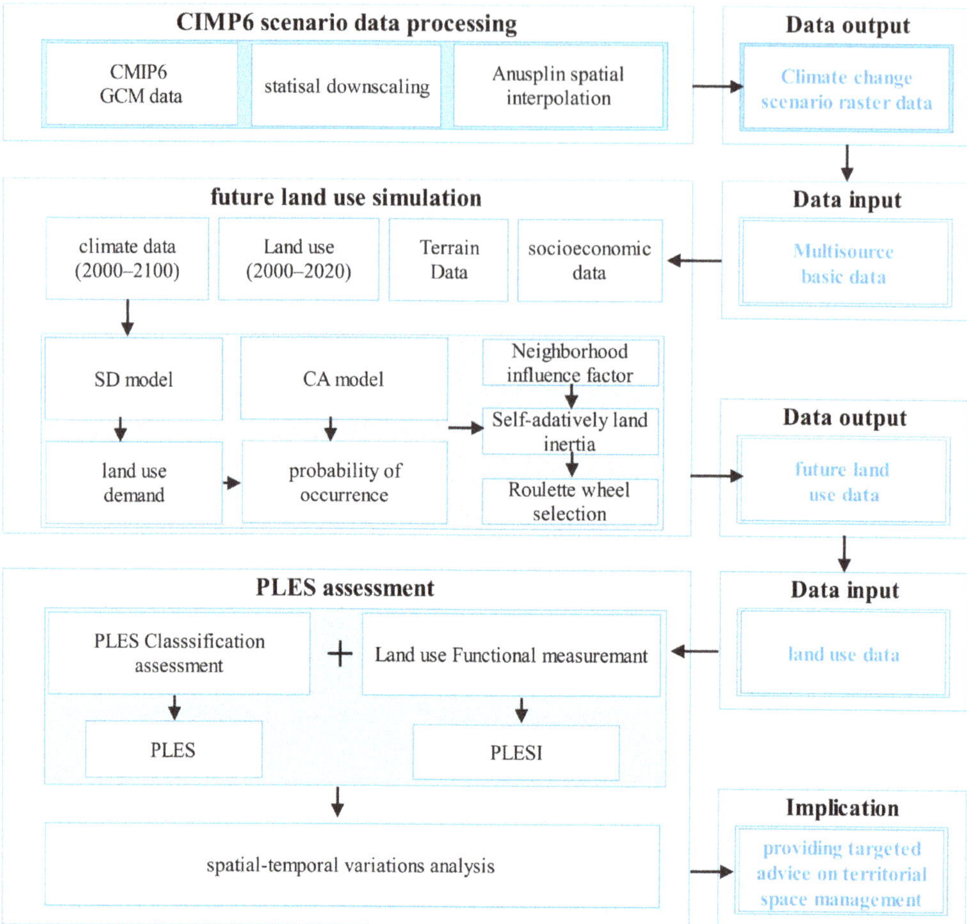

Figure 2. Research framework of the study.

2.3. Data Sources

In this study, multiple sources of historical data and future climate scenario data were used to process the simulation of future land use. Specific research data are in Table 1:

(1) Meteorological observation data of the national meteorological stations in the URHR and the surroundings during the historical period (2000–2020) were derived from the Chinese surface meteorological observation dataset released by China Meteorology Administration.

(2) CMIP 6 climate model data for the future period (2021–2100) were obtained from daily data of the global climate models of the World Climate Research Program under the SSP2-4.5 and SSP5-8.5 scenarios.

(3) Land use data in historical period (2000–2020) were from the global ESA Land Cover dataset ESA CCI Land Cover project released by the European Space Agency (ESA).

(4) In the simulation of land use in the future period, the main data are land use data in the base period, prediction data in the future climate scenarios, soil data, topographic data (DEM), and socio-economic data (raw data of population, GDP, city center and

transportation network elements). The slope and aspect data are obtained through DEM data processing. Resulting from the traffic network data, the driving factors including the distance data to places such as the city center, town center, expressway, airport, river, railway station, railway and traffic artery in the simulation of land use are generated through GIS.

Table 1. Summary information of data.

Data Category	Data Name	Time Resolution	Spatial Resolution	Data Sources
Land use	-	2000–2020	300 m	ESA CCI Land Cover project (https://cds.climate.copernicus.eu/cdsapp#!/dataset/satellite-land-cover?tab=overview accessed on 20 September 2021)
Climate data	temperature precipitation	2000–2020	Meteorological station	China Meteorology Administration (http://data.cma.cn accessed on 10 December 2021)
Climate scenarios	temperature precipitation	2021–2100	-	World Climate Research Programme (https://esgf-node.llnl.gov/search/cmip6/ accessed on 1 July 2021)
Terrain Data	DEM	2003	30 m	NASA Shuttle Radar Topographic Mission (https://srtm.csi.cgiar.org/ accessed on 15 June 2020)
Socioeconomic data	GDP population	2015	1 km	1 km Grid GDP, 1 km Grid population of Chinese Academy of Sciences (www.resdc.cn/ accessed on 20 June 2020)
	residential site traffic network	2020	vector data	National Geographic Information Data (www.webmap.cn/commres.do?method=dataDownload accessed on 3 September 2021)

All data were processed in TIF format with 1 km spatial resolution through the processes of vector-to-raster and resampling on the ArcGIS 10.6 software platform.

2.4. Methods

2.4.1. Future Climate Scenarios

Combining CMIP5 Representative Concentration Pathways (RCPs) and Shared Socioeconomic Pathways (SSPs), the CMIP6 constructs a new scenario matrix (RCP-SSP integrated scenario) to clearly reflect the simulation of greenhouse gas emissions and improve the precision of climate simulation [40–43]. Climate scenarios data were generated with global climate model output data and statistical downscaling method [44–46]. Specifically, the process of climate scenario data includes the following steps:

(1) The selection of climate scenarios:

The emission scenarios from CMIP6 include low emission scenarios (SSP1-1.9 and SSP1-2.6), a medium emission scenario (SSP2-4.5) and high emission scenarios (SSP3-7.0 and SSP5-8.5) [47,48]. The low-emission scenarios envisage a future in which greenhouse gas emissions are substantially reduced and socio-economic development is apt to be more sustainable. The medium emission scenario assumes that greenhouse emissions will maintain the current level and represent the closest to the current greenhouse emission scenarios. The high emission scenarios indicate rapid socio-economic growth and significant impacts on the global climate [49,50].

At present, the SSP2-4.5 and SSP5-8.5 scenarios have become common in the studies to simulate the response of climate change and related socio-economic scenarios, representing two scenarios of maintenance the status quo of development and maximum greenhouse gas emissions [48–53]. Therefore, these two scenarios are selected to drive the climate and land use models in this research.

(2) Download data in two scenarios of 27 models from the Global Climate Model (GCM) output of CMIP6.

(3) Based on the improved weather generator, GCM raster data are downscaled from spatial coarse resolution to meteorological station. For specific operating principles, please refer to the study of Liu et al. [44] and Wang et al. [45].
(4) The correlation coefficient between projected and observed data is evaluated through the quantitative index S of the Taylor Chart [54–56], the spatial skill score is used to evaluate the spatial correlation coefficient between projected and observed data [57], and the temporal skill score is quantified to assess the simulation efficiency of the projected value at each point in the space to simulate the inter-annual change of the observed value [58–60]. The two models, UKES and MIR2, were selected based on the ranking of the total scores. Therefore, with multi-model ensemble mean approaches [61,62], the average value of the two models was obtained as the prediction data in the future climate scenarios.

2.4.2. Land Use Simulation

In FLUS model applications, most studies take historical climate data as a driver of land use change, ignoring the future climate scenario [63], while the coupling of CMIP6 data and FLUS model in this study can increase the simulation accuracy of FLUS model.

The simulation processes of land use in the URHR are as follows: Firstly, the System Dynamics model (SD) is used to obtain land use demand data in the future based on climate change scenarios, socio-economic scenarios and historical land use. Secondly, a Cellular Automata model (CA) based on an Artificial Neural Network model (ANN) is used to estimate the change probability in different land categories, and the overall land adaptability probability of the cell is calculated. Thirdly, combining self-adaptive inertia coefficient and roulette wheel selection mechanism, multiple iterations are made based on the demand and the actual situation of land use, and then the future land use data were obtained [29].

In the process of data verification, with actual land utilization data in 2020, FLUS was applied to generate land utilization simulation data in 2020. Then, actual and simulated land utilization data in 2020 were input into FLUS, random sampling was selected in the sampling mode with the sampling proportion was 20%, and the Kappa coefficient was 0.96, indicating that FLUS model was in good consistency for the future land use simulation in the URHR.

2.4.3. Classification System and Functional Measurement of PLES

In terms of production space (PS), these land types can be divided according to three main industry types. Living space (LS) refers to areas that can provide living needs for human beings such as living, rest, entertainment and consumption. Ecological space (ES) is the foundation of PLES, which can provide the guarantee for sustainable land use [17] and supply ecosystem services such as supporting services and regulating services, and maintain the ecological environment for human being [18].

The production–living–ecological space (PLES) types of different land use were determined by the differences of land functions and land utilization types. With the first-level land classification, dryland (DL) and paddy field (PF) are defined as production space (PS), forest land (FL), grassland (GL) and water land (WL) as ecological space (ES), and building land as living space (LS) [64].

Based on the actual situation of the multifunctionality of different land utilization categories [15,19], the multi-functional measurement index of PLES (PLESI) was constructed to finely evaluate the functional level of PLES on a spatial scale in the URHR. The formula for evaluating is shown as

$$PLESI_n = 0.25 \times PFI_n + 0.5 \times EFI_n + 0.25 \times LFI_n \quad (1)$$

where $PLESI_n$ is the multi-function index of PLES in pixel n, PFI_n is the land production function index in pixel n, EFI_n is the land ecological function index in pixel n, and LFI_n is the land living function index in pixel n.

As for the actual situation of the diversity functions of different lands and the previous research results, the production, living and ecological function of different lands were quantitatively evaluated by four grades based on the scoring system [19,35]. Among them, the number 5 refers to the high function index (production function, ecological function or living function) of a land type, 3 means the medium land function, 1 is for the weak function, and 0 indicates that the function of a land type is missing (Table 2) [16,21,35].

Table 2. PLES classification and multi-functional assessment in the URHR.

PLES	Land Use	Production Function	Ecological Function	Living Function
production space	dryland	5	3	0
	paddy field	5	3	0
ecological space	forest land	0	5	0
	grassland	3	5	1
	water land	3	5	0
living space	building land	3	0	5

3. Results

3.1. Spatio-Temporal Variations in Land Use and PLES

3.1.1. Variations in Land Use and PLES in Historical Period

The results showed that dryland, paddy field, forest land and grassland in the URHR are the main land use categories, among which forest land occupies the largest proportion (78.87%) (Figure 3). Dryland was mainly distributed in the northeast, with an average annual distribution area of 6660 km^2. Paddy fields were gathered in the Hanjiang Valley, with an area of 5849 km^2. Forest land followed this, which was widely distributed in the study area, especially in the Qinling Mountains and Daba Mountains, with an area of 48,821 km^2. Grassland was concentrated in the border area between dry land and forest land, the area of which was about 1,138 km^2. Building land spread along the main stream of Hanjiang, mainly around Hanzhong City, Ankang City and Shangluo City with an area of approximate 152 km^2. Water land was primarily in the border area of ZR and ZHR with an area of 72 km^2.

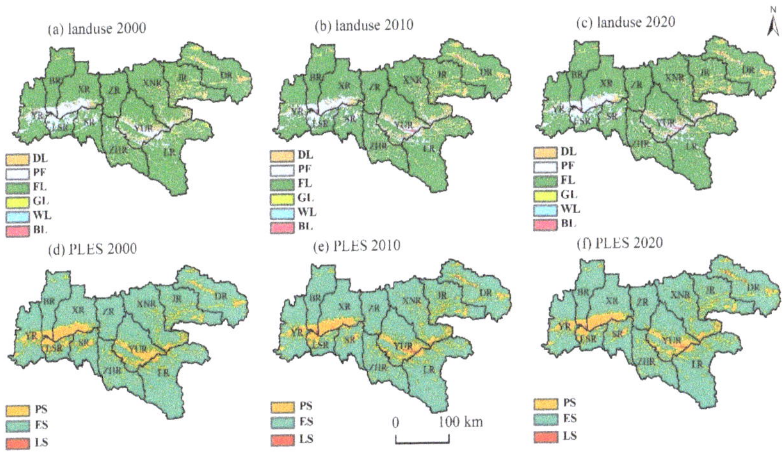

Figure 3. Spatial patterns of historical land use and PLES in the URHR.

The area of dryland and building land keeps growing from 2000 to 2010, meanwhile, the other land categories fall dramatically, indicating that the local social and economic development and urbanization process were significantly accelerated. In turn, the area of PS and LS was remarkably increased in this period, with the result of encroaching ES. From 2010 to 2020, paddy field, forest land and building land showed a growing trend, while dryland, grassland and water land were in a downward trend. The growth rate of building land had fell sharply, and other land types changed mildly, which indicated that the competition between local PLES tended to ease in this period.

The spatio-temporal simulation of PLES during the historical period showed that ES had the widest distribution area in the study area, followed by PS and LS. PS was concentrated in the Hanjiang Valley, and scattered in some areas of JR and DR, with an average area of 12,509 km^2, comprising 19.95% of the URHR. ES was widespread especially in the Qinling Mountains of the northern Hanjiang Valley, with an area of 50,031 km^2 over the years, comprising 79.80% of the URHR. LS was located near the central city, covering 152 km^2. This indicated that the land function of the URHR was dominated by ecological function, and the intensity of urban development and industrial production activities was small.

In 2000–2010, the change in LS was the most dramatic, followed by PS, which showed an increasing trend with a rate of 5.75%. In contrast, ES experienced a decreasing trend at a rate of 1.73%. In 2010–2020, PS had a reduction of 0.71%, ES showed a slight increase trend of 0.18%, and LS increased slowly at a rate of 0.18%. The results also showed that the competition for PLES in this area was more intense in the first decade, and the expansion of LS and PS had caused a greater encroachment on ES. In recent ten years, due to the effective conservation policies, the development of urban expansion and industrial activities tended to be slow, and ES showed an upward trend, which would better promote the ecological environment in the URHR.

3.1.2. Variations of Future PLES under Different Scenarios

The future period in this research consists of near-term, medium-term and long-term, in which 2030 represents the near-term, 2050 represents the medium-term, and 2100 represents the long-term, respectively. The results showed that ES occupied the widest distribution area, followed by PS and LS (Figure 4).

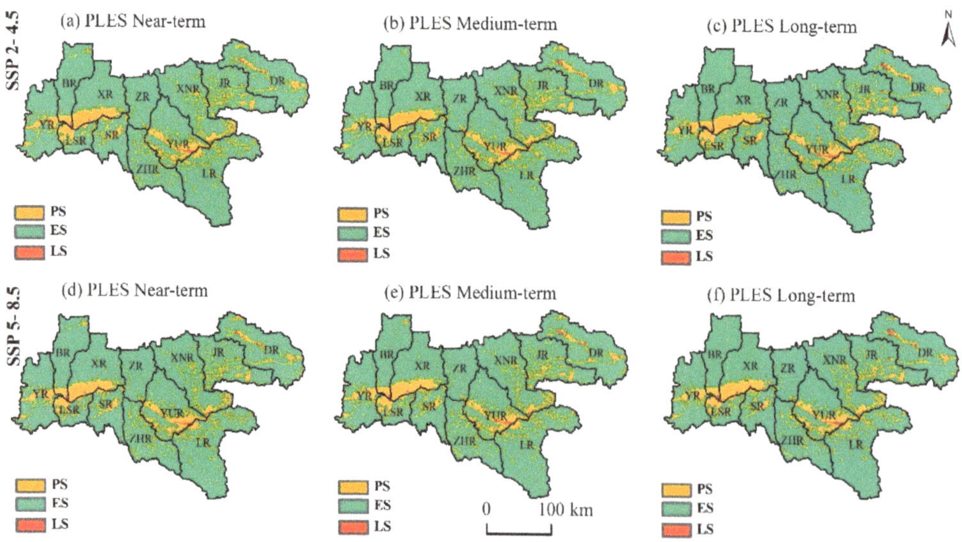

Figure 4. Spatial distribution of the future PLES under the SSP2-4.5 and SSP5-8.5 scenarios in the URHR.

Under the SSP2-4.5 scenario, the average projected distribution area of PS in the three periods is 12,301 km^2, 19.65% of the total area of URHR. The average projected distribution area of ES in the three periods is 50,124 km^2, comprising 80.07% of the URHR. The average projected distribution area of LS in the three periods is 237 km^2, only comprising 0.38% of the URHR. In terms of each stage, PS is projected to decrease at a rate of 0.71% from 2020 to 2030, while ES and LS are projected to increase at a rate of 0.16% and 4.25%, respectively. During 2030–2050, PS will decrease at a rate of 1.41%, and ES and LS will increase at a rate of 0.33% and 6.33%, respectively. During 2050–2100, PS will decrease with a rate of 4.13%, and ES and LS will grow at a rate of 0.81% and 8.51%, respectively. The outcomes reveal that the change amplitude of PLES in three periods is basically the same under the SSP5-8.5 and SSP2-4.5 scenarios.

In the three periods, PS is projected to decline while ES and LS are projected to increase under the two scenarios. The result indicates that local ES development is well guaranteed, and the ecological environment quality is well protected; however, the increase in ES and LS is at the expense of the decrease in PS. It should also be noted that the development of ES is accompanied by slight urban sprawl in the future, which will bring threats to ecological environment, and the decline of PS will also trigger challenges to agricultural production and food security.

3.2. *Transfer Evolution of Land Use and PLES under Different Scenarios*

3.2.1. Evolution Characteristics of Land Use in the URHR

The evolution of spatial pattern in PLES is directly affected by the change in land use structure. Given the small area of mutual conversion between paddy field, water land and building land, this research focused on the transfer pattern of forest land, dryland and grassland in the near-term (2020–2030), the medium-term (2030–2050) and the long-term (2050–2100) (Figure 5).

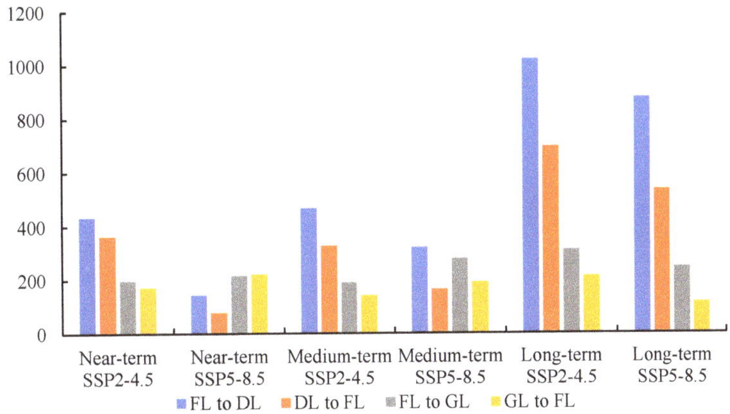

Figure 5. Transition matrix of future land use under SSP2-4.5 and SSP5-8.5.

In each period, the transfer of land use is mainly projected as the mutual transformation of forest land and dryland, forest land and grassland. Forest land is projected as the main type of ecological space and the main land class that provides ecological functions, so its change will inevitably have an impact on the structure and quality of PLES. Obvious differences are existing in the amplitude of land utilization transition during different periods. In 2050–2100, the range of land use transformation will be much larger than the other two stages, which would inevitably affect the mutual transformation of ES and PS. The findings showed that the mutual conversion between forest land and dryland under the SSP2-4.5 scenario is significantly larger than that under the SSP5-8.5 scenario, affecting the transformation of ES and PS, while it is the opposite in the case of forest land and grassland under two scenarios.

The transformation of forest land to dryland leads to ES transform into PS, and the function of regional ES will be weakened accordingly, which will further cause the changes in the multifunction level of land use. Both forest land and grassland belong to ES, the mutual transformation of forest land and grassland is only a type of change within ES, there are significant differences in the multifunctional characteristics between these two land use types. The results revealed that the mutual transformation of forest land and grassland will have certain impacts on the multi-functional pattern of land use, but will not change the pattern of PLES.

3.2.2. Evolution Characteristics of PLES in the URHR

The mutual transformation process and quantitative relationship of PS, ES and LS types of PLES were analyzed through the land transition matrix method and the spatial patterns of the three types of PLES were assessed in this research (Table 3). It is found that the mutual transformation of PLES is projected mainly in the mutual conversion of PS and ES.

Table 3. Transition matrix of future PLES in the URHR (km^2).

Research Period	PLES	SSP 2-4.5			SSP 5-8.5		
		PS	ES	LS	PS	ES	LS
2020–2030	PS	12,216	373	1	12,495	96	0
	ES	459	49,422	0	178	49,701	1
	LS	5	5	211	7	3	211
2030–2050	PS	12,056	356	1	12,203	208	3
	ES	521	49,522	1	380	49,661	2
	LS	13	3	219	8	11	216
2050–2100	PS	11,200	790	0	11,419	571	0
	ES	1201	49,245	1	981	49,465	1
	LS	12	9	234	14	7	234

In the SSP2-4.5 scenario in 2030, a total of 459 km^2 of PS will be converted to ES, and 373 km^2 of ES will be transferred into PS compared to 2020. In 2050, a total of 521 km^2 of PS will be converted to ES, and 356 km^2 of ES will be converted to PS from 2030. The change of PLES in 2100 is relatively larger than that in 2050, 1,201 km^2 of PS converted to ES and 790 km^2 of ES to PS. The transfer pattern of PLES under the SSP5-8.5 scenario will be consistent with that under the SSP2-4.5 scenario, except for the smaller amplitude of the PLES transfer in PLES. The results indicated that the change in the spatial pattern of PLES is mainly dominated by the mutual change of PS and ES in the future.

3.2.3. Evolution Characteristics of Land Use in Sub Basins

The spatial differences of the transition in the future land use were analyzed by further dividing the study area (Figure 6).

The results showed that the zones are projected with more frequent land use transformation including XNR, JR, BR and DR in the northwest of the URHR from 2020 to 2030. Forest land and dryland are projected to be mainly interconverted under SSP2-4.5, while under the SSP5-8.5 scenario, it will happen between forest land and grassland, which would trigger a wide range of mutual transformation between ES and PS and strong impact the spatial variations of PLES in these four regions in the SSP2-4.5 scenario.

In 2030–2050, a frequent land use transformation area appeared in XNR, JR, DR and BR. Transition occurs frequently between forest land and dryland in these regions under SSP2-4.5, while it is forest land and grassland under the SSP5-8.5 scenario. In 2050–2100, the transition between forest land and dryland is more frequent, and great changes are gathered in ZHR, DR, XNR and LR under the two scenarios.

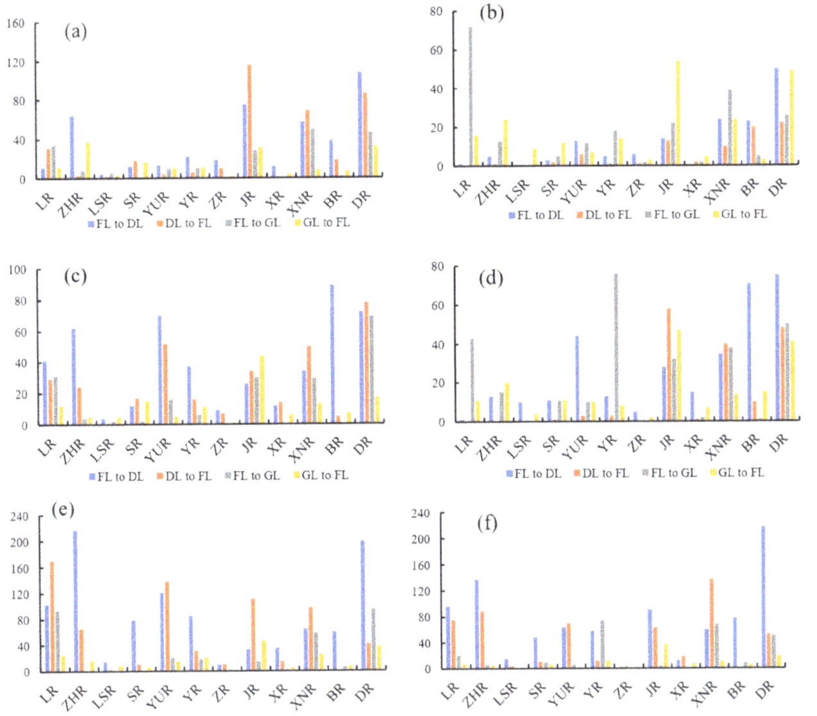

Figure 6. Transition matrix of future land use in near-term under (**a**) SSP2-4.5 and (**b**) SSP5-8.5, medium-term under (**c**) SSP2-4.5 and (**d**) SSP5-8.5 and long-term under (**e**) SSP2-4.5 and (**f**) SSP5-8.5 in sub-basins.

3.2.4. Evolution Characteristics of PLES in Sub Basins

From 2020 to 2030, the transfer of PS to ES are projected mainly in ZHR in the central URHR and XNR, JR and DR in the eastern URHR, while ES to PS transformation areas will be gathered in XNR, JR and DR under the SSP2-4.5 scenario (Figure 7). From 2030 to 2050, the transfer area of PS to ES will be concentrated in BR, and ES to PS transformation are projected mainly in YUR, XNR and DR. From 2050–2100, the regions of PS to ES transformation are projected mainly in ZHR, SR and DR, and the transformation of ES to PS are projected in the surrounding areas of Ankang City.

From 2020 to 2030, the regions of PS to ES are projected mainly in the north of BR and DR, which is basically the same as the region where ES is transformed into PS, indicating that ES and PS are frequently transformed into each other in these two regions during this period under the SSP5-8.5 scenario. From 2030 to 2050, the regions of PS to ES conversion are mainly in the northern part of BR, YUR and DR, while the regions of ES to PS conversion are mainly distributed in the border area of XHR and JR and the northern part of DR. From 2050 to 2100, the transformation area from PS to ES are projected mainly in the northern BR, ZHR and DR.

The findings showed that the BR, DR and XNR were the main concentration areas where the transition of PS and ES are projected frequently in the future, which would lead to great changes in spatial patterns of the PLES and PLESI in these regions in the future.

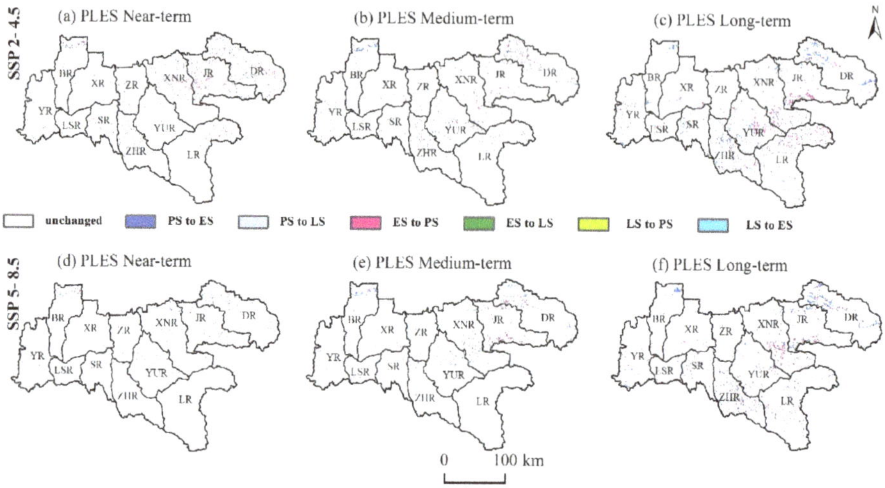

Figure 7. Space transition matrix of future PLES under the SSP2-4.5 and SSP5-8.5 scenarios.

3.3. Functional Measurement of PLES in Different Land Use and Climate Scenarios

3.3.1. Spatial Variations of PLESI

The spatial evolutions of PLESI in the URHR in historical period and two future climate scenarios were evaluated through the measurement model of PLES spatial function (PLESI) (Figure 8).

From 2000 to 2020, the PLESI is projected to decrease from 2.57 in 2000 to 2.56 in 2020. Spatially, the PLESI in the Hanjiang Valley in the west and the central YUR region has a moderate level (2.75), due to the wide distribution of paddy field and dryland in these areas. PLESI in Qinling Mountains, Micang Mountains and Daba Mountains is projected generally low, of which forest land is widely distributed. The urbanization level of Hanzhong City, Ankang City, and the center of Shangluo City is relatively higher than the peripheral areas, where the PLESI will be at the lowest level. The junction of YUR, ZHR and LR will have a higher PLESI, where Yinghu Lake is located. Yinghu Lake possesses excellent water quality and is an important water conservation area in Qinling mountains. The high value areas of PLESI are projected in the peripheral areas in the main stream. Under the future scenarios, the high value region of PLESI will generally migrate to the eastern DR, the border region of YUR, XNR and LR, and the western YR. And the junction region of ZR and ZHR in the Hanjiang River mainstream enjoys the higher value of PLESI (3.25), especially in the year 2100 under the SSP5-8.5 scenario.

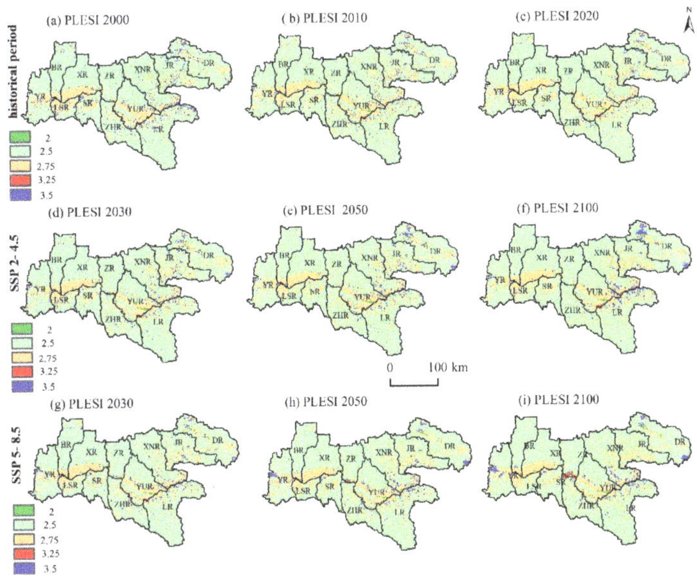

Figure 8. Spatial variations of future PLESI in the URHR from 2000 to 2100.

3.3.2. Temporal Variations of PLESI

The temporal variations in PLESI were evaluated through the PLES function measurement model (Figure 9). During 2000–2100, PLESI in the whole region is projected to descend under two scenarios. The results showed that PLESI had a large downward trend from 2000 to 2020, and the PLESI in the whole region decreased from 2.57 in 2000 to 2.56 in 2020. And after 2020, the PLESI changes under the two scenarios are projected significantly different. Under the SSP5-8.5 scenario, PLESI is projected a relatively obvious rising trend, while decreasing in the SSP2-4.5 scenario. For sub-basins, LSR, YUR and JR are the high-value regions of PLESI, while ZHR, ZR and BR are in low-value of PLESI.

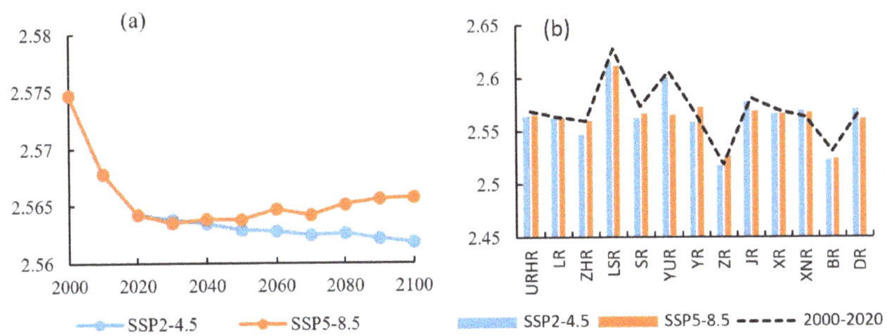

Figure 9. (**a**) Temporal variations of PLESI in the URHR from 2000–2100; (**b**) annual average values of PLESI in sub-basins in the historical period and under two scenarios.

The results showed that the change trends of PLESI in each sub-basin will differ greatly in the future (Figure 10). In ZHR, YR and ZR, the PLESI value is projected in higher upward trend under the SSP5-8.5 scenario than that in the SSP2-4.5 scenario. The PLESI in DR is projected in dramatic rise under the SSP2-4.5 scenario and the value is projected larger than that in the SSP5-8.5 scenario.

Similar change trends of PLESI are projected in the LSR, SR, YUR, JR, XR, XNR and BR under the two scenarios. In LSR, YUR and JR, the PLESI value in the SSP2-4.5 scenario will be larger than that under the SSP5-8.5 scenario, while in SR and BR, it is the opposite in the case.

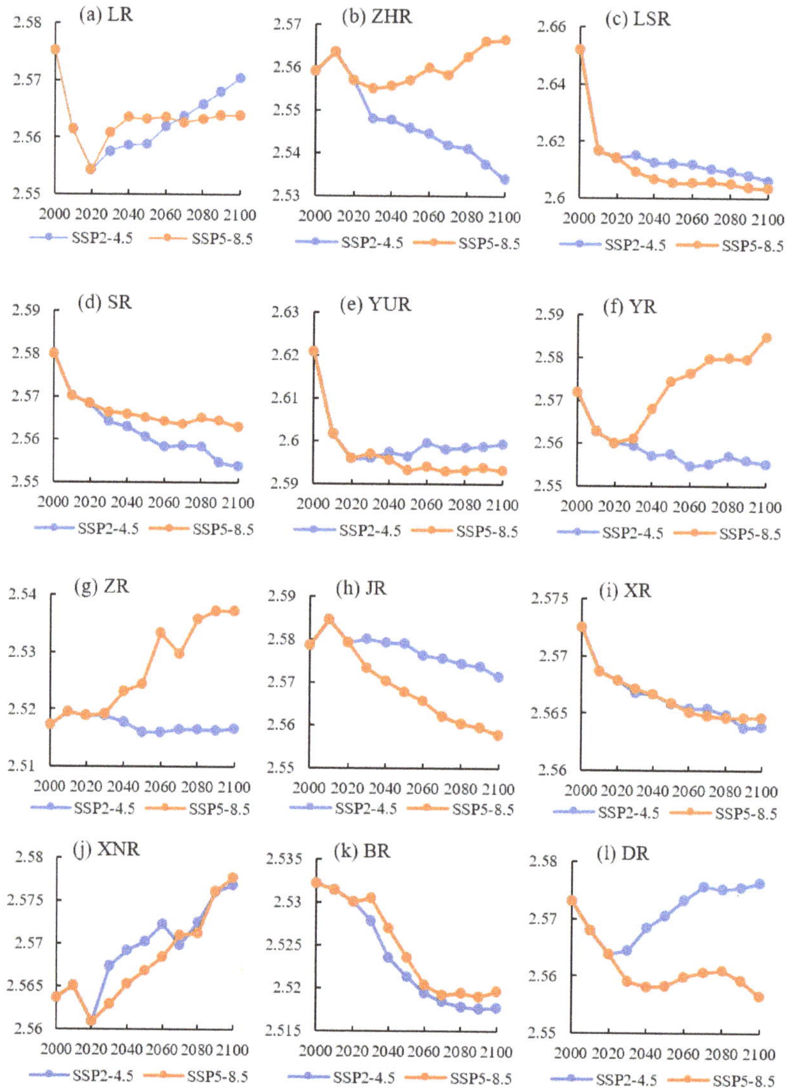

Figure 10. Temporal variations of PLESI in sub-basins (**a**–**l**) from 2000–2100 under SSP2-4.5 and SSP5-8.5.

4. Discussion

4.1. Implications of Land Use Multifunctionality Assessment

The results showed that forest land and dryland, forest land and grassland are the main types in the transfer of land use in the different periods. The transformation of forest land and dryland involves the mutual transformation of ES and PS. The areas of PLES type transfer in different periods are mainly BR, DR and XNR, which is consist with the distribution characteristics of the spatial transfer of land use. The alluvial plain spread over in the Hanjiang River valley area, with fertile soil low and flat terrain, of which dryland and building land are widely distributed. Due to intensive human activities, land use in these regions change frequently with more competition and conflicts in PLES space, and these regions becomes the main area for the mutual transformation of PLES types. The spatial and temporal changes in PLES types on a longer time scale were assessed through future climate scenarios and land use data, which has important scientific reference value for regional sustainable development [18,31].

The land system is a combination of economic, social and ecological subsystems [23], providing complex and comprehensive land functions such as environment, society and economy. In addition to dominant functions, land use also has secondary and tertiary functions [13]. Given this, it is necessary and imperative to conduct a comprehensive assessment of the land multifunctionality to guide the optimal allocation of land use scientifically and rationally.

The mutual transformation of forest land and grassland in different periods in this research represents the transformation of the ES type. This change has no impact on the spatial pattern of PLES, but bound to have certain impact on the land multifunctionality of PLES. The findings revealed that the comprehensive level of land function showed a continuous downward trend during 2000–2100 under the two scenarios. Meanwhile, PLESI had a large downward trend from 2000 to 2020, and PLESI presents a relatively obvious rising and dropping trend afterwards in the future. As for the zoning results, under the future scenario, PLESI is projected generally a significant increase trend in DR, YUR, XNR and LR border areas, and the western region of YR. Combined with the PLESI model, the multifunctionality of different land types in this research have been further clarified by systematic, comprehensive and quantitative methods [19]. It is also an important supplement to assessment of the spatiotemporal variations in PLES.

The ecosystem and land resources of the ecological protection areas are strictly managed and protected in the URHR. The study on PLES can clarify the spatial pattern of protected areas, then ES, LS and PS can be managed and controlled in different areas. Specifically, ES should be focused on in core protection areas and important protected areas, while LS and PS can be appropriately developed in general protected areas. The findings showed that PLESI can be taken to comprehensively evaluate land multifunctionality and monitor the coordination and conflict of PLES in these areas, ensuring reasonable protection and development of PLES in the study area. Besides, as an important ecological protection area in China, the evolution pattern of PLES and PLESI in Qinling Mountains have a certain reference effect on the territorial space development of similar ecological protection places. Meanwhile, it is noteworthy that the multifunctionality of land categories can be further clarified by systematic, comprehensive, and quantitative methods by the PLESI index. This research provides an insight for the comprehensive assessment of land functions in PLES at the regional and national scales.

4.2. Implications of Scenario Analysis for PLES

The superiority of CMIP6 scenarios lies in that the data are updated from previous climate models and deeply coupled with SSP data [65–67]. The simulation of land use in this research also involves regional social and economic development and policy formulation. The simulation of land use under future scenarios reflected the significant changes of major land types such as forest land, grassland and dryland, the significant changed areas such as DR, XNR, BR, and the significant differences in land use changes are existing in different years under SSP2-4.5 and SSP5-8.5.

From the perspective of land functional assessment, the PLESI decreased sharply from 2000 to 2020, and will moderate during 2030–2100. PLESI showed a relatively upward trend in LR, ZNR, DR, etc., indicating that the land functional level in these regions would be greatly improved in the future, probably due to the significant increase in precipitation in the study area under global warming [68,69].

The simulation of future PLES scenarios can supply spatially explicit assessment for land use change in different time scales and regions, different land categories and climate scenarios. The framework in this research would play an effective role in promoting social and economic development and ecological management at the regional and national scales.

4.3. PLES Management Implementation

As an important ecological barrier in central China, it is necessary to develop measures for territorial space protection and ecological management to ensure the stable development of ecological functions in the URHR. Firstly, in terms of the research on PLES, the ecological protection red line had been delimited within the region for spatial control [70]. The development activities of PS and LS should be banned in core and key protected areas, while the moderate development of PS and LS can proceed in general protected areas.

Secondly, considering the urgent needs of local economic development and ensuring people's livelihoods, the industrial structures need to be further optimized to improve the land use efficiency in production and life as well as the land function level of PS and LS. Additionally, ecological compensation could be received in the form of goods production in the surrounding areas by utilizing the advantages of a high-quality ecological environment, especially in water conservation areas. Moreover, the tertiary industries such as ecotourism and rural tourism could also be developed. Based on the research framework and methods in this research, the ultimate goal is to coordinate the spatial pattern of PLES structure on the basis of ensuring the stability of the ecosystem structure [31], thus promoting the balanced development of ecosystems and social economy.

Thirdly, the findings showed that forest land has been well protected as the major contributor to the regional ES. While as the main contributor to PS, building land is also limited by the policies of ecological protection in the study area, showing a weak expansion trend in the future. In terms of land multifunctionality, the ecological functions of dryland and paddy field are weaker than forest land and their residential functions are weaker than building land. In this context, cultivated land can be used to balance the changes of ES and PS and can be a key entry point for the coordinated development of PLES and the alleviation of regional land use in the URHR. The PLESI adopted in this research can be used as an effective tool to quantitatively evaluate the multifunctionality of different land use types in the area, and evaluate the spatial variations in the coordination and conflict of PLES, to ensure the stable development of land functionality in the whole region.

Lastly, with the time change and scenario analysis outcomes, the possible changes of PLES and PLESI in different climate scenarios and different periods were clarified, informing the long-term planning and protection of regional land resource management. In the long term, DR, XNR, BR and other areas could become the key areas for social sustainable development and ecosystem management. In addition, the mutual transformation between forest land and dryland, forest land and grassland under the two scenarios should be concerned further, which is in accord with the previous study [71].

4.4. Limitations and Future Research

The statistical downscaling method was employed to process the future climate scenarios, which can improve the regional applicability and simulation accuracy. Nevertheless, there are many driving factors affecting the simulation of land use, and the statistical scale and accuracy of data are different, resulting in a certain impact on the simulation effect. Finer resolutions of various forms of data collection in the further study are needed to effectively optimize and improve the simulation in the future scenarios, and inform reliable policy-making in the management of PLES. Additionally, the priority scenarios of PS, LS, ES and a comprehensive space optimization scenario would also be optimally designed for the regional PLES management.

The integrated approach of PLES and PLESI can be carried out in the management of regional PLES more comprehensively. However, the same land type in different regions also has great differences in its corresponding production, living and ecological functions due to the differences in soil environment, terrain and geomorphic conditions, climate and regional socio-economic conditions. The PLESI model applied in this study still needs to be improved in combination with regional actual conditions and socio-economic development indicators and ecological environment indicators, so as to improve the precision of PLESI assessment. Furthermore, how to clarify the competition and cooperation relationship, coordinate and balance the development of PLES are critical to advance the social sustainable development in the Qinling Mountains and other similar ecological function regions.

5. Conclusions

The spatial–temporal variations in land use and PLES in the URHR were simulated in the historical period and different scenarios through the combination of the newly released CMIP6 climate model, geographic ecology and socioeconomic data.

The findings indicated that the spatial pattern of land use and PLES in the future period will be almost consistent with that in the historical period. And the spatial transformations between forest land and dryland, forest land and grassland in each period resulted in the main transformation of PLES, which are reflected in the mutual conversions of PS and ES. Spatially, DR, XNR and BR are the main areas of the change in land use and PLES. As for the horizontal comparison of scenarios, a greater impact on the mutual transformation of forest land and grassland will be distinctive in the SSP5-8.5 scenario, while it will occur between FL and dryland under the SSP2-4.5 scenario, affecting the time-space transformation and distribution pattern of PS and ES. PLESI is projected trending downward under two different scenarios during 2000–2100. Meanwhile, PLESI was in a significant downward trend from 2000 to 2020, but in a relatively obvious upward trend under the SSP5-8.5 scenario and downward trend under the SSP2-4.5 scenario. This research is conducted with the aim to provide explicit reference for land resources planning, ecological environment governance and the socio-economic sustainability at the regional and national scales.

Author Contributions: All authors made significant contributions to the preparation of this manuscript. Conceptualization, P.W., X.L. and L.Z.; methodology, P.W. and X.L.; software, X.L. and Z.W.; formal analysis, P.W. and X.L.; resources, Z.W., J.B., Y.S. and T.Z.; writing—original draft preparation, P.W., X.L. and L.Z.; writing—review and editing, X.L., L.Z., H.H., T.Z., G.H. and J.Y.; funding acquisition, P.W. and J.Y. All authors have read and agreed to the published version of the manuscript.

Funding: This research was funded by the Natural Science Basic Research Program of Shaanxi Province of China (2021JQ-768), the Scientific Research Project of Shaanxi Provincial Education Department (21JK0306), the Special Research Project of Philosophy and Social Sciences of Shaanxi Province (2023HZ957), the Social Science Planning Fund Program of Xi'an city (23JX150).

Data Availability Statement: All data and materials are available upon request.

Conflicts of Interest: The authors declare no conflict of interest.

References

1. Verburg, P.H.; van de Steeg, J.; Veldkamp, A.; Willemen, L. From land cover change to land function dynamics: A major challenge to improve land characterization. *J. Environ. Manag.* **2009**, *90*, 1327–1335. [CrossRef] [PubMed]
2. Fu, B.; Zhang, L.; Xu, Z.; Zhao, Y.; Wei, Y.; Skinner, D. Ecosystem services in changing land use. *J. Soil Sediment.* **2015**, *15*, 833–843. [CrossRef]
3. Siddique, M.N.E.A.; Lobry de Bruyn, L.A.; Osanai, Y.; Guppy, C.N. Determining the role of land resource, cropping and management practices in soil organic carbon status of rice-based cropping systems. *Agric. Ecosyst. Environ.* **2023**, *344*, 108302. [CrossRef]
4. Lilburne, L.; Eger, A.; Mudge, P.; Ausseil, A.-G.; Stevenson, B.; Herzig, A.; Beare, M. The Land Resource Circle: Supporting land-use decision making with an ecosystem-service-based framework of soil functions. *Geoderma* **2020**, *363*, 114134. [CrossRef]
5. Lourenço, I.B.; Guimarães, L.F.; Alves, M.B.; Miguez, M.G. Land as a sustainable resource in city planning: The use of open spaces and drainage systems to structure environmental and urban needs. *J. Clean. Prod.* **2020**, *276*, 123096. [CrossRef]

6. Dorninger, C.; von Wehrden, H.; Krausmann, F.; Bruckner, M.; Feng, K.; Hubacek, K.; Erb, K.-H.; Abson, D.J. The effect of industrialization and globalization on domestic land-use: A global resource footprint perspective. *Glob. Environ. Chang.* **2021**, *69*, 102311. [CrossRef]
7. Furtado, I.S.; Martins, M.B. The impacts of land use intensification on the assembly of drosophilidae (Diptera). *Glob. Ecol. Conserv.* **2018**, *16*, e00432. [CrossRef]
8. Almulhim, A.I.; Cobbinah, P.B. Can rapid urbanization be sustainable? The case of Saudi Arabian cities. *Habitat Int.* **2023**, *139*, 102884. [CrossRef]
9. Thaweepworadej, P.; Evans, K.L. Urbanisation of a growing tropical mega-city during the 21st century—Landscape transformation and vegetation dynamics. *Landsc. Urban Plan.* **2023**, *238*, 104812. [CrossRef]
10. Kangas, K.; Brown, G.; Kivinen, M.; Tolvanen, A.; Tuulentie, S.; Karhu, J.; Markovaara-Koivisto, M.; Eilu, P.; Tarvainen, O.; Simila, J.; et al. Land use synergies and conflicts identification in the framework of compatibility analyses and spatial assessment of ecological, socio-cultural and economic values. *J. Environ. Manag.* **2022**, *316*, 115174. [CrossRef]
11. Baldini, C.; Marasas, M.E.; Tittonell, P.; Drozd, A.A. Urban, periurban and horticultural landscapes—Conflict and sustainable planning in La Plata district, Argentina. *Land Use Policy* **2022**, *117*, 106120. [CrossRef]
12. Zhang, K.; Wei, W.; Yin, L.; Zhou, J. Spatial-Temporal Evolution Characteristics and Mechanism Analysis of Urban Space in China's Three-River-Source Region: A Land Classification Governance Framework Based on "Three Zone Space". *Land* **2023**, *12*, 1380. [CrossRef]
13. Liu, J.; Liu, Y.; Li, Y. Classification evaluation and spatial-temporal analysis of "Production–Living–Ecological" spaces in China. *Acta Geol. Sin.* **2017**, *72*, 1290–1304.
14. Li, C.; Wu, J. Land use transformation and eco-environmental effects based on Production–Living–Ecological spatial synergy: Evidence from Shaanxi Province, China. *Environ. Sci. Pollut. Res. Int.* **2022**, *29*, 41492–41504. [CrossRef] [PubMed]
15. Ma, Q.; Wang, Z.; Zhao, Y. Evolution of Spatial-Temporal Pattern and Functional Measurement of "Production–Living–Ecological" Space in Xi'an, China. *Mt. Res.* **2021**, *39*, 722–733. [CrossRef]
16. Chen, J.; Fu, H.; Chen, S. Multi-Scenario Simulation and Assessment of Ecosystem Service Value at the City Level from the Perspective of Production–Living–Ecological-Spaces: A Case Study of Haikou, China. *Land* **2023**, *12*, 1021. [CrossRef]
17. Zou, L.; Liu, Y.; Yang, J.; Yang, S.; Wang, Y.; Cao, z.; Hu, X. Quantitative identification and spatial analysis of land use ecological-production-living functions in rural areas on China's southeast coast. *Habitat Int.* **2020**, *100*, 102182. [CrossRef]
18. Liu, X.; Wang, X.; Chen, K.; Li, D. Simulation and prediction of multi-scenario evolution of ecological space based on FLUS model: A case study of the Yangtze River Economic Belt, China. *J. Geogr. Sci.* **2023**, *33*, 373–391. [CrossRef]
19. Jin, X.; Lu, Y.; Lin, J.; Qi, X.; Hu, G.; Li, X. Research on the evolution of spatiotemporal patterns of Production–Living–Ecological space in an urban agglomeration in the Fujian Delta region, China. *Acta Ecol. Sin.* **2018**, *38*, 4286–4295.
20. Song, Y.; Xia, S.; Xue, D.; Luo, S.; Zhang, L.; Wang, D. Land Space Change Process and Its Eco-Environmental Effects in the Guanzhong Plain Urban Agglomeration of China. *Land* **2022**, *11*, 1547. [CrossRef]
21. Wang, A.; Liao, X.; Tong, Z.; Du, W.; Zhang, J.; Liu, X.; Liu, M. Spatial-temporal dynamic evaluation of the ecosystem service value from the perspective of "Production–Living–Ecological" spaces: A case study in Dongliao River Basin, China. *J. Clean. Prod.* **2022**, *333*, 130218. [CrossRef]
22. Zhao, J.; Zhao, Y. Synergy/trade-offs and differential optimization of production, living, and ecological functions in the Yangtze River economic Belt, China. *Ecol. Indic.* **2023**, *147*, 109925. [CrossRef]
23. Zhou, D.; Xu, J.; Lin, Z. Conflict or coordination? Assessing land use multi-functionalization using production-living-ecology analysis. *Sci. Total. Environ.* **2017**, *577*, 136–147. [CrossRef] [PubMed]
24. Turner, B.L.; Lambin, E.F.; Reenberg, A. Land Change Science Special Feature: The emergence of land change science for global environmental change and sustainability. *Proc. Natl. Acad. Sci. USA* **2008**, *105*, 2751. [CrossRef]
25. Musakwa, W.; Wang, S. Landscape change and its drivers: A Southern African perspective. *Curr. Opin. Environ. Sust.* **2018**, *33*, 80–86. [CrossRef]
26. Domingo, D.; Palka, G.; Hersperger, A.M. Effect of zoning plans on urban land-use change: A multi-scenario simulation for supporting sustainable urban growth. *Sustain. Cities Soc.* **2021**, *69*, 102833. [CrossRef]
27. Molinero-Parejo, R.; Aguilera-Benavente, F.; Gómez-Delgado, M.; Shurupov, N. Combining a land parcel cellular automata (LP-CA) model with participatory approaches in the simulation of disruptive future scenarios of urban land use change. *Comput. Environ. Urban Syst.* **2023**, *99*, 101895. [CrossRef]
28. Bacău, S.; Domingo, D.; Palka, G.; Pellissier, L.; Kienast, F. Integrating strategic planning intentions into land-change simulations: Designing and assessing scenarios for Bucharest. *Sustain. Cities Soc.* **2022**, *76*, 103446. [CrossRef]
29. Liu, X.; Liang, X.; Li, X.; Xu, X.; Ou, J.; Chen, Y.; Li, S.; Wang, S.; Pei, F. A future land use simulation model (FLUS) for simulating multiple land use scenarios by coupling human and natural effects. *Landsc. Urban Plan.* **2017**, *168*, 94–116. [CrossRef]
30. Liang, X.; Liu, X.; Li, X.; Chen, Y.; Tian, H.; Yao, Y. Delineating multi-scenario urban growth boundaries with a CA-based FLUS model and morphological method. *Landsc. Urban Plan.* **2018**, *177*, 47–63. [CrossRef]
31. Zhang, Y.; Zheng, M.; Qin, B. Optimization of spatial layout based on ESV-FLUS model from the perspective of "Production–Living–Ecological": A case study of Wuhan City. *Ecol. Model.* **2023**, *481*, 110356. [CrossRef]
32. Gebhardt, S.; van Dijk, J.; Wassen, M.J.; Bakker, M. Agricultural intensity interacts with landscape arrangement in driving ecosystem services. *Agric. Ecosyst. Environ.* **2023**, *357*, 108692. [CrossRef]

33. Bell, J.K.; Siciliano, S.D.; Lamb, E.G. Seasonality and bacterial community assembly processes dominate prairie ecosystem service disruption during invasion. *Soil Biol. Biochem.* **2023**, *184*, 109120. [CrossRef]
34. Wenzel, W.W.; Philipsen, F.N.; Herold, L.; Kingsland-Mengi, A.; Laux, M.; Golestanifard, A.; Strobel, B.W.; Duboc, O. Carbon sequestration potential and fractionation in soils after conversion of cultivated land to hedgerows. *Geoderma* **2023**, *435*, 116501. [CrossRef]
35. Tao, Y.; Wang, Q. Quantitative Recognition and Characteristic Analysis of Production–Living–Ecological Space Evolution for Five Resource-Based Cities: Zululand, Xuzhou, Lota, Surf Coast and Ruhr. *Remote Sens.* **2021**, *13*, 1563. [CrossRef]
36. Li, X.; Li, S.; Zhang, Y.; O'Connor, P.J.; Zhang, L.; Yan, J. Landscape Ecological Risk Assessment under Multiple Indicators. *Land* **2021**, *10*, 739. [CrossRef]
37. Wang, P.; Zhang, L.; Li, Y.; Jiao, L.; Wang, H.; Yan, J.; Lü, Y.; Fu, B. Spatio-temporal variations of the flood mitigation service of ecosystem under different climate scenarios in the Upper Reaches of Hanjiang River Basin, China. *J. Geogr. Sci.* **2018**, *28*, 1385–1398. [CrossRef]
38. Li, X.; Zhang, L.; J. O'Connor, P.; Yan, J.; Wang, B.; Liu, D.L.; Wang, P.; Wang, Z.; Wan, L.; Li, Y. Ecosystem Services under Climate Change Impact Water Infrastructure in a Highly Forested Basin. *Water* **2020**, *12*, 2825. [CrossRef]
39. Wang, P.; Zhang, L.; Li, Y.; Jiao, L.; Wang, H.; Yan, J.; Lv, Y.; Fu, B. Spatio-temporal characteristics of the trade-off and synergy relationships among multiple ecosystem services in the Upper Reaches of Hanjiang River Basin. *Acta Geol. Sin.* **2017**, *72*, 2064–2078.
40. Riahi, K.; van Vuuren, D.P.; Kriegler, E.; Edmonds, J.; O'Neill, B.C.; Fujimori, S.; Bauer, N.; Calvin, K.; Dellink, R.; Fricko, O.; et al. The Shared Socioeconomic Pathways and their energy, land use, and greenhouse gas emissions implications: An overview. *Glob. Environ. Chang.* **2017**, *42*, 153–168. [CrossRef]
41. Carvalho, D.; Rocha, A.; Costoya, X.; deCastro, M.; Gómez-Gesteira, M. Wind energy resource over Europe under CMIP6 future climate projections: What changes from CMIP5 to CMIP6. *Renew. Sustain. Energy Rev.* **2021**, *151*, 111594. [CrossRef]
42. Hamed, M.M.; Nashwan, M.S.; Shahid, S.; Ismail, T.b.; Wang, X.-j.; Dewan, A.; Asaduzzaman, M. Inconsistency in historical simulations and future projections of temperature and rainfall: A comparison of CMIP5 and CMIP6 models over Southeast Asia. *Atmos. Res.* **2022**, *265*, 105927. [CrossRef]
43. Bağçaci, S.Ç.; Yucel, I.; Duzenli, E.; Yilmaz, M.T. Intercomparison of the expected change in the temperature and the precipitation retrieved from CMIP6 and CMIP5 climate projections: A Mediterranean hot spot case, Turkey. *Atmos. Res.* **2021**, *256*, 105576. [CrossRef]
44. Liu, D.L.; Zuo, H. Statistical downscaling of daily climate variables for climate change impact assessment over New South Wales, Australia. *Clim. Chang.* **2012**, *115*, 629–666. [CrossRef]
45. Wang, B.; Liu, D.L.; Macadam, I.; Alexander, L.V.; Abramowitz, G.; Yu, Q. Multi-model ensemble projections of future extreme temperature change using a statistical downscaling method in South-Eastern Australia. *Clim. Chang.* **2016**, *138*, 85–98. [CrossRef]
46. Mohanty, M.P.; Simonovic, S.P. Changes in floodplain regimes over Canada due to climate change impacts: Observations from CMIP6 models. *Sci. Total Environ.* **2021**, *792*, 148323. [CrossRef]
47. Buhay Bucton, B.G.; Shrestha, S.; Kc, S.; Mohanasundaram, S.; Virdis, S.G.P.; Chaowiwat, W. Impacts of climate and land use change on groundwater recharge under shared socioeconomic pathways: A case of Siem Reap, Cambodia. *Environ. Res.* **2022**, *211*, 113070. [CrossRef]
48. Viseh, H.; Bristow, D.N. How climate change could affect different cities in Canada and what that means for the risks to the built-environment functions. *Urban Clim.* **2023**, *51*, 101639. [CrossRef]
49. Russo, M.A.; Carvalho, D.; Martins, N.; Monteiro, A. Future perspectives for wind and solar electricity production under high-resolution climate change scenarios. *J. Clean. Prod.* **2023**, *404*, 136997. [CrossRef]
50. Fournier, A.; Martinez, A.; Iglesias, G. Impacts of climate change on wind energy potential in Australasia and South-East Asia following the Shared Socioeconomic Pathways. *Sci. Total Environ.* **2023**, *882*, 163347. [CrossRef]
51. Seker, M.; Gumus, V. Projection of temperature and precipitation in the Mediterranean region through multi-model ensemble from CMIP6. *Atmos. Res.* **2022**, *280*, 106440. [CrossRef]
52. Rodríguez-Aguilar, O.; López-Collado, J.; Soto-Estrada, A.; Vargas-Mendoza, M.d.l.C.; García-Avila, C.d.J. Future spatial distribution of Diaphorina citri in Mexico under climate change models. *Ecol. Complex.* **2023**, *53*, 101041. [CrossRef]
53. Das, P.; Zhang, Z.; Ghosh, S.; Lu, J.; Ayugi, B.; Ojara, M.A.; Guo, X. Historical and projected changes in Extreme High Temperature events over East Africa and associated with meteorological conditions using CMIP6 models. *Glob. Planet. Chang.* **2023**, *222*, 104068. [CrossRef]
54. Taylor, K.E. Summarizing multiple aspects of model performance in a single diagram. *J. Geophys. Res. Atmos.* **2001**, *106*, 7183–7192. [CrossRef]
55. Grose, M.R.; Narsey, S.; Trancoso, R.; Mackallah, C.; Delage, F.; Dowdy, A.; Di Virgilio, G.; Watterson, I.; Dobrohotoff, P.; Rashid, H.A.; et al. A CMIP6-based multi-model downscaling ensemble to underpin climate change services in Australia. *Clim. Serv.* **2023**, *30*, 100368. [CrossRef]
56. Hersi, N.A.M.; Mulungu, D.M.M.; Nobert, J. Prediction of future climate in semi-arid catchment under CMIP6 scenarios: A case study of Bahi (Manyoni) catchment in Internal Drainage basin (IDB), Tanzania. *Phys. Chem. Earth Parts A/B/C* **2023**, *129*, 103309. [CrossRef]

57. Pierce, D.W.; Barnett, T.P.; Santer, B.D.; Gleckler, P.J. Selecting global climate models for regional climate change studies. *Proc. Natl. Acad. Sci. USA* **2009**, *106*, 8441–8446. [CrossRef]
58. Chen, W.; Jiang, Z.; Li, L. Probabilistic Projections of Climate Change over China under the SRES A1B Scenario Using 28 AOGCMs. *J. Clim.* **2011**, *24*, 4741–4756. [CrossRef]
59. Gleckler, P.J.; Taylor, K.E.; Doutriaux, C. Performance metrics for climate models. *J. Geophys. Res.* **2008**, *113*, D06104. [CrossRef]
60. Santer, B.D.; Taylor, K.E.; Gleckler, P.J.; Bonfils, C.; Barnett, T.P.; Pierce, D.W.; Wigley, T.M.; Mears, C.; Wentz, F.J.; Bruggemann, W.; et al. Incorporating model quality information in climate change detection and attribution studies. *Proc. Natl. Acad. Sci. USA* **2009**, *106*, 14778–14783. [CrossRef]
61. Camacho, A.M.; Perotto-Baldivieso, H.L.; Tanner, E.P.; Montemayor, A.L.; Gless, W.A.; Exum, J.; Yamashita, T.J.; Foley, A.M.; DeYoung, R.W.; Nelson, S.D. The broad scale impact of climate change on planning aerial wildlife surveys with drone-based thermal cameras. *Sci. Rep.* **2023**, *13*, 4455. [CrossRef] [PubMed]
62. Samuel, S.; Dosio, A.; Mphale, K.; Faka, D.N.; Wiston, M. Comparison of multi-model ensembles of global and regional climate model projections for daily characteristics of precipitation over four major river basins in southern Africa. Part II: Future changes under 1.5 °C, 2.0 °C and 3.0 °C warming levels. *Atmos. Res.* **2023**, *293*, 106921. [CrossRef]
63. Fan, Z. Simulation of land cover change in Beijing-Tianjin-Hebei region under different SSP-RCP scenarios. *Acta Geol. Sin.* **2022**, *77*, 228–244.
64. Yang, Y.; Bao, W.; Li, Y.; Wang, Y.; Chen, Z. Land Use Transition and Its Eco-Environmental Effects in the Beijing–Tianjin–Hebei Urban Agglomeration: A Production–Living–Ecological Perspective. *Land* **2020**, *9*, 285. [CrossRef]
65. Eyring, V.; Bony, S.; Meehl, G.A.; Senior, C.A.; Stevens, B.; Stouffer, R.J.; Taylor, K.E. Overview of the Coupled Model Intercomparison Project Phase 6 (CMIP6) experimental design and organization. *Geosci. Model. Dev.* **2016**, *9*, 1937–1958. [CrossRef]
66. Siabi, E.K.; Awafo, E.A.; Kabo-bah, A.T.; Derkyi, N.S.A.; Akpoti, K.; Mortey, E.M.; Yazdanie, M. Assessment of Shared Socioeconomic Pathway (SSP) climate scenarios and its impacts on the Greater Accra region. *Urban Clim.* **2023**, *49*, 101432. [CrossRef]
67. Mondal, S.K.; Huang, J.; Wang, Y.; Su, B.; Zhai, J.; Tao, H.; Wang, G.; Fischer, T.; Wen, S.; Jiang, T. Doubling of the population exposed to drought over South Asia: CMIP6 multi-model-based analysis. *Sci. Total. Environ.* **2021**, *771*, 145186. [CrossRef]
68. Zhu, X.; Lee, S.-Y.; Wen, X.; Ji, Z.; Lin, L.; Wei, Z.; Zheng, Z.; Xu, D.; Dong, W. Extreme climate changes over three major river basins in China as seen in CMIP5 and CMIP6. *Clim. Dynam.* **2021**, *57*, 1187–1205. [CrossRef]
69. Wu, X.; Wang, L.; Niu, Z.; Jiang, W.; Cao, Q. More extreme precipitation over the Yangtze River Basin, China: Insights from historical and projected perspectives. *Atmos. Res.* **2023**, *292*, 106883. [CrossRef]
70. Ma, Q.; Wang, P.; Yang, X.; Yuan, J.; Li, J.; Liu, W. Research on Delineation of Ecological Protection Red Line for Biodiversity Conservation in Qinling Mountains. *Resour. Environ. Yangtze Basin* **2020**, *29*, 634–642.
71. Jing, C.; Jiang, T.; Su, B.; Wang, Y.; Wang, G.; Huang, J.; Gao, M.; Lin, M.; Liu, S.; Zhai, J. Multiple application of shared socioeconomic pathways in land use, energy and carbon emission research. *Trans. Atmos. Sci.* **2022**, *45*, 397–413. [CrossRef]

land

MDPI

Article

Evaluation of Urban Commercial Land Use Intensification Based on Land Parcels: Taking Wuxi City as an Example

Haocong Wang [1,2], Kening Wu [1,*], Zhe Feng [1], Huafu Zhao [1], Hua Ai [3] and Chao Meng [3]

1 Faculty of Land Science and Technology, China University of Geosciences (Beijing), Beijing 100083, China; wanghaocong@cugb.edu.cn (H.W.); zhefeng@cugb.edu.cn (Z.F.); zhaohuafu@cugb.edu.cn (H.Z.)
2 Land and Water Division (NSL), Food and Agriculture Organization of the United Nations (FAO), 00153 Rome, Italy
3 Land Intensive Utilization Center, China Land Surveying and Planning Institute, Beijing 100035, China; aihua@mail.clspi.org.cn (H.A.); mengchao@mail.clspi.org.cn (C.M.)
* Correspondence: wukening@cugb.edu.cn

Abstract: Intensive land use assessment is a key research topic in urban land use, and most of the existing studies focus on macro-level assessment. There is a lack of research on the micro-level assessment of intensive urban land use, especially at the parcel level. The objective of this research is to propose a method for the parcel-based evaluation of urban commercial land intensification. The study uses a multidimensional evaluation framework and index system, comprehensive evaluation, and spatially exploratory analysis of urban commercial intensive land use based on "building intensity, use efficiency, compatibility, and diversity". The study finds that (1) the average value of intensive use of urban commercial land is 13.01, the standard deviation is 5.11, and the median value is 13, which generally indicate a medium level. (2) The degree of intensive use of commercial land has obvious characteristics of a high, medium, and low level. The study shows that when evaluating the degree of land use intensification at the parcel level, it is also necessary to consider the influence of the compatibility and diversity of external land use. The research results can provide a basis for spatial planning and the optimal design of urban land resources to improve urban vitality.

Keywords: evaluation of land use intensification; commercial land; land parcel; Wuxi City

1. Introduction

Intensive urban land use is an important strategy to ensure sustainable urban development and is an important component of the United Nations Sustainable Development Goals (SDGs) [1–3]. As one of the non-renewable resources, land in urban centers is under increasing pressure in terms of area, transport, and landscape. Only with a higher level of intensive land use is it possible to fulfill its core service function [4]. Therefore, urban centers place the greatest demands on high-density land use. However, during urban development, extensive land use patterns have led to urban sprawl [5]. With the acceleration of industrialization and urbanization, the contradiction between urban expansion and land use has intensified, leading to smart growth [6–8], compact development [9,10], infill development [11,12], multifunction intensive land use [13], and other ideas for sustainable development and modern urban planning [14–17]. Intensive urban land use is considered a key initiative to curb urban sprawl and promote sustainable urban development [18]. In addition, the extent of intensive urban land use has a significant impact on urban vitality [19]. The issue of intensive urban land use has gained increasing attention among scholars and practitioners [20].

Commercial land is the building land with the highest land price, the most obvious gain of extreme difference, and the most significant regional difference, and its intensive use has a great impact on the overall land use of the city [21]. As the population concentrates in cities, irrational land use regulations and planning lead to urban sprawl. Within cities, problems such as idleness and the inefficient use of commercial space are particularly prevalent. This has not only led to ineffective investment in urban public facilities and increased the burden on urban businesses [22]; the loss of urban vitality has also led to the great waste of natural resources and seriously affected the sustainable development of cities [23]. Thus, it is crucial to consider how to improve the level of the intensive use of urban commercial land scientifically and efficiently [24,25].

The evaluation of the intensive use of commercial areas is the basis of improving land use planning and optimizing the design of commercial areas. In recent years, some scholars have proposed diversified use characterized by three-dimensional and four-dimensional spatial use to achieve the intensification of spatial use, which has introduced new elements into the meaning of intensive urban land use [26]. From an assessment scale perspective, such assessment generally includes macro assessment (comprehensive assessment of national, regional, or urban land use intensification) [24,27–30], mid-range assessment (assessment of actual land use classification or thematic assessment) [31–33], and micro assessment (assessment of commercial land) [34,35]. The evaluation of land use intensification mainly consists of the evaluation of the current situation, policy discussion, and analysis of the factors influencing urban land use. At the macro level, studies have focused on the analysis and evaluation of urban land use policies, such as the feasibility analysis of urban intensification policies in England [36] and the evaluation of the compactness of urban development in the UK [37]. Other studies focus on examining land use intensification in a particular city or region, such as Singapore's Jurong Island [38], Tehran, Iran [39], Hamburg, Germany [40], Washtenaw, Michigan, USA [41], Gangnam, Seoul [42], and Hong Kong, China [26]. It is clear that research at the macro and meso levels has been very productive. Only a few studies have examined the evaluation of land use intensification from the micro perspective [34]. The studies on land use intensification at specific sites have analyzed horizontal and vertical forms regarding the input and output statuses of each floor to obtain a comprehensive overall picture of intensification [39]. It can be seen that there is still a lack of research evaluating the intensive use of urban land at the microscopic level. In particular, a systematic evaluation system based on the evaluation of intensive urban land use at the parcel level has not yet been developed and applied in practice.

The issue of the intensification of land use in cities is a central topic in urban research and management [43–45]. In recent years, commercial land use has gradually become the focus in evaluating land-intensive uses in cities [21]. However, most of the existing studies have focused on the evaluation of land use intensification at the city level. Because the characteristics of the sparing and intensive use of different spatial scales and different types of land can vary greatly, a single land use policy cannot be used to promote the intensive use of urban land [25,46]. The results of existing studies cannot be used to guide the behavior of specific parcel-based urban commercial land use at the micro level or city level [47]. Rather, the study of intensive commercial land use at the parcel level seeks to analyze and evaluate the use of each individual parcel. This can not only provide fundamental data for the design of urban business districts but also insights for the preparation of spatial planning as well as the optimization of the arrangement of urban land resources to increase urban vitality [19,48,49]. In addition, studies have found significant externalities of commercial land use [50]. The mixed use of different types of commercial land can have both positive and negative effects on urban vitality. When evaluating intensive commercial land use, it is important to consider not only the extent of per capita use but also the externalities of commercial land use. However, the impacts of the externalities of commercial land use have not been considered in previous assessment frameworks.

As one of the "1 + 7" cities in the Shanghai metropolitan area (Shanghai, Wuxi, Suzhou, Nantong, Ningbo, Jiaxing, Huzhou, Zhoushan) in the regional development pattern of the Yangtze River Delta, the city of Wuxi has a thriving industry and vibrant economy, accounting for 5.0 percent of the land area and 4.7 percent of the population, with a total economic output of 1.27 percent of the country. After Shanghai, Suzhou, Nanjing, and Hangzhou, Wuxi is the fifth city in the Yangtze River Delta with an economic output of over one trillion dollars. Wuxi has a high proportion of commercial land in the city center, and the use of commercial land has a significant impact on the intensive land use in the city. In addition, Wuxi has long faced resource scarcity, high-intensity land development, and a low ecological capacity [51]. Therefore, a scientific evaluation of the extent of intensive urban commercial land use in Wuxi can support the optimization of the urban construction land structure and the promotion of the new development of land with low utility value. It also serves as a reference for other highly urbanized areas to conduct parcel-based assessments of intensive commercial land use.

The objective of this research is to propose a method for the parcel-based evaluation of urban commercial land intensification. The result of this evaluation can provide the government and planners with a scientific basis for the optimal allocation and management of urban commercial land. Compared with previous studies, the contribution of this work lies in the following aspects: (1) the systematic discussion of the meaning and evaluation dimensions of the intensive use of urban commercial land; (2) the construction of a multidimensional evaluation framework and indicator system for the intensive use of urban commercial land based on construction intensity–use efficiency–sustainability–diversity.

2. Conceptual Framework and Methodology

2.1. Parcel-Based Index System for Evaluation of the Intensification of Urban Commercial Areas

In contemporary theories of urban planning, early urban planning concepts mainly emphasize function and the pursuit of an ideal urban condition [46]. With the rapid development of global urbanization and the use of a large amount of open space, the intensive use of urban land has become more important [52,53]. Planners promote the intensification of urban land use through zoning and measures such as use control. However, because urban land use functions are composite, it is not possible to increase the degree of intensive land use by simply increasing inputs [54]. The impact of urban land use externalities has been highlighted by planning scholars [50]. Commercial land use is particularly obvious. The agglomeration of commercial land uses can lead to complementary commercial functions, thereby increasing urban vitality. In contrast, the mutual exclusion of commercial forms can lead to the inefficient use of commercial land [55]. Academia has not yet developed a unified understanding of the economical and intensive uses of commercial land. This has also led to a lack of precise guidelines in planning practice to promote the intensive use of commercial land.

Based on the basic characteristics of agricultural land, the law of diminishing returns of land shows that the input and output of land do not always have a positive relationship, and, for parcels in different spatial locations in cities, the degree of intensive use does not increase with the increase in the input of productive resources [43]. For a parcel of land, there are many factors that characterize intensive use, such as the comprehensive floor area ratio, the building density, the occupancy rate of commercial properties, the degree of functional compatibility between the use of the land and surrounding parcels, and the diversity of commercial uses. As one of the forms of building land, commercial land plays an important role in the social and economic development of cities [21]. The intensive use of commercial land is an organic integration that considers the intensity of building use in the physical dimension, the input–output efficiency in the economic dimension, and the diversity and compatibility of commercial land use in terms of types of commercial activity [4]. Therefore, this study considers the factors affecting the effectiveness of the intensive use of urban commercial land from multiple dimensions and constructs a multidimensional evaluation framework and index system for the intensive use of urban commercial land based on "construction intensity, utilization benefits, compatibility, diversity" (Figure 1).

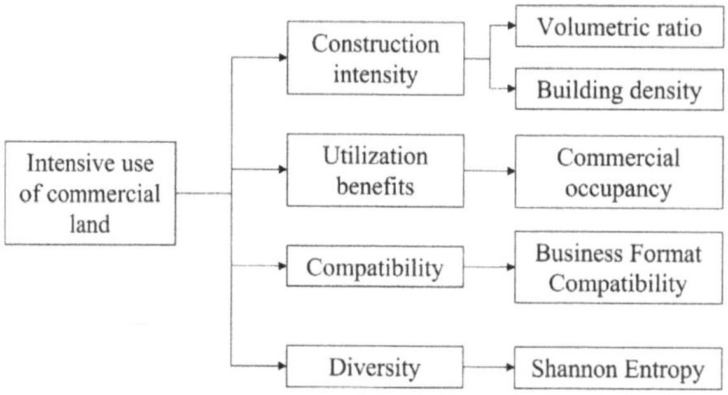

Figure 1. Evaluation framework.

2.1.1. Construction Intensity

For each type of land use within its possible intensity of use threshold, the optimal intensity of land use increases with the average productivity of society and the level of urban development, which means that the difference in land use intensity criteria reflects the difference in the level of intensive land use [35]. In essence, the floor area ratio of large functional buildings represents the degree of spatial concentration of urban functions and indicates the dominant development direction of urban functions in a city center [56]. Based on the results of existing studies, the combined floor area ratio and building density are chosen to characterize the building intensity of an individual commercial area.

$$VR = TAB/PS \tag{1}$$

$$BD = BBA/PS \tag{2}$$

VR denotes the combined floor area ratio, TAB represents the total building area within the commercial land parcel, PS represents the parcel size of the land, BD denotes the building density, and BBA represents the area of the building footprint within the parcel of commercial land.

2.1.2. Utilization Efficiency

Utilization efficiency can reflect the degree of land use intensification, and the higher the utilization efficiency per unit area, the higher the degree of land use intensification [26]. Therefore, the land use efficiency of individual commercial sites is also an important component of land use intensification. For commercial land, the degree of land use efficiency is characterized in various ways [21]. It is true that most studies have used the land output per unit area as a measure of the degree of intensive use of commercial land. However, due to the great differences in the degree of development and industrial structures, it is not possible to directly measure the degree of intensive use of commercial land in different cities using the economic output per unit area. For different cities, the degree of intensive land use of different properties can be objectively read by whether the property is vacant or not [46]. For this reason, this study chooses the occupancy rate of commercial properties as an indicator of the efficiency of commercial land use to objectively assess the degree of intensive use of commercial space.

$$RR = LCP/TCP \tag{3}$$

The occupancy rate (RR) represents the occupancy rate of commercial real estate, which is mainly determined by field study statistics. LCP represents the area of leased commercial properties, and TCP represents the total area of commercial properties.

2.1.3. Compatibility

Most of the existing assessments of land use intensification focus on the indicators of input–output aspects of land use. These assessments cannot effectively show the interdependent influence relationships between different land uses. Land use is characterized by significant externalities, and each commercial enterprise has a significant influence on the choice of business type of adjacent commercial land [57]. For this reason, compatibility among land parcels must be considered when assessing the effectiveness of land use intensification at the micro level, particularly on a parcel-by-parcel basis [4]. This study mainly considers the characteristics of externality arising from the process of commercial land use and constructs indices to evaluate the compatibility of commercial land use by relying on the measure of compatibility of mixed land use [50].

The commercial land use compatibility index (*COM*) is expressed as the ratio between the actual compatibility value and its theoretical maximum value among parcels in the area of influence.

$$COM = 1 - \frac{\sum_j^n c_{ij}}{n} \quad (4)$$

where c_{ij} is the compatibility relationship between parcels i and j; n is the number of parcels within the influence of parcel I, and 1 is the maximum score value of parcel incompatibility.

2.1.4. Diversity

The diversity index is mainly borrowed from the landscape index measure in landscape ecology, which is based on information theory and used to measure the complexity of the structural composition of a system [58]. For urban commercial land use, complex land use is a common form of land use. In order to accurately represent the degree of parcel-based intensification of commercial land use, the degree of diversity of commercial land use must be included in the evaluation index system. There are many measures of diversity, generally based on measures from landscape ecology, such as the richness index, the Shannon entropy index, the Simpson index, etc. From the perspective of commercial land use characteristics, this diversity index should, on the one hand, reflect the richness characteristics of the land use or functional types and, on the other hand, consider the area attributes of land use, i.e., reflect the unit characteristics among the different types [4]. With this in mind, we choose the Shannon entropy index as the measure because it can effectively account for both richness and uniformity characteristics, and thus better meets the requirements in characterizing land use diversity in commercial land use.

Thus, the indicators of commercial land use diversity based on the Shannon entropy index are as follows.

$$DIV = -\sum_{i=1}^{n} p_i \ln(p_i) \quad (5)$$

where n is the total number of land use or function types; p_i is the probability of occurrence of type i (usually estimated as the proportion of the number of raster or image elements of this type). For a given n, the DIV indicator reaches its maximum value (DIV_{max}) when the proportion of the area for each type of land use or function is the same (i.e., the uniformity is maximum, $p_i = 1/n$).

$$DIV_{max} = -\sum_{i=1}^{n} \left(\frac{1}{n}\right) \ln\left(\frac{1}{n}\right) = \ln(n) \quad (6)$$

2.1.5. Comprehensive Evaluation

After establishing the measurement index system for conservation and utilization, it is necessary to standardize these indicators because the value ranges and units of each indicator are not uniform. In this study, the index values are standardized to the interval $[0, 1]$ using the extreme difference standardization method [59].

After the standardization process is completed, these individual indicators must be aggregated into a comprehensive index, and, in this process, the appropriate distribution of weights is an important component. In previous studies, there are usually two methods used to determine the weights: the subjective assignment method and the objective assignment method. The former is highly subjective and the results obtained are easily influenced by the level of knowledge of decision makers, while the latter is a horizontal comparison between indicators and cannot reflect the degree of importance of different indicator dimensions [21]. Considering the respective advantages and disadvantages of these two types of methods, the authors opt for a combination of both, namely expert judgment and gray correlation, to determine the weights. This can compensate for the shortcomings of the above subjective and objective methods by fully utilizing the subjective information of the experts' empirical judgment results, while using mathematical models to objectively calculate the indicator weights [4]. The individual calculation steps are as follows.

(1) Judgement of weights by experts based on experience. There are m experts who make empirical judgements on the weights of n indicators at the same time, thus forming the empirical judgement data column for the indicator weights.

$$X = \begin{bmatrix} x_{11} & \cdots & x_{1m} \\ \vdots & \ddots & \vdots \\ x_{n1} & \cdots & x_{nm} \end{bmatrix} \tag{7}$$

(2) Determine the reference sequence. The largest weight value from X is selected as the "common" reference weight value, thus forming the reference data column X_0.

$$X_0 = (x_{01},\ x_{02},\ \ldots, x_{0m}) \tag{8}$$

(3) Find the distance between each indicator series and the reference data column X_0.

$$D_{0i} = \sum_{k=1}^{m} (x_{0k} - x_{ik})^2 \tag{9}$$

(4) Find the weights of each indicator and perform the normalization operation.

$$w_i = \frac{1}{1 + D_{0i}} \tag{10}$$

$$w_i^{\ *} = \frac{w_i}{\sum_{i=1}^{n} w_i} \tag{11}$$

where x_{0k} denotes the reference weight value of the kth indicator in the reference series, x_{ik} denotes the empirical judgment of the ith expert on the weight of the kth indicator, D_{0i} denotes the distance between the ith indicator series and the reference series, w_i is the weight value of the ith indicator, and $w_i^{\ *}$ is the normalized indicator weight value.

After the above process, the weight values of each individual indicator can be obtained. Afterwards, the individual indicators are weighted and summed according to the following formula to obtain a comprehensive indicator value for the intensive use of urban commercial land based on the parcel.

$$IU = f(VR, BD, RR, COM, DIV) = \sum_{i}^{n} IU_i \cdot w_i = VR \times w_{vr} + BD \times w_{bd} + RR \times w_{rr} + COM \times w_{com} + DIV \times w_{div} \tag{12}$$

where VR, BD, RR, COM, and DIV are the individual indicators for the evaluation of intensive urban commercial land use; w_{vr}, w_{bd}, w_{rr}, w_{com}, and w_{div} are the weights of each individual indicator, respectively; and IU refers to the integrated indicator value of intensive urban commercial land use based on the land parcel.

2.2. Exploratory Spatial Data Analysis (ESDA)

The ESDA method [60] is used to analyze the spatial agglomeration effect of commercial land intensification created by the spatial distribution characteristics of commercial land use behavior and the spatial spillover effects of the degree of commercial land intensification. The global spatial autocorrelation uses the global Moran's I index to test whether spatial dependence is positive in the distribution of spatial elements. Moran's I is defined as follows:

$$I = \frac{n\sum_{i=1}^{n}\sum_{j=1}^{n} W_{ij}|x_i - \overline{x}||x_j - \overline{x}|}{\sum_{i=1}^{n}\sum_{j=1}^{n} W_{ij}\sum_{i=1}^{n}|x_j - \overline{x}|^2} \tag{13}$$

where x_i and x_j are the observed values in parcel i and j, respectively; $\overline{x}$ is the mean of the observed values in the parcels; W_{ij} refers to the binary spatial weight matrix, and n is the total number of observed spatial units. For this analysis, the radius used for clustering is 3355 m, which is produced by the ArcGIS software platform.

Local Indicator of Spatial Association (LISA) reflects the extent to which commercial land parcels differ from or are consistent with their neighboring spatial units in terms of intensive use. The LISA model is expressed as follows:

$$I_i = \frac{n(x_i - \overline{x})\sum_i W_{ij}(x_j - \overline{x})}{\sum_i (x_i - \overline{x})^2} = Z_i\sum_j W_{ij}Z_j \tag{14}$$

where Z_i and Z_j are the normalized forms of x_i and x_j. A positive I_i indicates significant local clusters, where the similarity value in the local space tends to agglomerate (high–high or low–low clusters). A negative I_i indicates significant local spatial outliers, where the similar value of the local space presents a discrete distribution (high–low or low–high clusters).

2.3. Study Area

The city of Wuxi is located in the southeastern part of Jiangsu Province and belongs to the corridor between the rivers and lakes of the Yangtze River Delta. The geographical coordinates of the city are longitude 119°33′~120°38′ east and latitude 31°07′~32°02′ north. To the east is Suzhou, 128 km from Shanghai; to the south and southwest is the border with Zhejiang Province and Anhui Province; to the west is Changzhou, 183 km from Nanjing; to the north is the Yangtze River; on the other side of the river is the city of Jingjiang, which belongs to Taizhou City. The total area of the city is 4627.47 square kilometers. The study area is located in the central urban area of Wuxi, including Liangxi District, Huishan District (Yanqiao Street and Chang'an Street), Xishan District (Dongting Street and Dongdongtang City), Xinwu District (Jiangxi Street, Wangzhuang Street, and Meicun City), and Binhu District (Haili Street, Liyuan Street, Lihu Street, Rongxiang Street, and Taihu Street), with an area of 368.24 km^2 (Figure 2).

2.4. Data Collection and Processing

Commercial land data were extracted from the Wuxi municipal cadastral database to reflect the data available at the time of the assessment (including the area of buildings, number of floors, etc.). The map was overlaid with the current land use map in China's 2000 national geodetic coordinate system and checked for flatness. The basic data needed to evaluate the intensive use of commercial land, such as building intensity, use efficiency, compatibility, and diversity, were collected through visits to the Wuxi Development and Reform Commission, the Bureau of Industry and Information Technology, the Bureau of Natural Resources, and other departments, as well as through field research.

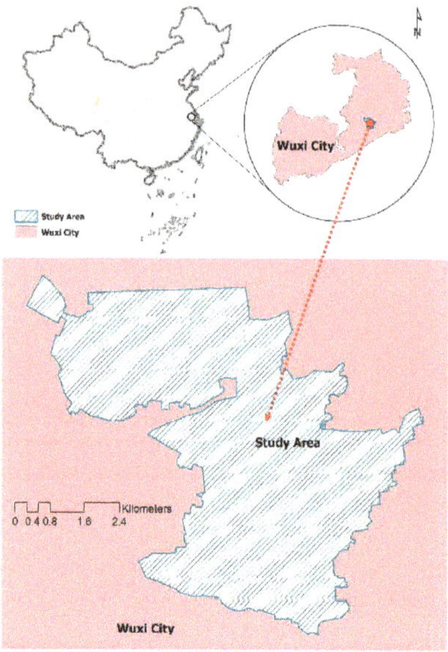

Figure 2. Study area.

3. Results

3.1. Statistical Information on Commercial Land Parcels

The study area is divided into 1664 commercial sites, with a total area of 18.12 km^2, which represents 8.91% of the perimeter of the study area (Table 1). In terms of number, Liangxi District is the most numerous, and in terms of area, Liangxi District and Xinwu District occupy the first two positions, indicating that the commercial sites in the old city are more mature and established but relatively small and fragmented in area, while the commercial areas in the new district are contiguous and larger in area, so that the site area is also larger (Figure 3).

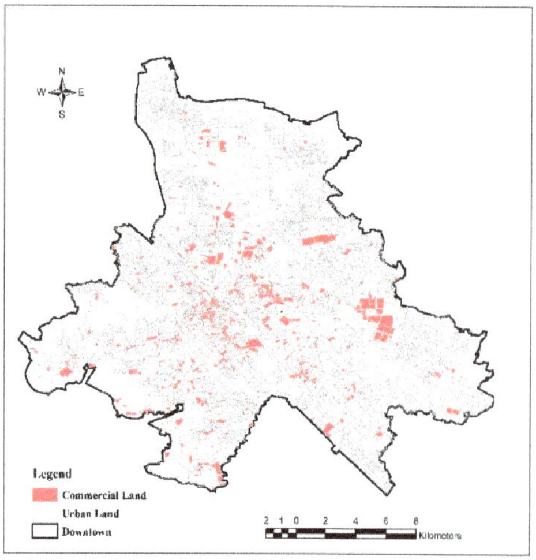

Figure 3. Commercial land distribution map.

Table 1. Summary of commercial land use statistics.

	Administrative District	Quantity (Pieces)	Area (km^2)	Proportion (%)
Commercial land	Binhu	190	2.12	11.68
	Huishan	179	1.57	8.67
	Jingkai	96	1.20	6.60
	Liangxi	743	5.56	30.71
	Xishan	217	3.16	17.45
	Xinwu	239	4.51	24.88
	SUM	1664	18.12	100.00

3.2. Evaluation Findings

3.2.1. Construction Intensity

The average plot ratio is 1.97 (Table 2). In terms of spatial distribution, the plot ratios show a gradient from the city center to the outskirts. The plot ratio of commercial land in the city center essentially reaches 5.0 or more, while the plot ratio of commercial land in the distant suburbs remains below 1.5 (Figure 4a). Moreover, the plot size and plot ratio show a significant negative correlation as far as the plot size is concerned. The larger the site area, the lower the plot ratio of commercial sites.

Table 2. Summary of evaluation findings.

	Max	Min	S.D.	Med	Aver
Volume ratio	15.37	0.04	1.81	1.40	1.97
Building density	100.02	3.84	25.11	56.92	59.10
Utilization efficiency	0.98	0.00	0.19	0.93	0.90
Compatibility	24.41	0.08	1.90	0.73	1.24
Diversity	12.46	0.04	1.61	1.37	1.86
Intensive urban commercial land use	24.00	2.00	5.11	13.00	13.01

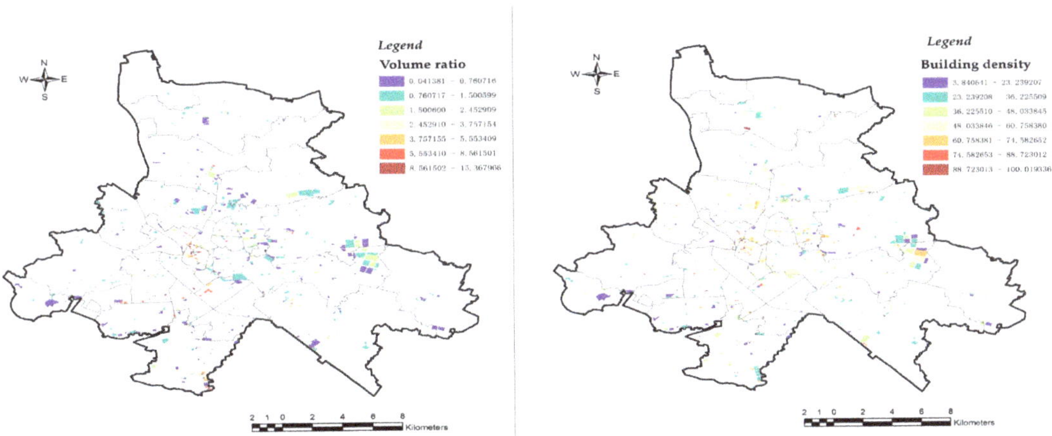

(**a**) Volume ratio (**b**) Building density

Figure 4. *Cont.*

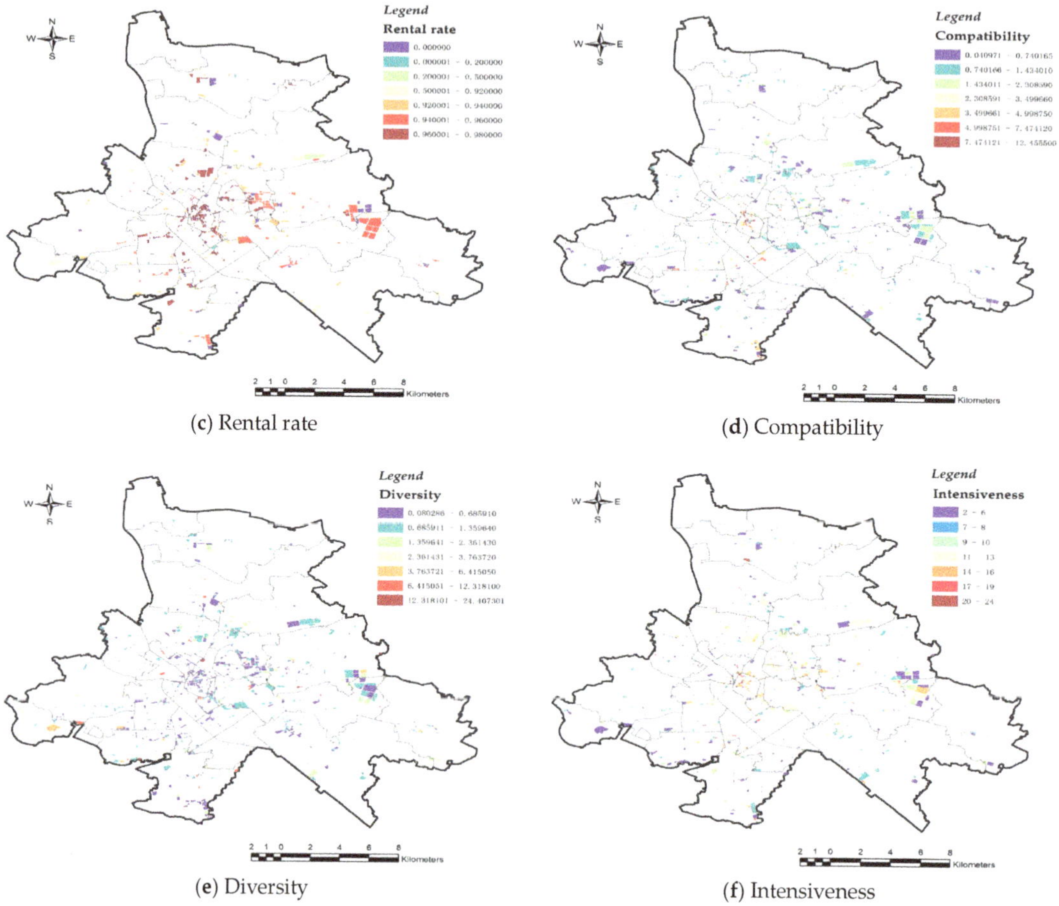

Figure 4. Evaluation findings.

The highest building density of commercial land in Wuxi is 100.02 and the lowest is 3.84, with a standard deviation of 25.11 (Table 2). Overall, the building density is also high in the city center and low in the suburbs. However, compared with the land ratio, the building density does not show a clear central concentration in space (Figure 4b).

3.2.2. Utilization Efficiency

The evaluation results show that commercial land use in Wuxi is highly effective. The highest commercial land occupancy rate is 0.98, while the median and average occupancy rates are both 0.9 (Table 2). This also reflects the developed commercial market in Wuxi and the high efficiency of commercial land use. However, the lowest occupancy rate of commercial land in Wuxi is 0, indicating that there is also unused commercial land in the city. In terms of spatial distribution, there is no obvious spatial clustering of commercial land rental rates in Wuxi. Most of the commercial sites have an occupancy rate of 92% or more, with the exception of a few scattered commercial sites in the north, where the occupancy rate is low (Figure 4c).

3.2.3. Compatibility

As can be seen from Equation (4), the higher the compatibility value, the less compatible the commercial sites are with each other. Based on the results summarized in Table

2, it can be seen that the mean value of commercial land compatibility in Wuxi is 1.24, with a median value of 0.73. This result indicates that there is a high level of compatibility between the various types of businesses carried by commercial land in Wuxi. In addition, Figure 4d shows that there is clear spatial clustering of the commercial land compatibility indices in Wuxi, with the degree of compatibility of commercial sites in the city center being much higher than the compatibility of sites in the suburbs.

3.2.4. Diversity

The results show that the highest commercial land use diversity index in Wuxi is 12.46 and the lowest is 0.04 (Table 2). The diversity of commercial land uses in Wuxi has significant differentiation characteristics. This result indicates that there is strong locational variability in the degree of composite use of commercial land in Wuxi. This is also evidenced in Figure 4e. Overall, the commercial land use diversity index in Wuxi is low, and the city center shows a depression in diversity.

3.2.5. Intensive Urban Commercial Land Use

The level of intensive use of commercial land based on the parcel is assessed according to Equation (12). As can be seen from Table 2, the average and median intensive use of commercial land in Wuxi is 13.01, indicating that the level of intensive use of commercial land in the city is comparable and at a medium level overall. Except for the high level of intensive commercial land use in the central business district, the level of intensive commercial land use in the periphery of the central city has declined significantly Figure 4f.

3.3. *Spatial Divergence Characteristics*

The spatial distribution and degree of intensive use of urban commercial land are significantly influenced by the spatial location of the city in which the parcel is located. The result shows that the degree of intensive use of commercial land has significant spatial differentiation characteristics. The results of the exploratory spatial analysis show that, with the exception of the indicators of occupancy and diversity, the indicators used to evaluate intensive use have significantly spatially divergent characteristics (Table 3).

Table 3. Summary of spatiotemporal differentiation.

	Moran's I	Z	*p*
Volume ratio	0.2134	2.0511	0.05
Building density	1.4144	13.5281	0.01
Rental rate	0.0461	0.4491	-
Compatibility	0.2359	2.2651	0.05
Diversity	0.0892	0.8721	-
Intensiveness	1.3164	12.5918	0.01

Firstly, the floor area ratio shows significant spatial clustering characteristics. From Figure 5a, it can be seen that the floor area ratio of commercial land in the central city area shows a significant high–high clustering feature. This result indicates that the city center's commercial area has a better location advantage, with high-rise commercial office buildings predominantly placed on commercial land. Moreover, the high floor area ratio of commercial land in the city center has spatial spillover characteristics, which can effectively enhance the intensive use of the surrounding land.

Secondly, the building density also has significant spatial clustering characteristics. However, unlike the floor area ratio, the building density is characterized by low clustering at the edge of the city center. The commercial sites away from the city center show a low-density sprawl. Moreover, as can be seen from Figure 5b, unlike the high–high agglomeration characteristic of a more concentrated volumetric ratio, the low–low agglomeration of low-density commercial land is scattered in multiple locations in urban areas away from the city center.

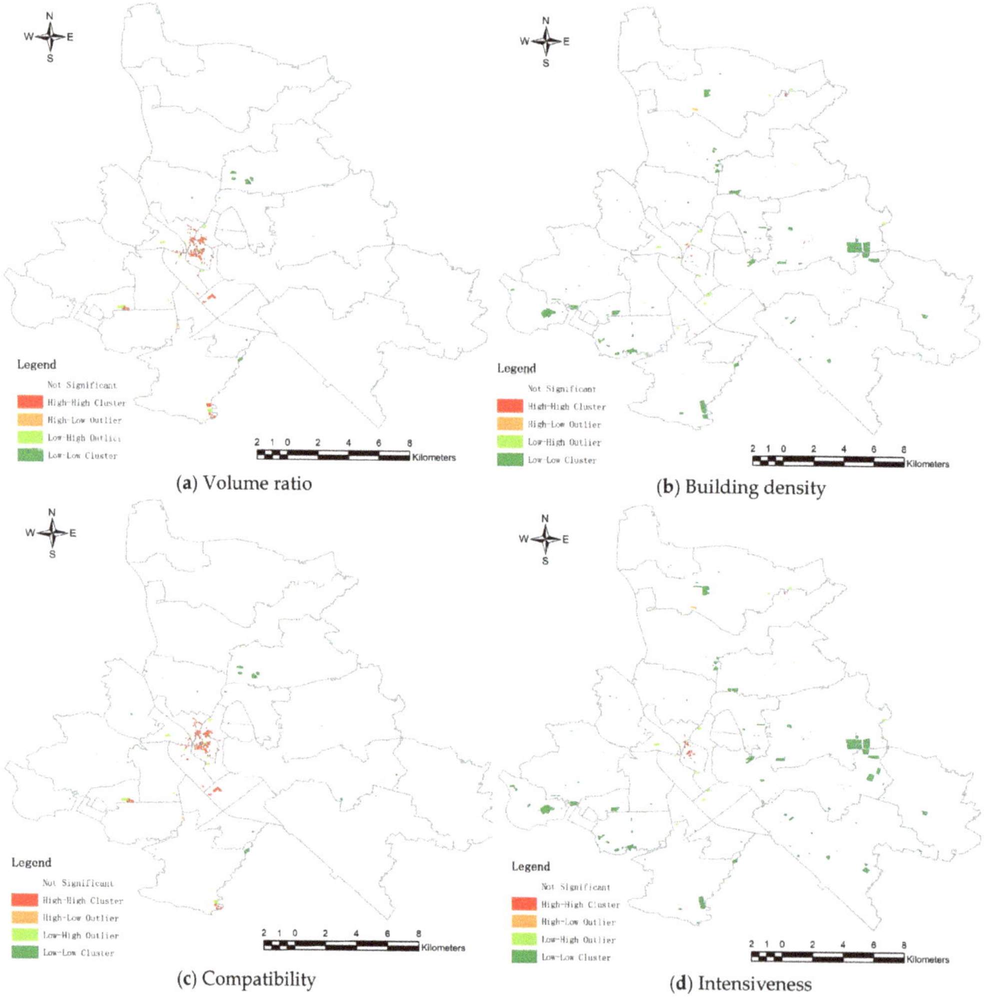

Figure 5. Spatial divergence characteristics.

The spatial clustering characteristics of compatibility are very similar to those of the volumetric ratios. The compatibility of commercial land uses in the city center area shows high–high agglomeration Figure 5c. This indicates that there is a high degree of compatibility between commercial businesses in the central city. The spatial spillover of the high compatibility of commercial land can help to further enhance the intensive use of commercial land.

Overall, the level of intensive land use of urban commercial land in Wuxi is characterized by significant spatial clustering Figure 5d. Firstly, the spatial distribution of the level of intensive land use of urban commercial land in Wuxi is relatively uneven. Commercial land in the city center is characterized by a high level of agglomeration. Commercial land use in the peripheral areas of the city is characterized by low concentration. Secondly, the number of parcels with high levels of intensive land use is small and the area is relatively small. The spatial spillover of commercial land parcels with high levels of intensive land use is limited. In contrast, there are many parcels with low levels of land intensive use, and they are scattered over a wide area. The spatial spillover of commercial land parcels with low levels of intensive land use is significant.

4. Discussion

4.1. Multidimensional Expression of Intensive Commercial Land Use

The input–output ratio is the core concept of land-intensive use evaluation. Most of the existing studies also consider the average input–output of the area to measure the degree of intensive use of urban land [26,56]. Without regard to spatial scale, similar assessment concepts and approaches are able to respond objectively to the degree of land use intensification and utilization [39,56]. The volume ratio and building density, as indicators of land use, can reflect the degree of land use intensification more objectively. However, the consideration of only these two indicators is not comprehensive [34]. The intensification of commercial land use takes into account the intensity of building use in the physical dimension, input–output efficiency in the economic dimension, and the organic integration of the diversity and compatibility of commercial land use from the commercial mode. The impact of land use externalities must also be included in the assessment of intensive commercial land use [50]. As noted in this study, a large area of commercial land on the edge of downtown has a relatively high floor area ratio and development density. However, due to the lack of reasonable commercial planning, these areas suffer from the problem of commercial compatibility [13]. As a result, the overall intensive use is not very high. For example, the downtown core area leads the region in terms of the floor area ratio and occupancy rate, but it is obvious that the diversity index of downtown is very low, which limits the intensive use of the downtown core area. From this, it can be seen that the evaluation of intensive land use based on parcels must consider not only the average land use of parcels, but also the relationships between parcels and the relationships between parcels and regions. When evaluating the intensification of commercial land use, compatibility and diversity between parcels must be considered [61].

4.2. Location Has a Significant Impact on Commercial Land Use Intensification

The influence of location on urban commercial land use is very large. Therefore, the degree of intensive commercial land use under the influence of location factors also shows clear spatial differentiation characteristics [21]. Overall, the degree of intensive commercial land use in Wuxi City shows a clear spatial downward trend from the core to the peripheral areas. In particular, it can be seen that the core area of the central city has a significantly high concentration of intensive commercial land use, while the peripheral area of the central city has a significantly low concentration. In addition to the rental rate and diversity index, the volume ratio, development density, and compatibility also exhibit significant spatial cluster characteristics. The core area of the central city is a more developed commercial area, where there are more commercial spaces and where a high volume ratio is more common. In contrast, the development density has significantly less clustering. Because areas distant from the city center tend to have less commercial development, the demand for commercial space is not as strong as in the core areas of the city center [43]. As a result, the density of commercial development in the outlying areas of downtown is relatively low. The study also shows that the volume ratio, development density, rental rate, compatibility, and diversity must be considered together when assessing the level of intensive commercial land use. However, the above indicators show similar spatial clustering characteristics. This also shows that location factors significantly affect the amount of commercial land use. Therefore, when planning commercial land use and developing optimization plans for intensive land use, it is necessary to consider the influence of location on commercial land use [46].

4.3. Implications for Spatial Planning

Spatial planning is the fundamental basis for all forms of territorial development, protection, and construction activities [48]. It is also an important means for the state to guide and monitor the development and use of urban land resources and to regulate and optimize the spatial design [49]. Therefore, in order to comprehensively improve the efficiency of urban land resource utilization, it is necessary to play the leading role of spatial planning scientifically [62]. Research has shown that the degree of intensive use of commercial land is affected by the floor area ratio, building density, compatibility, and diversity. Therefore, not only the floor area ratio, building density, and other controlling elements of each commercial land must be considered in the preparation of national land use planning. The study found that the lack of compatibility and the diversity of the spatial arrangement of commercial areas lead to a loss of urban vitality [50]. Therefore, the compatibility and diversity of commercial enterprises must also be taken into account when preparing land use plans or urban renewal programs. In addition, the spatial spillover effect of the degree of intensification of an individual commercial use must not be ignored [21]. Only when the per capita input index of an individual commercial site and its potential externalities are taken into account can the degree of intensification of commercial land use be effectively improved.

5. Conclusions

Assessing the degree of intensification of urban land use on the basis of the plot scale is a difficult problem in this field of study. When discussing the degree of intensification of commercial land use at the microscopic level, it is necessary to consider not only the indicators related to average land use but also other factors that affect the degree of intensification of land use. In this paper, we construct a system to evaluate the intensity of commercial land use that integrates the indicators of the volume ratio, building density, use efficiency, compatibility, and diversity. An empirical study was conducted in downtown Wuxi as a case study area. The research results show that the mean values of the volume ratio of commercial space, building density, use efficiency, compatibility, and diversity in Wuxi City are 1.97, 59.1, 0.9, 1.24, and 1.86, respectively. The mean value of the intensive use of commercial land in the city is 13.01, with a standard deviation of 5.11 and median value of 13, which generally indicate a medium level. The degree of intensive use of commercial land has obvious characteristics of high, medium, and low levels. Except for the high intensity of commercial land use in the central business district, the intensity of commercial land use decreases significantly at the periphery of the central city. The results of the study can help city managers or planners to optimize commercial land use in terms of land use per capita or in terms of the compatibility or diversity of commercial land types, based on the different intensities of commercial land use and its different indicators in different regions.

In addition, the impact of the externalities of commercial land use must be considered in the development of urban plans. In this study, it was also found that in addition to the average indicators of land consumption, such as the volume ratio, building density, and use efficiency, these can objectively respond to the degree of intensive commercial land use. When evaluating the intensification of land use at the parcel level, it is also necessary to consider the impact of external land use. The study shows that the degree of intensive commercial land use in Wuxi City is favored by the high compatibility among different types of enterprises located on commercial land. However, the diversity of commercial land use in Wuxi City is characterized by significant differentiation. The low diversity in the central business district affects the level of intensive commercial land use throughout the study area. In addition, the spatial clustering characteristics of the different indexes show significant differences. The development density is also high in the city center and low in the suburbs. However, compared to the volume ratio, the development density does not show a clear central concentration in the study area. The commercial land compatibility indexes are highly spatially clustered in Wuxi City, and the degree of commercial land compatibility in the central city is much higher than the compatibility of the land in the suburbs.

The contribution of this study is to develop an area-based evaluation approach for the intensive use of commercial land in the city. This evaluation approach is not only able to reflect the influence of the average area input of commercial land on intensive use. In addition, this system can also consider the external effects of neighboring commercial uses. However, a limitation of this work is that it only examines the evaluation of the intensive use of commercial land based on land size, without analyzing the applicability to other types of land. As with residential and industrial areas, it is necessary to determine which factors must be considered when evaluating compatibility and diversity. This will greatly affect the usefulness and accuracy of the assessment. Future research can further validate this evaluation system using different cities and different land use types.

Author Contributions: Conceptualization, H.W. and K.W.; methodology, H.W., Z.F. and H.Z.; validation, H.W., H.A. and C.M.; writing—original draft preparation, H.W.; writing—review and editing, H.W. and K.W. All authors have read and agreed to the published version of the manuscript.

Funding: This research was funded by the National Natural Science Foundation of China (NSFC No. 42171261).

Institutional Review Board Statement: Not applicable.

Informed Consent Statement: Not applicable.

Data Availability Statement: The data are not publicly available due to privacy.

Conflicts of Interest: The authors declare no conflict of interest.

References

1. Schiavina, M.; Melchiorri, M.; Freire, S.; Florio, P.; Ehrlich, D.; Tommasi, P.; Pesaresi, M.; Kemper, T. Land use efficiency of functional urban areas: Global pattern and evolution of development trajectories. *Habitat Int.* **2022**, *123*, 102543. [CrossRef]
2. Xia, L.M.; Chen, S.P. Evaluation on Land Intensive Use Based on the Sustainable Development Theory in Tianjin. In Proceedings of the 2010 International Conference on Logistics Systems and Intelligent Management, Harbin, China, 9–10 January 2010; Volume 1–3, pp. 395–398.
3. Cai, G.; Zhang, J.; Du, M.; Li, C.; Peng, S. Identification of urban land use efficiency by indicator-SDG 11.3.1. *PLoS ONE* **2021**, *15*, e0244318. [CrossRef]
4. Zhuo, Y.; Zheng, H.; Wu, C.; Xu, Z.; Li, G.; Yu, Z. Compatibility mix degree index: A novel measure to characterize urban land use mix pattern. *Comput. Environ. Urban Syst.* **2019**, *75*, 49–60. [CrossRef]
5. Kazak, J.K.; Błasik, M.; Świąder, M. Land use change in suburban zone: European context of urban sprawl. *J. Water Land Dev.* **2022**, 92–98. [CrossRef]
6. Miller, J.S.; Hoel, L.A. The "smart growth" debate: Best practices for urban transportation planning. *Socio-Econ. Plan. Sci.* **2002**, *36*, 1–24. [CrossRef]
7. Gabriel, S.A.; Faria, J.A.; Moglen, G.E. A multiobjective optimization approach to smart growth in land development. *Socio-Econ. Plan. Sci.* **2006**, *40*, 212–248. [CrossRef]
8. Howell-Moroney, M. Studying the effects of the intensity of US state growth management approaches on land development outcomes. *Urban Stud.* **2007**, *44*, 2163–2178. [CrossRef]
9. Stevens, M.R. Does Compact Development Make People Drive Less? *J. Am. Plan. Assoc.* **2017**, *83*, 7–18. [CrossRef]
10. Neuman, M. The Compact City Fallacy. *J. Plan. Educ. Res.* **2005**, *25*, 11–26. [CrossRef]
11. Farris, J.T. The barriers to using urban infill development to achieve smart growth. *Hous. Policy Debate* **2001**, *12*, 1–30. [CrossRef]
12. Kim, J.; Larsen, K. Can new urbanism infill development contribute to social sustainability? The case of Orlando, Florida. *Urban Stud.* **2016**, *54*, 3843–3862. [CrossRef]
13. Eom, S.; Suzuki, T.; Lee, M.-H. A land-use mix allocation model considering adjacency, intensity, and proximity. *Int. J. Geogr. Inf. Sci.* **2019**, *34*, 899–923. [CrossRef]
14. Echenique, M.H.; Hargreaves, A.J.; Mitchell, G.; Namdeo, A. Growing Cities Sustainably. *J. Am. Plan. Assoc.* **2012**, *78*, 121–137. [CrossRef]
15. Cen, X.; Wu, C.; Xing, X.; Fang, M.; Garang, Z.; Wu, Y. Coupling Intensive Land Use and Landscape Ecological Security for Urban Sustainability: An Integrated Socioeconomic Data and Spatial Metrics Analysis in Hangzhou City. *Sustainability* **2015**, *7*, 1459–1482. [CrossRef]
16. Qian, Q.; Liu, H.; Zheng, X. A Regional Sustainable Intensive Land Use Evaluation Based on Ecological Constraints: A Case Study in Jinan City. *Sustainability* **2019**, *11*, 1434. [CrossRef]

17. Shao, J.; Ge, J. Investigation into Relationship between Intensive Land Use and Urban Heat Island Effect in Shijiazhuang City Based on the Tapio Decoupling Theory. *J. Urban Plan. Dev.* **2020**, *146*, 04020043. [CrossRef]
18. Wang, X.; Shi, R.; Zhou, Y. Dynamics of urban sprawl and sustainable development in China. *Socio-Econ. Plan. Sci.* **2020**, *70*, 100736. [CrossRef]
19. Xia, C.; Yeh, A.G.-O.; Zhang, A. Analyzing spatial relationships between urban land use intensity and urban vitality at street block level: A case study of five Chinese megacities. *Landsc. Urban Plan.* **2020**, *193*, 103669. [CrossRef]
20. Cui, X.; Huang, S.; Liu, C.; Zhou, T.; Shan, L.; Zhang, F.; Chen, M.; Li, F.; de Vries, W.T. Applying SBM-GPA Model to Explore Urban Land Use Efficiency Considering Ecological Development in China. *Land* **2021**, *10*, 912. [CrossRef]
21. Garang, Z.; Wu, C.; Li, G.; Zhuo, Y.; Xu, Z. Spatio-Temporal Non-Stationarity and Its Influencing Factors of Commercial Land Price: A Case Study of Hangzhou, China. *Land* **2021**, *10*, 317. [CrossRef]
22. Wu, Y.; Fan, P.; You, H. Spatial Evolution of Producer Service Sectors and Its Influencing Factors in Cities: A Case Study of Hangzhou, China. *Sustainability* **2018**, *10*, 975. [CrossRef]
23. Blazy, R.; Łabuz, R. Spatial Distribution and Land Development Parameters of Shopping Centers Based on GIS Analysis: A Case Study on Kraków, Poland. *Sustainability* **2022**, *14*, 7539. [CrossRef]
24. Zhang, P.; Yang, D.; Qin, M.; Jing, W. Spatial heterogeneity analysis and driving forces exploring of built-up land development intensity in Chinese prefecture-level cities and implications for future Urban Land intensive use. *Land Use Policy* **2020**, *99*, 104958. [CrossRef]
25. Wu, Y.; Hui, E.C.M.; Zhao, P.; Long, H. Land use policy for urbanization in China. *Habitat Int.* **2018**, *77*, 40–42. [CrossRef]
26. Lau, S.S.Y.; Giridharan, R.; Ganesan, S. Multiple and intensive land use: Case studies in Hong Kong. *Habitat International* **2005**, *29*, 527–546. [CrossRef]
27. Chen, L.; Yang, X.; Li, L.; Chen, L.; Zhang, Y. The Natural and Socioeconomic Influences on Land-Use Intensity: Evidence from China. *Land* **2021**, *10*, 1254. [CrossRef]
28. Cheng, X.; Shao, H.; Li, Y.; Shen, C.; Liang, P. Urban Land Intensive Use Evaluation Study Based on Nighttime Light—A Case Study of the Yangtze River Economic Belt. *Sustainability* **2019**, *11*, 675. [CrossRef]
29. Zeng, C.; Zhang, A.L.; Liu, L.; Liu, Y. Administrative restructuring and land-use intensity-A spatial explicit perspective. *Land Use Policy* **2017**, *67*, 190–199. [CrossRef]
30. Luo, X.; Qin, J.; Cheng, C.; Pan, Y.; Yang, T. Spatial effects and influencing factors of urban land intensive use in the Yangtze River Delta under high-quality development. *Front. Environ. Sci.* **2022**, *10*, 1270. [CrossRef]
31. Borchers, A.; Pieler, T. Programming pluripotent precursor cells derived from Xenopus embryos to generate specific tissues and organs. *Genes* **2010**, *1*, 413–426. [CrossRef]
32. Zhou, L.; Shi, Y.; Cao, X. Evaluation of Land Intensive Use in Shanghai Pilot Free Trade Zone. *Land* **2019**, *8*, 87. [CrossRef]
33. He, W.; Kuang, F.M.; Wang, P. Evaluation of Intensive Land Use in Shuikoushan Economic Development Zone, Hunan Province, China. *IOP Conf. Ser. Earth Environ. Sci.* **2017**, *86*, 012013. [CrossRef]
34. Yang, J.; Yang, Y.; Tang, W. Development of evaluation model for intensive land use in urban centers. *Front. Archit. Res.* **2012**, *1*, 405–410. [CrossRef]
35. Yang, Z.; Li, S.; Sun, D.; Li, C.; Wu, J. Intensive Evaluation and High-Quality Redevelopment of Enterprise Land Use: A Case Study in China. *Land* **2022**, *11*, 432. [CrossRef]
36. Williams, K. Urban intensification policies in England: Problems and contradictions. *Land Use Policy* **1999**, *16*, 167–178. [CrossRef]
37. Burton, E. Measuring Urban Compactness in UK Towns and Cities. *Environ. Plan. B Plan. Des.* **2002**, *29*, 219–250. [CrossRef]
38. Yang, P.P.-J.; Lay, O.B. Applying ecosystem concepts to the planning of industrial areas: A case study of Singapore's Jurong Island. *J. Clean. Prod.* **2004**, *12*, 1011–1023. [CrossRef]
39. Taleai, M.; Sharifi, A.; Sliuzas, R.; Mesgari, M. Evaluating the compatibility of multi-functional and intensive urban land uses. *Int. J. Appl. Earth Obs. Geoinf.* **2007**, *9*, 375–391. [CrossRef]
40. Thinh, N.X.; Arlt, G.; Heber, B.; Hennersdorf, J.; Lehmann, I. Evaluation of urban land-use structures with a view to sustainable development. *Environ. Impact Assess. Rev.* **2002**, *22*, 475–492. [CrossRef]
41. Lewis, G.M.; Brabec, E. Regional land pattern assessment: Development of a resource efficiency measurement method. *Landsc. Urban Plan.* **2005**, *72*, 281–296. [CrossRef]
42. Oh, K.; Jeong, Y.; Lee, D.; Lee, W.; Choi, J. Determining development density using the Urban Carrying Capacity Assessment System. *Landsc. Urban Plan.* **2005**, *73*, 1–15. [CrossRef]
43. Li, H.; Chen, K.; Yan, L.; Zhu, Y.; Liao, L.; Chen, Y. Urban Land Use Transitions and the Economic Spatial Spillovers of Central Cities in China's Urban Agglomerations. *Land* **2021**, *10*, 644. [CrossRef]
44. Lin, G.C.S.; Yi, F. Urbanization of Capital or Capitalization on Urban Land? Land Development and Local Public Finance in Urbanizing China. *Urban Geogr.* **2013**, *32*, 50–79. [CrossRef]
45. Wu, Y.; Luo, J.; Zhang, X.; Skitmore, M. Urban growth dilemmas and solutions in China: Looking forward to 2030. *Habitat International* **2016**, *56*, 42–51. [CrossRef]
46. Kosow, H.; Wassermann, S.; Bartke, S.; Goede, P.; Grimski, D.; Imbert, I.; Jenssen, T.; Laukel, O.; Proske, M.; Protzer, J.; et al. Addressing Goal Conflicts: New Policy Mixes for Commercial Land Use Management. *Land* **2022**, *11*, 795. [CrossRef]
47. Abolhasani, S.; Taleai, M.; Karimi, M.; Rezaee Node, A. Simulating urban growth under planning policies through parcel-based cellular automata (ParCA) model. *Int. J. Geogr. Inf. Sci.* **2016**, *30*, 2276–2301. [CrossRef]

48. Yanbo, Q.; Shilei, W.; Yaya, T.; Guanghui, J.; Tao, Z.; Liang, M. Territorial spatial planning for regional high-quality development—An analytical framework for the identification, mediation and transmission of potential land utilization conflicts in the Yellow River Delta. *Land Use Policy* **2023**, *125*, 106462. [CrossRef]
49. Hersperger, A.M.; Oliveira, E.; Pagliarin, S.; Palka, G.; Verburg, P.; Bolliger, J.; Grădinaru, S. Urban land-use change: The role of strategic spatial planning. *Glob. Environ. Change* **2018**, *51*, 32–42. [CrossRef]
50. Yang, H.J.; Song, J.; Choi, M.J. Measuring the Externality Effects of Commercial Land Use on Residential Land Value: A Case Study of Seoul. *Sustainability* **2016**, *8*, 432. [CrossRef]
51. Sun, W.; Chen, W.; Jin, Z. Spatial Function Regionalization Based on an Ecological-economic Analysis in Wuxi City, China. *Chin. Geogr. Sci.* **2019**, *29*, 352–362. [CrossRef]
52. Liu, Y.L.; Wei, X.J.; Jiao, L.M.; Wang, H.M. Relationships between Street Centrality and Land Use Intensity in Wuhan, China. *J. Urban Plan. Dev.* **2016**, *142*, 05015001. [CrossRef]
53. Luo, J.; Wu, Y.; Choguill, C.L.; Zhang, X. A study on promoting the intensive use of industrial land in China through governance: A game theoretical approach. *J. Urban Manag.* **2022**, *11*, 298–309. [CrossRef]
54. Esfandi, S.; Nourian, F. Urban carrying capacity assessment framework for mega mall development. A case study of Tehran's 22 municipal districts. *Land Use Policy* **2021**, *109*, 105628. [CrossRef]
55. Peng, C.; Song, M.; Han, F. Urban economic structure, technological externalities, and intensive land use in China. *J. Clean. Prod.* **2017**, *152*, 47–62. [CrossRef]
56. Su, H.M.; He, A.X.; Fang, G. Study of Urban Land Intensive Use Dynamic and Spatial Difference in Anhui Province. *Adv. Mater. Res.* **2011**, *347–353*, 3597–3602. [CrossRef]
57. Irwin, E.G.; Bockstael, N.E. Land use externalities, open space preservation, and urban sprawl. *Reg. Sci. Urban Econ.* **2004**, *34*, 705–725. [CrossRef]
58. Millard, J.; Outhwaite, C.L.; Kinnersley, R.; Freeman, R.; Gregory, R.D.; Adedoja, O.; Gavini, S.; Kioko, E.; Kuhlmann, M.; Ollerton, J.; et al. Global effects of land-use intensity on local pollinator biodiversity. *Nat. Commun.* **2021**, *12*, 2902. [CrossRef]
59. Wang, F. *Quantitative Methods and Socio-Economic Applications in GIS*; CRC Press, Inc.: Boca Raton, FL, USA, 2014.
60. Anselin, L. Local Indicators of Spatial Association. *Geogr. Anal.* **1995**, *27*, 93–115. [CrossRef]
61. Gao, J.; Song, J.; Wu, L. A new methodology to measure the urban construction land-use efficiency based on the two-stage DEA model. *Land Use Policy* **2022**, *112*, 105799. [CrossRef]
62. Wang, L.Y.; Zhang, S.Y.; Tang, L.P.; Lu, Y.C.; Liu, Y.F.; Liu, Y.L. Optimizing distribution of urban land on the basis of urban land use intensity at prefectural city scale in mainland China. *Land Use Policy* **2022**, *115*, 106037. [CrossRef]

MDPI AG
Grosspeteranlage 5
4052 Basel
Switzerland
Tel.: +41 61 683 77 34

Land Editorial Office
E-mail: land@mdpi.com
www.mdpi.com/journal/land

www.ingramcontent.com/pod-product-compliance
Lightning Source LLC
Chambersburg PA
CBHW040819250926
51850CB00007BA/1214